Space Commercialization: Satellite Technology

Edited by
F. Shahrokhi
The University of Tennessee Space Institute
Tullahoma, Tennessee

N. Jasentuliyana
United Nations
New York, New York

N. Tarabzouni
King Abdulaziz City
for Science and Technology
Riyadh, Saudi Arabia

Volume 128
**PROGRESS IN
ASTRONAUTICS AND AERONAUTICS**

A. Richard Seebass, Editor-in-Chief
University of Colorado at Boulder
Boulder, Colorado

Technical papers selected from the Symposium on Space Commercialization: Roles of Developing Countries, Nashville, Tennessee, USA, March 1989, and subsequently revised for this volume.

Published by the American Institute of Aeronautics and Astronautics, Inc.
370 L'Enfant Promenade, SW, Washington, DC 20024-2518.

American Institute of Aeronautics and Astronautics, Inc.
Washington, DC

Library of Congress Cataloging in Publication Data

Symposium on Space Commercialization: Roles of Developing Countries
  (1989: Nashville, Tenn.)
Space commercialization. Satellite technology/edited by
  F. Shahrokhi, N. Jasentuliyana, N. Tarabzouni.
  p. cm. – (progress in astronautics and aeronautics; v. 128)
  "Technical papers selected from the Symposium on Space
commercialization: Roles of Developing Countries, Nashville,
Tennessee, U.S.A., March 1989, and subsequently revised for this
volume."
  1. Artificial satellites – Congresses.   2. Remote sensing –
Congresses   3. Space industrialization – Developing countries –
Congresses.   I. Shahrokhi, F.   II. Jasentuliyana, Nandasiri, 1938-.
III. Tarabzouni, N.   IV. Title.   V. Series.
TL507.P75   vol. 128   90-39563
[TL796.A1]
629.1 s – dc20
[384.5'1]
ISBN 0-930403-77-0

Copyright © 1990 by the American Institute of Aeronautics and Astronautics, Inc. All rights reserved. Reproduction or translation of any part of this work beyond that permitted by Sections 107 and 108 of the U.S. Copyright Law without the permission of the copyright owner is unlawful. The code following this statement indicates the copyright owner's consent that copies of articles in this volume may be made for personal or internal use, on condition that the copier pay the per-copy fee ($2.00) plus the per-page fee ($0.50) through the Copyright Clearance Center, Inc., 21 Congress Street, Salem, Mass. 01970. This consent does not extend to other kinds of copying, for which permission requests should be addressed to the publisher. Users should employ the following code when reporting copying from this volume to the Copyright Clearance Center:

0-930403-77-0/90 $2.00 + .50

Progress in Astronautics and Aeronautics

Editor-in-Chief
A. Richard Seebass
University of Colorado at Boulder

Editorial Board

Richard G. Bradley
General Dynamics

John R. Casani
*California Institute of Technology
Jet Propulsion Laboratory*

Allen E. Fuhs
Carmel, California

George J. Gleghorn
*TRW Space
and Technology Group*

Dale B. Henderson
Los Alamos National Laboratory

Carolyn L. Huntoon
NASA Johnson Space Center

Reid R. June
Boeing Military Airplane Company

John L. Junkins
Texas A&M University

John E. Keigler
*General Electric Company
Astro-Space Division*

Daniel P. Raymer
*Lockheed Aeronautical Systems
Company*

Joseph F. Shea
*Massachusetts Institute
of Technology*

Martin Summerfield
*Princeton Combustion Research
Laboratories, Inc.*

Charles E. Treanor
*Arvin/Calspan
Advanced Technology Center*

Norma J. Brennan
Director, Editorial Department
AIAA

Jeanne Godette
Series Managing Editor
AIAA

Table of Contents

Preface vii

Present and Future Imaging Radar Systems 1
    R. Keith Raney, *RADARSAT Project, Ottawa, Ontario, Canada*

**SPOT: Commercial Policies for an International Remote
Sensing System** 11
    Pierre Bescond, *SPOT Image Corporation, Reston, Virginia*

**Satellite Remote Sensing of Water Resources in the Yangtze
and Yellow Rivers of China Based on Infrared Imagery
of Cloud Distributions**...18
    R. J. Hung, *University of Alabama in Huntsville, Huntsville, Alabama*,
    and James C. Dodge, *NASA Headquarters, Washington, D.C.*

**Earth-Orbiting Satellite Imageries for Geodetic Data:
A Simulation Study** 32
    Sanjib K. Ghosh and Zhengdong Shi, *Laval University, Quebec, Canada*

Remote Sensing Applications to Tectonism in West Tennessee 44
    D. P. Argialas, *Louisiana State University, Baton Rouge, Louisiana*,
    and F. Shahrokhi, *The University of Tennessee Space Institute,
Tullahoma, Tennessee*

**Satellite Technology in the African Center of Meteorological
Applications for Development (ACMAD)** 54
    John A. Leese, *World Meteorological Organization (WMO), Geneva,
Switzerland*

**Hydrologic Assessment of Critical Erosion Areas Using Satellite Data
and a Geographic Information System**..........................70
    K. M. Morgan and L. W. Newland, *Texas Christian University, Ft. Worth,
Texas*, and S. A. Hayes, *Water Utilities, City of Weatherford, Texas*

Applications of High-Resolution Remote Sensing Image Data.........77
    W. M. Strome, D. Leckie, J. Miller, and R. Buxton, *PCI Inc.,
Richmond Hill, Ontario, Canada*

**Remote Sensing Applications to Earth Resources Survey
in Pakistan**..94
    Salim Mehmud, *Pakistan Space and Upper Atmosphere Research
Commission, Karachi, Pakistan*

Use of the Spectroradiometer LI-1800 to Solve Problems of Preservation of the Environment 105
Miloslav Krizek, *Remote Sensing Center, Prague, Czechoslovakia*

Chinese Very Small Aperture Terminal System for Ministries 109
Sen Dan, *China Broadcasting Satellite Corporation, Beijing, China*

Use of Satellite Communication for Technology Development and Transfer in Developing Countries 117
W. T. Weerakoon, *Center for Space Science and Technology, Kelaniya, Sri Lanka*

Low Traffic Density, Small Terminal Network, and Satellite Antenna Design for Communications in the Rural Areas 132
L. Bardelli, F. Martinino, and F. Rispoli, *Selenia Spazio, Rome, Italy*

Payload, Bus, and Launcher Compatibility for Multibeam Mobile Communication Satellite Systems 150
Nizar Sultan, *Canadian Astronautics Limited, Ottawa, Ontario, Canada*

Rupture of the Spit of Sangomar – Estuary of the Saalum, Senegal 170
Amadou Tahirou Diaw, *University of Dakar, Senegal, West Africa*, Nouhoum Diop, *Dakar Port, Senegal, West Africa*, and Yves-Francois Thomas, *Ecole Normale Superieure, Montrouge, France*

Saudi Arabia's Experience in Solar Energy Applications 181
Fahad S. Huraib, *King Abdulaziz City for Science and Technology, Riyadh, Saudi Arabia*

The Saudi Center for Remote Sensing 191
Muhammad A. Tarabzouni, *King Abdulaziz City for Science and Technology, Riyadh, Saudi Arabia*

Agricultural Applications of Remote Sensing in Hungary 205
G. Csornai, *FÖMI Remote Sensing Center, Budapest, Hungary*

Yield Prognosis by the Productivity Criteria Using Spectral Signatures in the VIS, NIR, and TIR Ranges 214
H. Barsch, *Pedagogical University Potsdam, Potsdam, GDR*, K.-H. Marek, *Academy of Sciences of the GDR*, H. Weichelt, *Central Institute for Physics of the Earth, Potsdam, GDR*, and A. Gebhardt, *Academy of Agricultural Sciences of the GDR, Muencheberg, GDR*

Measures for Minimizing Radiation Hazardous to the Environment in the Advent of Large-Scale Space Commercialization 227
S. Nataraja Murthy, *Indian Space Research Organization, Bangalore, India*

Remote Sensing Activities in Japan 240
Keiji Maruo, *Remote Sensing Technology Center of Japan, Tokyo, Japan*

Communications and Broadcasting Satellites in Japan 259
    H. Dobashi, E. Kimura, and K. Aikyo, *Telecommunications Satellite Corporation of Japan, Tokyo, Japan*

Space Research Satellite Program of Japan 271
    Tomonao Hayashi, *Institute of Space and Astronautical Science (ISAS), Sagamihara, Kanagawa, Japan*

Mobile Satellite Communications: Applications for Developing Countries 282
    Wolf D. von Noorden, *International Maritime Satellite Organization, London, United Kingdom*

Remote Sensing Program of the Federal Republic of Germany 296
    A. Langner, *DFVLR, Köln, Federal Republic of Germany*, and H. Schüssler, *Dornier GmbH, Friedrichshafen, Federal Republic of Germany*

Author Index for Volume 128 318

List of Series Volumes 319

Table of Contents for Companion Volume 126

Preface xi

Long March Launch Vehicles in the 1990s 1
    Zuwei Huang, *Ministry of Aerospace Industry, Beijing, China*

Advent of Commercial Space 7
    William B. Wirin, *The Space Commerce Corporation, Colorado Springs, Colorado*

ISAS's New Launch Vehicle for Science Use 24
    R. Akiba, H. Matsuo, and Y. Matogawat, *The Institute of Space and Astronautical Science, Yosinodai, Sagamihara, Kanagawa, Japan*, and T. Hosomurat, *Nissan Motor Company, Ltd., Momoi, Suginami-ku, Tokyo, Japan*

NASA Directions in Space Propulsion for 2000 and Beyond 35
    Gregory M. Reck, *NASA Headquarters, Washington, D.C.*

Establishing a Center for Advanced Space Propulsion 50
    George W. Garrison, *The University of Tennessee Space Institute, Tullahoma, Tennessee*

Possible Uses of the External Tank in Orbit 60
    L. F. Ronquillo and F. F. Baillif, *Martin Marietta Manned Space Systems, New Orleans, Louisiana*

Pegasus: Key to Low-Cost Space Applications 72
    Marty R. Mosier and David A. Steffy, *Orbital Sciences Corporation, Fairfax, Virginia*

Power from the Space Shuttle 90
    William M. Speier, *NASA Johnson Space Center, Houston, Texas*

SP-100 Nuclear Space Power Systems with Application to Space Commercialization .. 106
    J. M. Smith, *NASA Lewis Research Center, Cleveland, Ohio*

Power from Space for Use on Earth: An Emerging Global Option 121
    Peter E. Glaser, *Arthur D. Little, Inc., Cambridge, Massachusetts*

Legal Problems of Developing Countries' Access to Space Launch Vehicles 132
    Stephen E. Doyle, *Aerojet TechSystems, Sacramento, California*

International Law of Responsibility for Remote Sensing 140
    Carl Q. Christol, *University of Southern California, Los Angeles, California*

Obstacles to Space Commercialization in the Developing World: Lessons from Landsat ... 159
    Christine Specter and Robert Amann, *Florida International University, Miami, Florida*

Hermes Ground Segment: Support for European Orbital Infrastructure Operations ... 176
    Michel Cadé, *Centre National d'Etudes Spatiales, Toulouse, France*

Raumkurier: The German Recovery Program 193
    Th. G. Müller and K. J. Jäger, *Dornier GmbH, Friedrichshafen, Federal Republic of Germany*

Joint DoD/NASA Advanced Launch System: Pathway to Low-Cost, Highly Operable Space Transportation 202
    Malcolm G. Wolfe, *The Aerospace Corporation, Los Angeles, California*

Broad-Based Space Education: Prerequisite for Space Commercialization 217
    R. N. Singh, *Institute of Technology, Banaras Hindu University, Varanasi, India*

Low-Cost Satellites and Satellite Launch Vehicles 227
    Sikandar Zaman, *Pakistan Space and Upper Atmosphere Research*
    Commission (SUPARCO), Karachi, Pakistan

Satellite Launch and TT&C Systems of China and Their Roles
   in International Cooperation 235
    Xia Nanyin, *Luoyang Institute of Tracking and Telecommunications Technology,*
    Luoyang, China, and Chen Laixing, *China Satellite Launch and TT&C General,*
    Luoyang, China

Ariane Transfer Vehicle: Logistic Support to Space Station Freedom 255
    C. Cougnet and C. Ricaud, *Matra Espace, Toulouse, France,* and
    N. Deutscher, *MBB/ERNO, Bremen, Federal Republic of Germany*

Author Index for Volume 126 270

List of Series Volumes 271

Table of Contents for Companion Volume 127

Preface xiii

OUTPOST CONCEPT: A Transportation and Service Platform in Low-Earth Orbit 1
Thomas C. Taylor, John D. Hodge, and William A. Good,
GLOBAL OUTPOST Inc., Alexandria, Virginia

Columbus Polar Platform: Concept Evolution and Current Status 19
Romano Barbera, *European Space Agency, Paris, France*

User Accommodations on Space Station Freedom 36
Thomas L. Moser, *NASA Headquarters, Washington, D.C.*

Planning for Space Station Freedom Laboratory Payload Integration 61
Harvey J. Willenberg and Larry P. Torre, *Boeing Aerospace Company, Huntsville, Alabama*

Space Station Application of Lessons Learned from Space Shuttle Integrated Operational Prototypes 69
Michael J. Wiskerchen, *DYSE Corporation, Stanford, California*

Low-Gravity Materials Experiments in the Space Station Freedom 84
Roger P. Chassay, *NASA George C. Marshall Space Flight Center, Huntsville, Alabama*

Preparation of Synthetic Polymer Membranes in a Microgravity Environment 94
Ivan A. Vera, *Venezuelan Electric Company (CADAFE), Caracas, Venezuela*

Multiple Experiment Processing Furnace – Crystal Growth Facility 103
R. Srinivas and S. Walker, *Teledyne Brown Engineering, Huntsville, Alabama,*
and D. Schaefer, *NASA George C. Marshall Space Flight Center, Huntsville, Alabama*

Modular Containerless Processing Facility 112
Andrew D. Morrison, *California Institute of Technology, Pasadena, California*

Dynamics of Surface Tension in Microgravity Environment 124
R. J. Hung, Y. D. Tsao, and B. B. Hong, *University of Alabama in Huntsville, Huntsville, Alabama,* and F. W. Leslie, *NASA George C. Marshall Space Flight Center, Huntsville, Alabama*

Containerless Processing Using Electromagnetic Levitation 151
A. B. Gokhale and R. Abbaschian, *University of Florida, Gainesville, Florida*

Review of Drop Tube and Drop Tower Facilities and Research 179
Michael B. Robinson, *NASA George C. Marshall Space Flight Center, Huntsville, Alabama,* and Robert J. Bayuzick and William H. Hofmeister, *Vanderbilt University, Nashville, Tennessee*

Low-Cost Low-Volume Carrier (Minilab) for Biotechnology and Fluids Experiments in Low Gravity 199
John M. Cassanto, Walter Holemans, and Ted Moller, *Instrumentation Technology Associates, Malvern, Pennsylvania,* Paul Todd and Robin M. Stewart, *National Institute of Standards and Technology, Boulder, Colorado,* and Z. Richard Korszun, *University of Wisconsin-Parkside, Kenosha, Wisconsin*

Cell Separation and Electrofusion in Space 214
    D. R. Morrison, *NASA Johnson Space Center, Houston, Texas,*
    and G. A. Hofmann, *BTX, Inc., San Diego, California*

Red Cell Membrane Under Zero Gravity: Interpretation of ARC Experiment on STS51-C 235
    L. Dintenfass, *University of Sydney, Sydney, N.S.W., Australia*

Glass Preparation Under Microgravity 251
    Masaki Makihara, Junji Hayakawa, and Toru Komiyama, *Government Industrial Research Institute, Osaka, Japan*

Acoustic Levitation for High Temperature Containerless Processing in Space 270
    C. A. Rey, R. Sisler, D. R. Merkley, and T. J. Danley, *Intersonics, Inc., Northbrook, Illinois*

Containerless Processing of Fluoride Glass 286
    Robert H. Doremus, *Rensselaer Polytechnic Institute, Troy, New York*

China Can Conduct Materials Processing and Experiments in Space Microgravity 292
    Da Daoan, *Chinese Academy of Space Technology (CAST), Lanzhou, China*

Japanese Approach to the Space Station 296
    Yasushi Horikawa, *National Space Development Agency of Japan, Tokyo, Japan*

Japan's Space Development Activities for the Practical Application Field 316
    Ryu-i-chi Nagashima and Tadahico Inada, *National Space Development Agency of Japan, Tokyo, Japan*

Space Station Freedom – Optimized to Support Microgravity Research and Earth Observations 336
    Vincent J. Bilardo Jr., *NASA Headquarters, Reston, Virginia,* and Daniel J. Herman, *NASA Headquarters, Washington, D.C.*

Opportunities for the Small Space Entrepreneur: A Guide to Strategic Planning 353
    Peter L. Portanova, *Los Angeles, California*

ORBITEC: Orbital Technology Demonstration Program 366
    Peter W. Sharp, *MBB/ERNO, Bremen, Federal Republic of Germany,* and Gerd Goelz, *German Aerospace Research Establishment, Cologne, Federal Republic of Germany*

Development of a Microgravity Experiment: Experiences of a Scientist from a Developing Country 374
    Ivan A. Vera, *Venezuelan Electric Company (CADAFE), Caracas, Venezuela*

Author Index for Volume 127 379

List of Series Volumes 381

Preface

Satellite technology and its Earth-oriented applications have evolved enormously since the early days of the space age. In the early 1960s, the potential of satellites to contribute to international communications and national and regional weather forecasting was quickly recognized, and the first experimental satellites were launched. The benefits of the early experiments were sufficiently convincing that operational communication and meteorological satellite systems were functioning by the mid-1960s. Remote sensing, which posed more difficult technological problems, began experimentally in the early 1970s and quickly became technologically operational, although there are still organizational questions concerning operational satellite remote sensing that need to be resolved.

In the field of communications, satellites have fundamentally changed intercontinental communications by introducing live television transmissions and steadily reducing the cost of voice and data circuits, leading to a steadily growing international communications traffic. As communication satellite technology developed and the costs per circuit came down, satellites became competitive for regional and national communications, particularly for television distribution. As the satellites became more powerful, receivers became smaller and cheaper, making mobile satellite systems and direct television reception possible.

In the field of satellite meteorology, satellite technology evolved from simple cloud pictures to cloud temperature observations, temperature and water vapor profiling, and cloud level wind measurements. The information that could be extracted from the satellite data evolved from basic qualitative information to voluminous quantitative measurements for input into numerical weather forecasting models. More recently, satellite observations of such climatic factors as solar radiation, stratospheric ozone, and aerosol concentration have contributed to our growing understanding of the dynamics of the global climate.

In the field of remote sensing, satellite images of the Earth's surface have become more detailed and more frequent, with increased spectral resolution. Images have been obtained in the thermal infrared and microwave regions, and special sensors have been developed to study the ocean surface, providing information on wave heights, wave length and direction, surface temperature and wind speed, and large-scale ocean topography.

As satellite technology and applications have been advancing, mainly through work in the developed countries, people in developing countries have been creating applications to meet the specific needs of their societies. Work has been done on low-cost Earth stations for communications in remote and rural areas, on national education television networks, on tropical meteorology and storm warning systems, and on remote sensing for agricultural development in subsistence agricultural areas. Some developing countries have designed and built their own satellites for communications, meteorology, and remote sensing.

The papers in this volume describe work currently underway in the further development of satellite technology and Earth-oriented applications. They include developments in communications, meteorology, and remote sensing in a variety of developed and developing countries. The field of satellite technology and applications is so vast today that such a collection of papers cannot begin to cover the full range of activities, but can only offer some highlights of current work.

Nonetheless, the collection as a whole does accurately reflect a number of aspects of the international structure of technological development. Most of the developments of advanced space technology are taking place in the developed countries, with the western countries particularly active in promoting international use of this technology. A few developed countries, notably China, India, and Pakistan, are developing their own technological capabilities, initially based largely on western models, to meet their specific needs and to ensure a substantial degree of self-sufficiency and independence in these increasingly essential activities. However, the great majority of developing countries remain users of space technology, working on developing economic applications to meet their developmental needs in such basic areas as agriculture, water resources, energy, and communications. These countries are dependent on international cooperation for access to the technology.

The historical international division of labor, with the developed countries as suppliers of technology and the developing countries as users, still dominates the field and is changing very slowly. Although a few developing countries have made substantial steps in developing national space capabilities, they have only started to explore the possibility of entering the international market for commercial products and services.

The areas in which developing countries can begin to compete internationally are indicated by a few of the papers in this volume. Interestingly, it appears from these and other works that the best opportunities may be both at the low technology end, such as mass production of low-cost Earth station antennas, and at the high end, such as custom design and construction of low to medium capacity meteorological and communication satellites. In both of these cases, the low cost of labor can make developing countries that have substantial numbers of engineers and technicians highly competitive. In the midlevel technology, such as data processing systems or satellite components that require both volume production and advanced technology, entry into the international market will probably be more difficult.

Perhaps the greatest obstacle developing countries face in participating in the international commercialization of space technology and applications is the difficulty in establishing the international connections necessary to conduct commercial operations in other countries. Multinational companies with existing operations, agents, and cooperative arrangements around the world can easily promote, deliver, and service equipment in different countries. Small national operations, even if highly competitive on a quality and price basis, have a hard time selling their products and services to foreign clients. These problems are perhaps most severe for companies producing components that must function as an integral part of larger systems. These problems also limit the marketing of complete systems, many of which require substantial training and servicing. In addition, of course, companies in developed countries generally have easier access to communication and transportation facilities.

As the papers presented here indicate, satellite technology and applications are continuing to develop rapidly, both in terms of technical innovation and in terms of the number of countries involved. Whereas the developed countries are developing ever new technologies and applications, the developing countries are learning the new skills, adapting the technology to their needs, and beginning to explore ways to gain a share of this growing international market using their advantage in lower-cost labor. Space technology is generally still in a developmental phase, changing rapidly, and with often uncertain commercial possibilities. Although there certainly are opportunities for developing countries, they will frequently be difficult to discover and develop. The papers in this volume provide a valuable look at the current activities being pursued in both developed and developing countries and indicate the directions in which international exchange of knowledge, skills, and technology are likely to develop in the near future.

Highlights of the Volume

Large Format Camera

In 1977, NASA provided funding for the development and construction of the Large Format Camera (LFC) in response to the recommendations of the Geodesy Cartography Panel established by the National Research Council as part of its efforts to examine the applications of Earth observation satellites. In 1984, the LFC was flown during Space Shuttle Mission 41-G aboard the Challenger. During this mission more than 4000 ft of assorted color, color infrared, and black-and-white films were used to test the LFC.

Following the 1984 flight, more LFC missions were planned, but the subsequent problems of the Space Shuttle program have canceled or delayed these LFC missions. During the six years since the maiden voyage, however, more than 40 technical papers have been published on the 41-G data. Although further use of the LFC has been delayed, there has been an increasing amount of interest created by these papers and other works. Currently, the LFC is being proposed as the primary mapping sensor/camera on the planned space station scheduled for the mid-1990s.

Although limited in its remote sensing applications by the film return requirement, the LFC has nevertheless added a new dimension to civil remote sensing technology. Each frame from the 41-G mission, at an approximate scale of 1:750,000, covered about 63,368 square kilometers. With regard to spatial resolution, color products provided 20 m and black-and-white 10 m. Although the LFC does not produce digital data, the color film products can be digitized to simulate multispectral bands of digital imagery.

SPOT System

The SPOT System was conceived as the world's first commercial remote sensing system, allowing the industry and the applications to develop and grow based on the economic value of the data as determined by the users. The design of the SPOT system involves several key elements that are geared specifically toward meeting the needs of the international user community. Among these are 1) design of the sensors and satellite operating systems; 2) design of the ground receiving stations and data processing operations; 3) a worldwide commercial distribution system responsive to the needs of the users; 4) a range of products and services developed in response to users' needs; and 5) a commitment to long-term data availability.

The SPOT system is proving the value of this approach by providing open access to its advanced remote sensing capabilities. SPOT's commercial approach is important to remote sensing's further development, as it is user responsive and provides services to support application development.

The family of European Earth observation satellites started with the first generation multimission platform, referred to as SPOT Mk 1. This development, undertaken in the late 1970s with Matra as the prime industrial contractor, has been used for SPOT 1, 2, and 3, and for the European Space Agency's (ESA) ERS 1 remote sensing satellite, demonstrating the versatility of the design. The same service module has been mated with completely different payloads for these two programs. SPOT 1 has performed well in orbit since its launch in February 1986; SPOT 2 was launched in early 1990; and SPOT 3 is currently under construction. Integration and tests of ERS 1 are in process for a launch planned in mid-1990.

The second generation platform (Mk 2) is an evolution of the Mk 1 with a number of improvements in its subsystems to achieve better performance and a 4-yr life time, compared to the planned 2.5 years for SPOT 1. The most significant improvements are found in the electrical and control systems: the solar array delivers 25% more power and has more growth potential than the Mk 1 design, the four NiCd batteries have a larger capacity, and the power electronics have a capability of up to 4.4 Kw of power. The central computer is now based on a Fairchild 9450 radiation hardened microprocessor providing full immunity to heavy ions and the memory capacity has been increased from 64 to 128 Kwords, of which 88 Kwords are available for the payload. The Mk 2 platform also allows a greater altitude range and better attitude control. The four hydrazine tanks provide a total capacity of 300 kg. Attitude sensors have been improved and high-accuracy star trackers will considerably enhance attitude measurement and pointing accuracy for those missions that require better performance.

SPOT Mk 2 is a state-of-the-art development designed to meet Earth observation requirements over the next twenty years. The Mk 2 multimission platform will be used for the SPOT 4 program and the HELIOS reconnaissance satellite program.

Radar Remote Sensing

Airborne and satellite synthetic aperture radar (SAR) systems can be used by planners in developed and developing nations to meet resource observation requirements.

Important airborne SAR systems currently available include the STAR 1 (Intera, Canada) and the Varan-S (France). Planned SAR satellites include ERS-1 (European Space Agency, 1991), J-ERS-1 (Japan, 1990), and RADARSAT (Canada, 1994). There may be a few short-duration (one week) Shuttle-based SAR missions after 1991. A SAR system on the Earth Observing System Polar Orbiter is planned for the late 1990s. Airborne SAR surveys are now available to all nations through international arrangements, and satellite data will become routinely available in the future. Countries interested in using satellite radar data should begin planning now for the new systems.

Low-Cost Airborne Remote Sensing

Frequently, the high cost of remote sensing data acquisition makes projects impractical. A combined multispectral video and small-format photographic system has been developed for use in airborne data acquisition. The sensor configuration includes three Sony XC-37 CCD video cameras, one Sony G-5 CCD color camera, four Sony EV-C8 video cassette recorders and two Nikon F-250 (35 mm) reconnaissance cameras. Medium (50 nm) and broad band optical filters can be used, and both the video and photographic systems cover the photographic spectrum (330-1100 nm).

Metric Camera

Topographic maps are proven and essential tools for the planning and management of Earth resources. Maps at 1:50,000 scale can meet most planning requirements, but according to a United Nation (UN) survey, such maps are available for only 25% of the world and their current production rate falls well below that needed to complete and periodically update world coverage.

Currently, almost all maps are produced from aerial photographs. The technology employed for this purpose has been operational for several decades, and the cameras, aircraft, and analysis techniques are universally used. These mapping tools are also available in developing countries at reasonable cost.

To extend the use of photogrammetric aerial survey cameras to space platforms, the German Ministry for Space Technology and the European Space Agency jointly sponsored the Metric Camera Experiment onboard the first Spacelab Mission from November 28 to December 8, 1983. From an altitude of 250 km, the Metric Camera obtained 1000 images at a scale of about 1:800,000 on a format of 23 x 23 cm. Although the illumination conditions were bad during the mission, the evaluation of the images showed that the experiment was successful. Photogrammetric images from space promise to become an efficient tool for the cartography of developing countries. In order to improve the image quality, the camera has been equipped with a forward motion compensation system and is now being prepared for a second mission in 1990.

Rural and Remote Communications

The last two decades have seen extensive efforts by the UN and other agencies, in particular the International Telecommunication Union (ITU), to examine the disparity in the distribution of telecommunications facilities between the developed and the developing world, and to identify activities whereby improved telecommunications and greater access to space technology can contribute to economic and social development.

Mobile satellite services, in particular, can contribute to the special communications needs of the developing world by offering greater flexibility in a variety of applications, and by complementing terrestrial and fixed satellite systems without the need for expensive supporting infrastructures.

Consistent with principles formulated by the United Nations General Assembly, INMARSAT's Convention requires it to provide a global mobile satellite communications system to serve all nations and all regions, on a nondiscriminatory basis. INMARSAT's initial mandate to serve the maritime community is being expanded to cover aeronautical and land-based mobile communications services. The developing countries are well represented in the management and activities of INMARSAT. The organization is a commercial, nonprofit entity, which, through international ownership and control, ensures economies of scale, enabling small, less affluent countries to benefit from modern technology. INMARSAT cooperates with international agencies concerned with telecommunications development, and its role was recognized in the recommendations of UNISPACE 82 and the work of the ITU.

From the inception of its services in 1982, INMARSAT's mobile system has been made available for use in developing countries at locations that lack satisfactory communications facilities. The system has been used for humanitarian purposes, such as disaster relief and emergency medical treatment, for oil exploration, and uses in deserts and other remote areas as well as for maritime uses. The evolution of low-cost technology is leading rapidly to the introduction of small, less expensive, transportable data and voice terminals, which will be available for land transport, rural communications, health, education, environmental monitoring, safety, emergency, and many other applications of importance to developing nations.

For low-volume communications for rural and remote areas, two techniques for satellite multiple access are appropriate: SCPC/DAMA and CDMA/SS. Two examples of the applications of these techniques are the CDMA/SS network for 9600 b/s voice communication and television program distribution in African countries, and an onboard, reconfigurable, multispot antenna to optimize the coverage of African regions.

African Meteorology Program

The proposed African Center of Meterological Applications for Development (ACMAD), to be located at Niamey, Niger, is designed to enhance the capabilities of national meteorological services in Africa. The basic activities of ACMAD include the collection, processing, and analysis of meteorological data and the routine dissemination of information applicable to a broad spectrum of human activities to national meteorological services and other national and international organizations of information applicable to a broad spectrum of human activities.

Satellite Communications in the Federal Republic of Germany

Communication satellites now form an essential part of the telecommunication system of the Federal Republic of Germany (FRG), supplementing cables and terrestrial transmitters. In the field of fixed satellite services (FSS), of INTELSAT IV and EUTELSAT ECS1, transponder channels are used for relaying telephone calls and for distributing television signals such as SAT1, 3-SAT, 1 PLUS, and RTL-PLUS.

Most of this television distribution traffic was transferred to the KOPERNICUS (DFS) telecommunication satellite in 1989. KOPERNICUS is the first national communication satellite, with 10 active transponders in the 14/12 GHz and 1 transponder in the 30/20 GHz range, providing a signal strength of about 50 dBW EIRP over the national territory, including Berlin.

Direct broadcasting satellite development was commissioned as long ago as 1978 by the West Germany Federal Ministry of Research and Technology. Since 1980 the governments of France and West Germany initiated the development and construction of the German TV-SAT and the French TDF satellites.

The first TV-SAT, launched in November 1987, failed due to a non-deployed solar array. The second TV-SAT, launched in August 1989, was augmented to provide the full capacity of five high-power channels, as was allocated by the WARC regulations. Four channels are being used for television broadcasting and one channel for the transmission of 16 high-quality radio programs. The TV-SAT signals can be received within West Germany by small terminals equipped with antennas of 0.4-m diam for the television signals and of 0.2-m diam for the digital radio signals. The signals will conform to the new D2-MAC norm accompanied by four digital audio signals.

In the future, the consumer will demand a better picture quality on a larger screen. This demand will be fulfilled by direct broadcasting satellites radiating about 100 W per channel to national and/or multinational coverage zones. The modulation method could be HDMAC to accommodate the high-definition television-signal structure.

The state-of-the-art technology now available for the space and ground segments will allow 1) a high number of mixed low/medium/high power channels per spacecraft; 2) a multimission/multinational type of system (FSS + DBS) on one common spacecraft; and 3) flexible adaptation of different spot coverage areas by use of reconfigurable onboard antennas.

Remote Sensing in the Federal Republic of Germany

The national remote sensing activities in the FRG are complementary to the European Earth Observation Program of the ESA, in which the FRG takes a substantial part. The activities sponsored by the Federal Ministry for Research and Technology cover preparatory scientific and technological studies, development of new space-born instruments, and demonstration projects to promote usage of remotely sensed data for geoscientific applications. The main emphasis is given to instrument development, in particular to active microwave sensors.

Development is underway on microwave sensors including the Microwave Remote Sensing Experiment (MRSE), the X-Band Imaging Radar (X-SAR), and the Millimeterwave Atmospheric Sounder (MAS), as well as on optical imaging instruments such as the Metric Camera, the Modular Optoelectronic Multispectral Scanner (MOMS) and instruments for atmospheric research such as the Michelson Interferometer (MIPAS) and LIDAR. Finally, the Precise Range and Range Rate System, which will be flown on ERS-1 and ERS-2 in the 1990s, will provide precise range and range rate information to support radar altimeter measurements.

Satellite Communications in Japan

For the casual observer, it might be hard to understand the necessity of communication satellites in Japan, where the terrestrial telephone network as well as the television broadcasting network are well established as fundamental infrastructures. However, there are at least two factors that justify the use of satellites, mainly reflecting geographical conditions in Japan. The first is the risk of natural disasters such as earthquakes and typhoons, which paralyze almost all terrestrial networks. The second factor is remote areas and islands like Bonin, where microwave or cable services are not economically feasible.

The development of communications and broadcasting satellites was initiated by the government in 1972. As a first stage, medium-sized satellites of about 350 kg for experimental purpose, CSE and BSE, were built by the National Space Development Agency (NASDA) and launched in 1977 and 1978 by a Delta rocket of NASA. Experiments with CSE and BSE were conducted for several years, mainly by the Ministry of Posts and Telecommunications at the Kashima Earth Station of the Radio Research Laboratories with the participation of Nippon Telephone and Telegraph Corporation (NTT) and Japan Broadcasting Corporation (NHK).

Following the experiments with CSE and BSE, it was decided, as a second stage, to build operational systems. Two communications satellites (CS-2 a and b) and two broadcasting satellites (BS-2 a and b) were built and launched by NASDA using its N-II launch vehicle between 1983 and 1986. Design of the CS-2 and BS-2 satellites was quite similar to that of CSE and BSE, respectively. However, the design life was extended to five years from three years, and some changes were implemented as a result of the flight experience obtained with the experimental satellites. At present, the CS-2's are mainly used by NTT as a supplement to its nationwide network. Two out of six Ka-band transponders onboard the CS-2's, however, are shared by several small-capacity users such as the government authorities, police, railway companies, and power companies. Two Ku-band 100 W transmitters onboard the BS-2s are used for NHKs television service, with one channel for conventional service directed at remote areas and the other for a new service called "around-the-clock broadcasting."

In 1979, the Ministry of Posts and Telecommunications established a new organization called Telecommunications Satellite Corporation of Japan (TSCJ) in cooperation with NTT, KDD, and NHK. One of the main tasks of TSCJ is station-keeping and house-keeping for the communications and broadcasting satellites through its Kimitsu Satellite Control Center located 50 miles southeast of Tokyo. The center was completed in August 1982. At present, four satellites (CS-2's and BS-2's) are under control of the center. The ground facilities of the center were designed in such a way that satellites can be monitored and controlled by a single station. For economic and operational availability, the equipment for CS-2 and BS-2 control have been kept compatible with each other and shared as much as possible. The center has a total of eight antennas.

A new generation CS-3a communications satellite was successfully launched by NASDA using its H-I launch vehicle in February 1988. The weight of CS-3a is about 550 kg and the life is expected to be more than seven years. The number of Ka-band transponders was increased to 15, including 5 spares, as compared to 6 for CS-2. Also, full eclipse operation is possible. A newly developed GaAs solar cell is used for the first time in space as the main power source for the CS-3 satellite. The number of Ka-band transponders on board has also increased to 14 from 7 for CS-2. The follow-on broadcasting satellites BS-3 also weigh approximately 550 kg and are equipped with 3-channel 130 W Ku-band transmitters that will be used by NHK and the Japan Satellite Broadcasting Company, which is a newly organized private broadcasting company. Pay television service is now under consideration for distribution by BS-3.

Having achieved the objectives set for the initial stage of communications satellites and broadcasting satellites, a satellite mobile communication system is envisaged as an important target for the years up to the late 1990s, particularly for maritime communications. For that purpose, NASDA has already launched the experimental satellite, ETS-V. For future broadcasting satellites, high definition television (HDTV) is now drawing much attention as a prospective service. Recently, TSCJ has been authorized to use a 130 W transponder on BS-3b for experiments with HDTV.

Space Science in Japan

Space science has played a pioneering role in the space development of Japan. The first major Japanese satellite was the Shinsei scientific satellite launched in 1971 by the Institute of Space and Aeronautical Science (ISAS) of the University of Tokyo, preceded by the launches of two small test satellites by the same institute. Since then 11 scientific satellites have been launched. They are dedicated mainly to studies of X-ray astrophysics and solar-terrestrial physics. ISAS spacecraft took part in the Halley armada that took a close look at Halley's comet in March 1986. The eleventh satellite, GINGA, detected X-ray emissions from the supernova of 1987. ISAS has also launched rockets and balloons for scientific purposes, and a number of launches have been made overseas. These include balloon observations of the X-ray star ScoX-1 and cosmic rays from Hyderabad India, and of the F'corona from Indonesia. Rocket observations of X-ray astronomy and the upper atmosphere were made from Thumba, India, and of the ionosphere from Peru.

Satellite Remote Sensing in Japan

Japan has several remote sensing satellite programs: the Marine Observation Satellite-1 (MOS-1) Program; the Earth Resources Satellite-1 (ERS-1) Program, and the Advanced Earth Observing Satellite (ADEOS) Program.

The MOS-1, the first Japanese Earth observation satellite, is an experimental satellite to establish basic technologies for and to make practical use of Earth observation. The preliminary design was carried out in 1979, and the critical design was completed in June 1983. MOS-1 was launched by an N-II launch vehicle from Tanegashima Space Center on February 19, 1987. MOS-1b, the successor to MOS-1, was launched in early 1990. MOS-1 has three sensors and a Data Collection System Transponder (DCST). The sensors are the Multispectral Electronic Self Scanning Radiometer (MESSR), the Visible and Thermal Infrared Radiometer (VTIR) and the Microwave Scanning Radiometer (MSR).

The primary purpose of Earth Resources Satellite-1 (ERS-1), which is the second Japanese Earth observation satellite, is to establish the basic technology of remote sensing from space by SAR and optical sensors, for surveying and monitoring natural resources, land use, agriculture, and forestry, as well as for environmental protection, prevention of natural disasters, and surveillance of coastal region. The development of the satellite will be shared with STA and the Ministry of International Trade and Industry (MITI). NASDA conducted the system studies of ERS-1 in 1980 and 1981, and started research on sensors and other devices in 1981. At present, ERS-1 is in the development phase, and is scheduled for launch in 1991. ERS-1 will carry the following primary mission instruments: SAR; Optical Sensor (OPS) using CCD system; Visible and Near Infrared Radiometer (VNIR); and Short Wave Infrared Radiometer (SWIR).

The main objectives of Advanced Earth Observing Satellite (ADEOS), which is the next generation Japanese Earth observation satellite, are to continue and advance the Earth observation technology developed in the MOS-1 and ERS-1 Programs. The preliminary design for ADEOS includes two advanced optical sensors, Ocean Color and Temperature Scanner (OCTS), and Advanced Visible and Near Infrared Radiometer (AVNIR). In addition to these two sensors, other announcement of opportunity (AO) sensors can be installed to provide opportunities for the international community. The AO for participation in the ADEOS program was issued in January 1988.

In October 1978, NASDA completed construction of the Earth Observation Center (EOC) for reception and preprocessing of Landsat data, and the center started operation in January 1979. The EOC is located in Jatoyama, Saitama prefecture. EOC receives and processes the data from Landsat-5, MOS-1, and SPOT-1.

Comprehensive operation of the EOC was transferred from NASDA to RESTEC in May 1987, and RESTEC was designated by NASDA as the data distribution office to general users in Japan and foreign customers. RESTEC commenced data distribution in July 1979 and began distribution of Landsat MSS floppy disk data for personal computer users in the spring of 1984. As an exclusive distributor for the French SPOT IMAGE, RESTEC commenced distribution of SPOT data received by EOC in May 1986. RESTEC also distributes MOS-1 data to general users in Japan and to foreign customers, they began in August 1987 on a test basis, in November 1987 on a routine basis, and since April 1988 they have distributed information in floppy disk form.

Of the applications of remote sensing data in Japan, land applications have had the largest share of the market. The major users of the data are industry, government organizations, nonprofit foundations, and universities and schools. Foreign customers sharply increased in 1983 but subsequently decreased again.

Since 1977, a "Group Training Course in Remote Sensing Technology" sponsored by the Japan International Cooperation Agency (JICA) has been held every year in Tokyo for people from developing countries. The purpose of this training course is to introduce participants to recent developments in equipment and techniques for data processing and analysis. A total of 112 participants from 23 countries have participated in these courses.

Agriculture Remote Sensing in Hungary

Agriculture plays an important role in Hungary's economy, and timely information on major crops is important. For crop inventory, Landsat MSS data can be used effectively because of the large fields (70-150 hectares). Computer classification accuracies achieved were in the 80-97% range, making possible a 92-98% area estimation accuracy. Accuracies depend on the uniformity of the area. Landsat data seem to be adequate for inventory and crop survey and for use in geographic information systems under Hungarian circumstances.

Agricultural Remote Sensing in Poland

The Polish Remote Sensing Center has been using remote sensing techniques for agriculture, concentrating on crop inventory within large administrative units, erosion hazard assessment, and soil moisture determination.

Large area crop inventories have been based on a two-stage stratification technique using satellite images, land evaluation maps, and information about crop structure collected from sampling areas with the use of aerial photographs. With the technique, crop acreage within large administrative units can be estimated with high accuracy. Erosion hazard assessment is based on land use information derived from aerial and satellite images, including slope data and soil-type information. These three types of data are combined with meteorological data in a mathematical model, which indicate erosion hazard and soil loss.

For soil moisture determination, research was based on aerial and satellite data, and indicated relations between the radiation temperature of the soil and vegetation and its water content. The research is continuing within an FAO project aimed at elaborating a method for dynamic grassland soil moisture assessment using satellite data.

Environment Remote Sensing in Czechoslovakia

Aerospace monitoring of the environment in Czechoslovakia is being conducted using a spectroradiometer LI-1800. This device measures spectral reflectance in the range of wavelengths from 300 nm to 110 nm. In cooperation with the Slovak Academy of Sciences, spectral reflectance of the different soil types is measured on the ground simultaneously with a Landsat overflight. For purposes of observing the negative effects of open-cast coal mining on the surrounding environment, spectral reflectance of the forests has been measured. There are great differences between healthy and damaged trees, there are great differences especially in the near IR spectral region. Surveys are conducted for dumps with petroleum waste and its effect on water. In cases of ecological damage caused by leaks of petroleum products, the reflectance of the vegetation is measured. The spectroradiometer is also used from a helicopter.

Experimental investigations to characterize the behavior of crop geosystems has been done through intercosmos cooperation for several years. In 1985, experiments were carried out on a test site in the forest steppe of the USSR near Kursk under relatively homogeneous soil conditions of black loess. In 1986, the experiment GEOEX-86 was conducted in the glacial lowland region of the GDR to extend the results of the KURSK-85 experiment to the typical heterogeneous soil conditions of glacial lowlands, and to extend the methodology for determining agricultural productivity and yield. This work was based on earlier investigations of test sites in the GDR to derive crop parameters from their spectral signatures and to develop a methodology for the automatic derivation of input data for a computer-assisted agricultural crop-monitoring system.

Solar Energy Research in Saudi Arabia

In Saudi Arabia, the King Abdulaziz City for Science and Technology (KACST), a government research entity, has undertaken several major research projects with the objective of advancing the development of solar energy technology as a viable cost-competitive energy alternative, introducing the solar applications industry to the Kingdom, and developing Saudi human resources in the solar field. The projects include: converting solar energy into electricity for everyday use by the inhabitants of several rural villages (such as the Saudi Solar Village); testing solar power as an energy source for space cooling and desalinization in Saudi Universities and the Yanbu water desalinization project; developing agricultural systems powered by solar energy to control the growing environment; undertaking fundamental research in photovoltaic and solar thermal power; and establishing high-technology laboratories for advanced solar energy research at the universities.

KACST solar programs are also engaged in the production of hydrogen gas using solar energy. This involves fundamental research and design and construction of 10 kW and 350 kW photovoltaic/electrolytic hydrogen generation plants.

Remote Sensing in Saudi Arabia

Following the launch of Landsat-4 with its new Thematic Mapper (TM) system with 30-meter ground resolution, Saudi Arabia decided to receive the satellite data directly. Therefore, the government signed an agreement with the United States National Oceanic and Atmospheric Administration (NOAA) to access the data from Landsat and SPOT, and built a receiving station with image processing and photo processing systems. The government also signed a preliminary agreement with SPOT IMAGE to access data from the French SPOT satellite. The Saudi Center for Remote Sensing (SCRS) consists of a Data Acquisition System, a Data Processing System and a Photo Processing Laboratory.

Satellite Observation of Desert Landscapes

Wind streaks, by far the most numerous features on the Syrian landscape, have been observed and studied using TM, SPOT, RBV, SIR-A and LFC images. The streaks have been categorized as involving either the deposition of dust or the deflation of a deposit of wind transportable material. The study area in Syria is about 50 km east of Damascus (Tloll Al-Safa). Lava flows and volcanic cones occupy most of this area, which extends toward the southeast for about 450 km, through Jordan to the Al Harrah region of northern Saudi Arabia. The lava flows on the TM and SPOT images range from medium to light grey tones. The TM images show an alignment of cones towards the northwest, a feature which is evident though less clear on the Landsat 2 and 3 images in the nearer wavelength. Very bright returns represent rougher, probably younger flows, whereas darker patterns indicate smoother, probably older flows.

China is greatly affected by the monsoon circulation of Southeast Asia. The onset, advance, break, and withdrawal of the summer monsoons determines the main characteristics and distribution of precipitation. Chinese meteorologists ascribe the activity of the summer monsoon in China to the effects of the Tibet Plateau. An understanding of the initiation and development of heavy rainfall and severe storms in China relies heavily upon an understanding of the convective storm systems on the Tibet Plateau. Geostationary meteorological satellite observations are a unique and powerful tool for studying short-lived mesoscale convective storm systems, but observations at 15 minute intervals are urgently needed for this study.

Infrared imagery from the GOES-1 satellite shows that heavy rainfall in the Tibet Plateau area is usually preceded by a high growth rate of convective clouds followed by a rapid collapse of the cloud top. Comparison of satellite observations and ground-based rainfall observations shows good agreement between the collapsing of the cloud as observed from the satellite, and the beginning of the rainfall observed by the ground stations, and between the dissipation of the cloud, observed from the satellite infrared imagery, and the ending of the rainfall observed by the ground station. A comparison of the dissipation of clouds over a ground station location, based on satellite infrared imagery, with rainfall recorded at that station shows a linear relationship between the volumetric dissipation of clouds per unit area and the amount of rainfall for amounts exceeding 8 mm. These studies indicate that the analysis of the dissipation of clouds using rapid-scan infrared imagery from a geostationary satellite can be valuable for quantitative estimation of rainfall.

China VSAT System

China's VSAT (Very Small Aperture Terminal) data communications system is primarily designed for governmental ministries, such as the Ministry of Railways, Ministry of Coal, Ministry of Forests, Seismic Bureau, Civil Aviation Administration, etc.

To meet the needs for management and information collecting, and for low initial investment from ministries, a system with a centralized processing facility and a number of groups of remote terminals has been designed. Either INTELSAT or China's domestic communications satellites can be used for the space segment, although it was initially designed for INTELSAT. The network is built in a star configuration because of simplicity and the nature of the ministries' management structure. VSAT-to-VSAT communications, however, can be provided by double hopping, but at the expense of band width and/or power, depending on satellite characteristics and system design. In addition to data transmission, voice and freeze-frame picture services are also under consideration. The system can be used not only for business purpose, but also for professional training as well.

The general characteristics of the VSAT system are as follows:

- Frequency bank   C band
- Hub Station   13-m-diam antenna
                          3 KW HPA
                          G/T (Clear sky) 16.5 dB/K
- Information rate   2.4 - 9.6 Kb/s inroute
                          57.6 kb/s outroute
- Error control 10^{-7} at Eb/No of 8.7 dB
                          FEC with code rate 1/2
- Channel configuration   SCPC inroute
                          TDM outroute
- Terminal interface   RS-232c

The CBSC (China Broadcasting Satellite Corporation) has integrated the system with facilities and equipment provided by international and Chinese suppliers. Performance demonstrations with a small number of VSATs were conducted successfully in 1987. Field trials of the entire system for the first users were scheduled for summer 1988.

Chinese FY-1 Meteorological Satellite

The first Chinese Meteorological satellite (FY-1) was launched into a near circular orbit at 900 km altitude and 99 deg inclination. With a period of 102 minutes, the satellite makes 14 orbits of the Earth every day. The spacecraft was designed and developed in China and constructed of Chinese-made components. It comprises 7 subsystems: remote sensing, image transmission, structure, attitude control, TT&C, thermal control, and power subsystems. The net weight of the spacecraft is 750 kg. Its main body is a cube 1.2 x 1.4 x 1.4 m, and the deployable solar panels are symmetrically connected to the left and right side of the body. The length of the spacecraft, including the solar panels, is 8.6 m. There are two sets of high-resolution spin-scanning radiometers (AVHRR), one for redundance. The radiometer has five channels, one infrared and four visible channels. The information provided includes day and night cloud images, surface temperatures of the land and ocean, a vegetation index, and a measure of suspended sediment in water. The image signal from the AVHRR is transmitted to the Earth continuously in real time by the automatic picture transmission system (HRPT). Image data from other parts of the world are stored on tape recorders and transmitted by the playback transmission system. There is also a space environment monitor to detect solar-wind protons. The satellite is oriented to the Earth by a three-axis stabilization system, including three reaction wheels, desaturation jet nozzles, gyros, and a computer. The FY-1 satellite can make effective contributions to weather forecasting, crop estimation, forest monitoring, storm prediction, ocean observation, and solar activity monitoring. In order to facilitate reception of the transmitted signals by world-wide ground stations, the transmission parameters are the same as those of NOAA satellites. The automatic picture transmission subsystem of FY-1 is especially intended for the developing countries, considering that a large number of low-cost APT receiving stations have been established in many developing countries.

Pakistan Space Program

The Pakistan Space and Upper Atmosphere Research Committee (SUPARCO) was established in 1961 to acquire a capability for peaceful uses and applications of space science and space technology. SUPARCO was reorganized through a Presidential decree in 1981 as an independent national organization, with the status of a full-fledged Commission with substantial autonomy.

SUPARCO has to its credit a number of achievements, which include launching of sounding rockets for neutral atmosphere research, operation of ground receiving station for ionospheric research, reception of cloud cover pictures from meteorological satellites, applications of satellite remote sensing data, completion of a feasibility study for a domestic direct-broadcasting and communication satellite system called PAKSAT, and development of television receive only (TVRO) terminals for reception of TV signals from direct broadcasting satellites. To support these programs, SUPARCO has established facilities for the production of sounding rockets and for the design, development, and limited production of instrumentation for rockets and satellites. Further, it has established a number of laboratories, libraries and a computer center for its scientists and engineers.

Some of the major new projects expected to be completed by the end of 1989 include: establishment of a ground receiving station for reception and processing of remote sensing data from Landsat, SPOT and NOAA satellites; collection of data from unattended platforms via the Argos system and data reception and processing in Pakistan, reception and processing of COSPAS-SARSAT transmissions for search and rescue; and development of a facility for airborne remote sensing.

In the future, SUPARCO would like to implement the PAKSAT project, establish a capability in the design and development of light-weight experimental satellites for remote sensing and communication purposes, and ultimately launch these into near-earth orbits. SUPARCO is at present training scientists and engineers in the fields of space science and space technology and has also hosted a number of international seminars in remote sensing applications. For advanced training of its scientists and engineers in space science and space technology, an Aerospace Institute is being established.

SUPARCO has also made noteworthy contributions in international bodies such as the UN Committee on Peaceful Uses of Outer Space (COPUOS) and its subcommittees, the Committee on Space Research (COSPAR), and the International Astronautical Federation (IAF).

Domestic Communication Satellite for Pakistan

In recognition of the need to augment the existing telecommunications network in Pakistan for nationwide television coverage and with a view to bringing the majority of the population into the mainstream of national activity in the shortest possible time, SUPARCO has carried out a prefeasibility study for a domestic communication satellite system. The study was conducted with assistance from ITU and with active participation from various user agencies including the Telegraph and Telephone Department, the Pakistan Television Corporation, the Pakistan Broadcasting Corporation, the Civil Aviation Authority, and the Ministry of Education. The prefeasibility study (while recommending use of Ku-band for fixed point-to-point telecommunication as well as broadcasting satellite services), identified preferred slots for positioning of PAKSAT-I and II in the geostationary orbit.

The system definition study providing performance specifications of two PAKSAT satellites and the TTC&M station was completed by an international consultant, on the basis of the requirements of various users agencies responsible for telephone, telex, data communication, emergency services for disaster relief operation, air traffic control, TV broadcasting and distribution, and radio networking for mass literacy education, health, and agricultural extension. The TTC&M system and the ground network have also been specified as a basis for inviting international tenders from satellite system manufacturers. In view of the operational nature of the project, the manufacturing, integration, launching, installation of TTC&M station, and in-orbit test and evaluation of the satellites will be carried out by international firms with the necessary expertise in the field. However, SUPARCO is expected to be actively involved in fabrication, integration, and installation of satellite ground stations, especially TVROs, DBSRs and SCPC terminals for rural applications.

The two satellites are proposed for placement at 38 deg E and 41 deg E in the geostationary orbit. The system will use Ku-band (11/14 GHz) for fixed as well as broadcasting satellite services and UHF for data collection. Pakistan has already been allotted an orbital slot at 38 deg E for Broadcasting Satellite Services (BSS) and has applied to IFRB for another position at 41 deg E for telecommunication and broadcasting satellite services.

Remote Sensing in Pakistan

Ever since the launching of Landsat in July 1972, the beginning of civilian satellite remote sensing, the Pakistan Space and Upper Atmosphere Research Commission (SUPARCO) has accorded high priority to satellite remote sensing applications, reflecting its firm belief that this field can make important contributions to the national development process through identification of new natural resources and better management of those already known, particularly for developing countries such as Pakistan. A Remote Sensing Applications Center (RESACENT) was set up in 1973, in order to acquire and analyze remote sensing data, conduct studies, provide data, studies, and facilities to interested user agencies and scientists, and train SUPARCO scientists and those of other institutions. The data analysis facilities include both analog and digital instruments, including microcomputers and a mainframe computer. More than 60 scientific studies dealing with problems of special interest to Pakistan in diverse fields such as agriculture, forestry, water resources, geology, land use, and environment have been completed using satellite remote sensing data. Some studies, including identification of land use features, demarcation of plantation areas, and mapping of snow and glaciers have also been initiated using high resolution SPOT (HRV) data.

SUPARCO is now in the process of establishing a satellite ground receiving station near Islamabad for direct reception and precision processing of Landsat MSS and TM, SPOT HRV and NOAA AVHRR data. Preparations are also being made to purchase a suitable aircraft and sensors for airborne remote sensing, and SUPARCO is conducting a regular training program in remote sensing techniques. It also encourages and sponsors national scientists to make use of training facilities in remote sensing technology offered by foreign institutions. SUPARCO has also played host to a number of international seminars and training workshops dealing with different aspects of remote sensing technology.

Future of Landsat

Since the early 1970s, the United States has been recognized as a global leader in supplying remotely sensed data and related technology for civilian applications through the Landsat system. This has been called "one of the most successful technological programs in the history of the United States," and a major part of this success is linked to the transfer of this technology to developing countries. However, the U.S. lead position in this technology is being eroded. Competing systems are being offered by the French (SPOT), the Japanese (MOS-1), and others. Now international users can choose the system that best meets their needs. In 1988 the Landsat program was at the threshold of a new era, as it had been repositioned from its research and development stage under U.S. government agencies to an operational system within the U.S. private sector. Congress appropriated funds in fiscal year 1988 for a technical study of a U.S. Advanced Civil Earth Remote Sensing System. A major component of this study is to ensure that the needs of international users are identified and met in order to increase their usage of U.S. remote sensing technology as it moves beyond Landsat 6.

United Nations Principles on Remote Sensing

On December 11, 1986 the General Assembly of the United Nations gave its unanimous approval to Resolution 41/65. This resolution consisted of 15 "Principles on Remote Sensing," that had been under consideration since 1970.

Many views were advanced in the negotiation of the Principles, and only as a result of major accommodations of outlook was it possible to obtain the required consensus. As agreed, the Principles enable and authorize States, nongovernmental entities, and international intergovernmental organizations to engage in remote sensing activities, in accordance with the 1967 Treaty on Principles Governing the Activities of States in the Exploration and Use of Outer Space, including the Moon and other Celestial Bodies. International cooperation rather than of unlimited national privacy, based on claims of national sovereignty, is the central theme of the agreed principles.

The "open skies" approach was conditioned by the legal doctrine of responsibility. Countries, and particularly developing countries that will be the object of remote sensing activities by the space resource states, sought and obtained assurances that they would be protected against certain kinds of harm resulting from disclosures of facts and information.

Principle 14 contains the separately stated norms of "international" and "state" responsibility, indicating that there are situations where each of the expressions is relevant, although in the final legal analysis they convey essentially the same substantive meaning. This may appear to be an unusual conclusion, since it might be expected that the separate and distinct terms would only have found their way into the Principles if unique legal significance were to be accorded to each expression.

Liability for damages can result from a breach of the international law of responsibility, and several situations where this law could be invoked are identified and analyzed. The international law of responsibility will facilitate remote sensing for peaceful and scientific purposes.

F. Shahrokhi
N. Jasentuliyana
N. Tarabzouni
June 1990

Present and Future Imaging Radar Systems

R. Keith Raney*
RADARSAT Project, Ottawa, Ontario, Canada

Abstract

This paper provides an encapsulated summary of synthetic aperture imaging radars (SAR) systems for the use of planners in developed and developing nations in anticipation of meeting future resource observation requirements. Nominal image parameters are described. System deployment globally and access to data are described. Key points of contact are listed for data requests. Important airborne SAR systems available for civilian use include the STAR 1 (Intera, Canada) and the Varan-S (France). Expected SAR-carrying spacecraft include ERS-1 (European Space Agency, 1990), J-ERS-1 (Japan, 1992), and RADARSAT (Canada, 1994), and research-oriented SAR systems will be flown on the Shuttle (1992-1994) and on polar-orbiting platforms (after 1997). It is concluded that SAR data are available to all nations at present under terms agreed through international arrangements, and that from satellites such data will become routinely available in the future. Planning is encouraged to best take advantage of such data for the needs of particular nations.

Introduction

More than a decade has passed since Seasat made its debut. In operation for only three months in the summer of 1978, the synthetic aperture radar (SAR) imagery from that satellite system has proven to be of unique value for many Earth resource applications. Since radar systems are able to observe the Earth's surface with rather little interference from darkness and cloud cover, such information had been anticipated to be of special interest for environmentally important regions from Canada's north to the tropics. Seasat data, reinforced by the Shuttle imaging radar experiments in November 1981 and October 1984, exceeded expectations, for there were some pleasant surprises awaiting the SAR scientists. Satellite SAR data proved to be of higher quality than expected from earlier aircraft-based SAR trials and contained unique information on ice, oceans, forestry, geology, and other important application areas.

In spite of the encouraging scientific content of Seasat data, however, for many years there were no concrete plans for SAR satellite systems to provide data on a

---

Canadian Crown Copyright © 1989. Published by the American Institute of Aeronautics and Astronautics with permission.

*Chief Radar Scientist.

global and routine basis. Fortunately, there are now several approved SAR satellite systems, and beginning in 1991, SAR data will be provided to users world-wide much as Landsat and SPOT data are now available. The purpose of this paper is to give an overview of the SAR satellites now planned and to outline the availability of current airborne SAR systems for users who wish to prepare for the satellite SAR data.

Synthetic Aperture Radar

Whereas many users of image data in Earth resource applications need not know the details of the instrument itself, in the case of SAR systems there are certain features of the sensor that deserve consideration. A SAR system differs from all other satellite imaging systems in two respects: It is an *active system*, in the sense that it provides its own illumination; and it requires extensive *processing* of the data delivered to ground stations in order to form the basic image product.

SAR as Active System

Like other types of radar, SAR systems transmit their own energy, which, after reflection from the Earth's surface, is received and stored in memory, either in the radar system or ground facilities. The SAR systems of interest in Earth resource applications use microwaves, chosen among other considerations for their ability to penetrate fog, cloud, and modest precipitation, and for their properties of probing surface features of interest. The popular microwave frequency bands are known by the following designations: L-band, about 23 cm wavelength; C-band, about 5 cm wavelength; and X-band, about 3 cm wavelength. (All three spacecraft systems flown to date have been at L-band.) Each of these bands has advantages for specific applications, such as better surface penetration by L-band, particularly in arid regions, and discrimination of crop, foliage, and ice type for C-band. X-band has advantages as well, but it is more difficult technically to achieve acceptable performance from spacecraft altitudes, so that none of the operational systems now planned use that frequency.

As a corollary to active microwave operation, radar systems may be built with the capability for the antenna to form the radiating and receiving beams with some flexibility, which enables different modes of operation to be available from a single system. This feature is important to users since image information content depends to a large degree on the incidence angle at which the radar illumination intercepts the Earth's surface. For example, ocean waves are best viewed near 20 ° incidence, agricultural and forestry applications are often best served by incidence angles in the neighborhood of 35 °, and detection of objects such as icebergs and ships is improved as the incidence angle moves out beyond 45 °. No single mode serves all purposes equally well.

SAR Processing

Unlike satellite-based optical imaging systems, which might be unkindly described as very expensive cameras, SAR imaging systems do not yield imagery without extra effort. The output signals of the radar itself are very *unlike* an image. SAR sensor data must be subjected to intense processing in order to form the image from the radar signals. It is this processing that gives rise to the name "synthetic aperture" radar, for the computer operates on the ensemble of received signals to focus them into the desired image, thus "synthesizing" the result that would have been obtained directly from a "large" aperture. By optical standards, an antenna 15 m long might seem large, but for radar wavelengths and satellite altitudes, an *effective* aperture size of several *kilometers* is required to form imagery of resolution on the order of 25 meters. Although requiring care, it is far easier to synthesize such an aperture in computer memory than it would be to actually construct one in space!

For most modern SAR systems, the image processing step (known as pre-processing in Europe) is essentially invisible to users. However, it is important in the sense that several image properties are determined to a large degree by the processor, rather than the radar. A key parameter is the spatial resolution, which determines the spacing between two small objects that allows them to be observed as two rather than one reflector. (The concept of resolution is fundamental, and should be kept distinct from the pixel, the size of each spatial element in a digital image.) Small resolution, also known as fine or high resolution, is generally considered to be "better", although not always with sufficient justification! Within limits set by the radar, the data may be processed to yield fine resolution or to achieve more modest resolution while simultaneously reducing the noise inherent to SAR imagery known as "speckle". The number of "looks" employed in the processor is an indication of the speckle reduction, with more looks corresponding to less speckle noise. For most Earth resource applications, the use of several looks rather than only one is preferred, even though some of the inherent fine resolution in the data is sacrificed. As with other imaging systems, the use of fine resolution always implies smaller image width, if we assume of course that there are given limits on the amount of data that can be handled by the supporting digital systems.

Airborne SAR Systems

Both for the purposes of SAR surveys in their own right, and for gathering SAR data in preparation for satellite SAR systems, consideration should be given to the use of airborne SAR systems. Table 1 lists the principal systems that are now available. Each of these systems has demonstrated its ability to provide SAR data of use in applications for a variety of users. Some have seen service in many parts of the world in projects ranging from local experiments to large mapping surveys.

Airborne SAR imagery typically yields resolutions on the order of 10 m or less. Whereas fine resolution may be required for certain applications, it has a special advantage when seen in the context of future satellite SAR systems. There are now available computer programs that prepare SAR imagery to simulate the expected output of a satellite SAR. The input to the best of these simulation packages is fine-resolution airborne SAR imagery, which when suitably processed, emulates rather well the reflectivity, resolution, and speckle properties that one day will actually be delivered by the particular satellite system being simulated. For users who wish to explore the utility of satellite SAR data as used on their own resource information requirements, an airborne SAR survey of the region of interest coupled with simulation processing will provide a basis for much of the required insight. Several of the airborne systems listed in Table 1 are candidates for generating the required SAR data-base.

Table 1 Airborne SAR systems (civilian) available world-wide

STAR-1	Operational X-band (HH) Real-time o/b processor, D/L Resource surveys, ice mapping, etc. (Intera, Calgary, Alberta, Canada)
STAR-2 (IRIS)	Operational X-band (HH) Production prototype (MDA, Canada) Improved STAR-1 (Intera, Calgary, Alberta, Canada)
SAR System	Operational X-band (HH) (ex-Goodyear system) Terrain mapping, etc. (Aero Service, Houston, Texas)
VARAN-S	Operational two channel X-band HH + HV or VV + VH o/b and ground digital processing (CNES, Toulouse, France)
X/C SAR	Experimental two/four channels o/b real-time processor, recorders Offline precision processing (CCRS, Ottawa, Canada)
DC-8	Experimental C, L, and P bands o/b and/or ground processing Supports interferometry, quadrature polarimetry. (JPL, Pasadena, California, USA)
ERIM	Experimental X, C, and L bands Six channels, o/b r/t processing Quadrature polarimetry (ERIM, or NADC, U.S.A.)
CASSAR	Experimental X-band (four channels) r/t downlink; polarizations Optical processing (ground-based) (Chinese Academy of Science, Beijing)

SAR Satellites

Overview

Given the absence of SAR satellites since 1978, the expected satellite missions including Earth resource imaging radars during the 1990's is significant (Table 2). ERS-1, J-ERS-1, and RADARSAT are discussed below. NASA has approved the Shuttle Imaging Radar mission, known as SIR-C, to fly on three occasions in the 1992-1994 time frame. Being on the Shuttle, each of these flights will have a duration of only about one week, and will be able to observe only a limited amount of the earth's surface. They are very interesting opportunities from an experimental and scientific point of view. SAR systems on polar-orbiting platforms are also planned, and would fly after about 1997, although their characteristics and data availability have yet to be finalized. The polar orbiters are being planned in the context of the Earth Observation System, led by NASA, and with substantial participation by Europe and Japan.

ERS-1

The European Space Agency is completing final preparations for the launch of the European remote sensing satellite ERS-1. The principal sensor on-board is the AMI, the advanced microwave instrumentation, an amalgamation of a scatterometer and SAR, with parallel operation of a microwave altimeter. The scatterometer, designed as a nonimaging sensor to estimate wind speed and direction based on observation of the (microwave) roughness of the ocean's surface, shares major electronic systems with the SAR. As a consequence, operation of the two instruments is mutually exclusive. Although the "on-time" budget of the SAR is limited, it is

Table 2 Summary of SAR satellites for the decade of the 1990 s

	ERS-1	J-ERS-1	(SIR-C)	ERS-2[1]	RADARSAT	EOS[2]
Operation	91-93	92-94	92- -	94-96	94-99	97(?)-
Radar Bands	C	L	C,L,X	C	C	C,L (X)
Polarimetry			yes			yes
Swath width (km)	80	75	15-90	80	50-500	50-500
Resolution (m)	30	18	~30	30	10-100	10-100
Incidence angle (°)	23	39	15-55	23	20-55	15-55
o/b recorder		yes	(yes)		yes	(yes)

[1]For purposes of this discussion, ERS-2 parameters are assumed to be similar to those for ERS-1.

[2]Current NASA planning has deleted the SAR as a facility instrument on EOS-2, and is seeking approval for a separate but related SAR satellite. System parameters of the proposed SAR are not established, but are likely to be rather close to those listed above.

expected that substantial SAR data will be gathered over those parts of the world that lie within radio visibility of satellite-receiving ground stations taking part in the ERS-1 mission. The SAR should provide data generally available following a three-month initialization period for the satellite and support systems so that users may have access to ERS-1 SAR data starting early in 1991. More information may be obtained from the European Space Agency, Paris, France.

An outline of the ERS-1 SAR parameters is given in Table 3. This radar will comprise the first operational SAR satellite and will operate at C-band. Since previous space SAR's operated at L-band, there is considerable interest in the results to be gained from ERS-1; a similar interest exists in preparatory missions by C-band airborne SAR systems. The nominal spatial parameters of ERS-1 are rather like those of Seasat.

The normal SAR data product from ERS-1 is known as "quick look". Processing will be done by ESA, and national centers devoted to this purpose. There are special SAR products planned, including the so-called "precision" or fully processed SAR image, available at special request from ESA. Terms of data availability have yet to be finalized, but it is expected that purchase rates will be comparable to those for Landsat and SPOT imagery. Rights of data access are global, but with priority given to nations participating in the mission. There are no data recorders on board the spacecraft for the SAR, so that there will be regions of the world not observable by ERS-1. Coverage maps are available from ESA.

Table 3 Radar parameters for ERS-1

Resolution @ looks	
> Quick look processing	30 m az, 30 m r @ 4 looks
> "Precision"	30 m az, 26.3 r @ 6 looks
Frequency	5.3 GHz (5.6 cm wavelength)
Antenna size	10 m az × 1 m h
Swath width	80.4 Km (99 Km processed)
Avg. min/orbit	7.5 (2 min in eclipse)
Noise eq. sigma$_0$	-18 dB
Dynamic range	21 dB (mean reflectivity)
Incidence angle	23 ° (swath center)
Bits/sample (I & Q)	5 (fixed-point)
Downlink	100 mb/s, direct, X-band

J-ERS-1

Pertinent parameters of the Japanese Earth resources satellite, known popularly as J-ERS-1, are given in Table 4. This SAR is very much like Seasat in many regards (L-band and solid-state implementation), but differs from Seasat in significant respects. Having somewhat more effective power than its predecessor, the J-ERS-1 SAR design is aimed at a slightly longer range, thus gaining in two respects. The incidence angle is about 39° (as opposed to 22° for Seasat), and the range resolution is 18 m rather than the 23 m typical of Seasat, an improvement that results from the more favourable imaging geometry. The primary data product therefore has been specified to have symmetrical resolution, thus 18 m in the second (along track or azimuth) direction, with the number of looks reduced to three as a consequence. (These numbers are subject to processor configuration, as previously remarked.)

It should be noted that J-ERS-1 includes onboard recording capability so that the entire globe (except the polar regions) may be observed by the SAR. Japan has announced an open data access policy, although the details and conditions are yet to be finalized. More information may be obtained from NASDA.

RADARSAT

The only other operational SAR mission for the next decade is RADARSAT, Canada's entry into satellite-based earth observation. Table 5 lists the main

Table 4 Radar parameters for J-ERS-1

Resolution @ looks	18 m az, 18 m r @ 3 looks
Frequency	1.275 Ghz (23.5 cm)
Antenna size	12 m az × 2.2 m h
Swath width	75 km
Avg. min/orbit	20
Mission duration	2 years
Noise eq. sigma$_O$	-20.5 dB
Dynamic range	15 dB (mean reflectivity)
Incidence angle	38.7° (swath center)
Bits/sample (I & Q)	3 (floating-point)
Downlink	2 × 32 Mb/s, X-band (direct and recorder)

Table 5 Parameters for RADARSAT

Resolution	(variable in MODES)
Swath	(variable in MODES)
Frequency	5.3 GHz (5.6 cm wavelength)
Antenna size	15 m az × 1.5 m h phased array in elevation
Max. min/orbit	28
Orbit	dawn-dusk, 24 day repeat (nominal 800 km altitude)
Mission duration	5 years
Launch date	December, 1994 (NASA)
Noise eq. sigma$_O$	-21 dB (nominal)
Dynamic range	30 dB (mean reflectivity)
Incidence angle	20° to 59° (swath edges) (MODE dependent)
Bits/sample (I & Q)	4 (floating-point)
Downlink	100 Mb/s, X-band (direct and recorder)

characteristics of RADARSAT. It is to be a dedicated SAR satellite having no other sensing systems on board. It will operate at C-band, and use a variety of imaging modes made possible by the programmable beam-pointing flexibility of its phased-array antenna.

RADARSAT has been designed to serve the needs of a broad spectrum of user requirements. Chief among these is that RADARSAT data products are meant to match the timeliness requirements of different applications. For example, for perishable purposes such as ship navigation, the elapsed time from SAR observation to delivery of a finished chart-compatible product to the bridge of a ship is planned to be within four hours. Of necessity, the supporting image processing, annotation, database merge, and communication links to accomplish this have been included from the beginning in the system design. Not all users have such tight delivery requirements, so that somewhat more time is allotted to agricultural and geological data sets. It is planned, however, that RADARSAT will be operated on a "zero backlog" basis so that no radar data are to be delayed in the image processing stage. Delivery of data products to users world wide is designed to be satisfactory for operational purposes.

Table 6 Image parameters for RADARSAT MODES

Basic:
    Any one of seven available beams
    Beams span 500 km, 20° to 50° incidence
    Each swath nominal 100 Km width
    28 m × 25 m × 4 looks

Fine-Resolution:
    Any one of five (nominal) beams
    Narrow swaths (40 Km nominal)
    Incidence angle > 35° preferred
    10 m × 10 m × 1 look (nominal)

Extended:
    Incidence angle < 20°; > 50°
    Resolution and swath variable

Wideswath:
    Nominal 150 Km
    Incidence angle < 35°
    28 m × 40 m × 4 looks (nominal)

ScanSAR:
    Resolution/swath width trade-off
    500 Km preferred (300 Km available)
    Incidence angles 20° to 49° imaged
    100 m × 100 m × 8 looks (nominal)

RADARSAT will be able to observe the entire globe, a consequence of its beam-pointing flexibility, and the onboard recorders. The satellite is to be placed in a so-called dawn-dusk orbit so that its solar panels provide nearly constant power to the radar, thus lengthening available operation time per orbit. It has been designed with a long lifetime so that the operational use of its data is encouraged.

The imaging modes available for the SAR are outlined in Table 6. These modes have been designed following extensive user analysis of SAR data from Seasat, the experimental Shuttle missions, and airborne SAR data. An interesting innovation is the ScanSAR mode that will provide wide-swath SAR imagery having 100 m resolution. This mode requires that the radar be designed so as to make such wide-swath imaging possible, but the subtleties in design should not be obvious to the user. Those who wish ScanSAR imagery need only specify it in their request, as the modes are selected and operated by electronic command through the normal mission operations sequence.

Access to RADARSAT data for most users will be through a distribution agent for which the Canadian partner is RADARSAT International, Limited. Terms of access will be competitive with other sources of Earth satellite remote sensing data, and consistent with international standards although the details are not yet confirmed. More information may be obtained through the RADARSAT Project Office.

CONCLUSION

The routine and reliable availability of satellite-based SAR imagery for use in a wide variety of Earth resource applications will become a reality early in the next decade. SAR data are well suited to the quantitative observation of critical national and global resources such as tropical forests and, as such, will be a primary source of information for many resource monitoring and analysis responsibilities. Developing countries together with all other nations should be prepared for the constructive use of such data, and plan for its reception and analysis. The information summarized in this paper has been presented to allow those agencies unfamiliar with these developments to be able to take the first steps toward taking advantage of satellite SAR data.

SPOT: Commercial Policies for an International Remote Sensing System

Pierre Bescond*
SPOT Image Corporation, Reston, Virginia

Abstract

Satellite remote sensing has long been recognized as a valuable technology for planning and managing our Earth's resources. The early systems, developed as government research and development programs, served to introduce the potential of this technology, though the inherent nature of these programs restricted further growth and development. The concept of a commercial system such as SPOT introduces a realistic approach that allows the industry and use of the technology to develop and grow based on its true economic value as determined by the users. The design of such a system involves several key elements such as technological and commercial operations that are geared specifically toward meeting the needs of the international user community. SPOT's commercial policies and activities serve as a model for further development of the remote sensing industry. The commercial approach is essential to the technology's further evolution and to ensuring that the benefits of this technology are increasingly available to meet the needs of the international user community.

Introduction

Since 1972, research and development efforts were the guiding force behind the development and launch of satellite-based remote sensing systems. Several applications of this unique technology, such as geologic exploration and environmental planning, became well established using these early noncommercial experimental systems. However, realizing the true potential of this powerful tool was well beyond the reach of these early programs.

In February 1986, the SPOT 1 satellite was launched from an Ariane rocket into sunsynchronous orbit, 517 miles above the Earth's surface. This event marked the beginning of a new era in satellite remote sensing. The SPOT program introduced two very important concepts: 1) an advanced technology that was designed to be user-

---

Copyright © 1990 by the American Institute of Aeronautics and Astronautics, Inc. All rights reserved.
*President.

oriented instead of research- and development-oriented; 2) a worldwide commercial data distribution system designed to support its own operation and to eventually pay for the launch of future satellites.

The commercial policies implemented by the SPOT system, and particularly by SPOT Image Corporation in the United States, are setting the standards for the future of the remote sensing industry and are increasing the accessibility of this valuable technology.

SPOT System Background

The SPOT system has taken a relatively obscure technology and has turned it into a powerful, accessible tool that is increasingly known around the world. Two identical high-resolution visible (HRV) sensors aboard SPOT 1 produce 10 m resolution panchromatic and 20 m resolution multispectral imagery. Ten m resolution is the most detailed imagery available from civilian satellite systems, detailed enough to detect objects the size of half a tennis court. Adjustable mirrors in each HRV direct the viewing angle over a range from vertical to 27 deg. This capability allows the satellite to capture images from anywhere within a 950 km swath and to acquire repeat images anywhere on Earth on the average of once every two and one-half days, a capability referred to as "rapid revisit." This off-nadir viewing also allows SPOT to acquire stereoscopic imagery that can be viewed and analyzed in three dimensions.

Image data acquired by SPOT are either transmitted directly to a ground receiving station or are recorded by an onboard tape recorder for later transmission. Once processed on the ground, the image data are made available in photographic or computer-compatible tape (CCT) formats. SPOT data are available through a worldwide network of commercial data distributors. SPOT Image Corporation is the exclusive distributor of SPOT products and services in the United States.

SPOT's Commercial Background

Through its commercial perspective, SPOT is making remote sensing technology openly available and more beneficial than ever before. Commercial success is achieved by providing not only image data, but also service and support to users. This includes the timely and reliable delivery of image data to clients. As part of user support, SPOT is working with all components of the remote sensing industry to market the entire technology, including the hardware/software systems required for digital image data processing.

One of the cornerstones of SPOT's commercial approach is the plan to launch a minimum of five satellites, thus ensuring a continuous supply of image data into the 21st century. This will provide a secure background upon which users and other industry components can reliably develop applications, operational programs, and the image/information processing equipment needed for industry growth.

The existing user community formed during remote sensing's early days provides a significant base upon which the commercial industry is developing. For these users, SPOT provides a more powerful, more accessible data source that allows them to expand their applications and to work toward developing operational systems based on SPOT's reliable and timely data delivery, and the long-term availability of data.

SPOT's advanced capabilities also lead the way for the development of whole new groups of commercial applications that were previously not possible because of lack of image detail, the long turnaround times between data acquisition and delivery, and the lack of a user-oriented perspective. These new applications are found in such disciplines as agriculture, urban planning, engineering, forestry, land management, and news gathering.

As the world's first commercial remote sensing venture, SPOT is setting the standards for the future of remote sensing as a commercial industry. The true value of remote sensing is becoming recognized worldwide because of SPOT and concurrent developments throughout other sectors of the industry. This is particularly true as the higher-quality, more detailed image data continue to inspire the exploration and development of new applications. With advanced technology, aggressive marketing, and a dedication to the commercial approach, SPOT is providing a new impetus for unlocking the potential of this powerful technology.

With the largest and most advanced user community in the world, the United States provides a unique perspective on the development of new commercial applications and an increasing integration of remote sensing into operational systems that are of benefit to the global user community. The commercial policies and experience of SPOT Image Corporation in the United States provide a model for ongoing worldwide industry development and the expansion of the user community.

SPOT Image Corporation Commercial Policies: A Model for Global Development

The launch of the SPOT 1 satellite two years ago marked the operational beginning of "commercialized" remote sensing and the transition from a primarily government-supported research and development program to a commercially based, market-driven industry. SPOT Image Corporation's business-oriented goals and direction have resulted in progress in technological development throughout the industry, growth of commercial applications, and a strong foundation for the future.

Initial Commercialization

SPOT began with the assumption that the foundation for a financially successful industry demanded a commercial source for satellite data. This assumption has been proven correct both by sales revenue and the increasing number of new users turning to SPOT. We have demonstrated that when satellite data are of high technical quality and delivery occurs on a reliable and commercial basis, users will have confidence in relying on our technologies and our industry as a valuable source of information in a variety of applications.

Users' "operational" information requirements are currently being met with efficient and economical remote sensing solutions. Now that this initial stage has been successfully completed, the remote sensing industry is prepared to move on to the next "phase" of business development: working with customers to establish the value of remote sensing solutions and delivering those solutions in a commercially acceptable and reliable manner.

SPOT is a service-oriented business dedicated to providing SPOT data as a reliable, useful, and economical information tool. This involves working closely with customers, both previous users of satellite data and new clients, to establish the business of SPOT as a data supplier and solution provider. A sound commercial foundation has been established and the user community continually demonstrates increasing acceptance of and confidence in the future of this commercial industry.

The remote sensing industry is taking on its commercial definition. The identity and role of the major firms shaping the industry are now known. Leading "value-added" service and system companies have responded to the commercialized environment by aggressively pursuing new opportunities that emphasize the use of remote sensing as an operational information tool rather than simply as a research and development project opportunity. SPOT has worked closely with many of these value-added firms to determine how best to identify and meet users' requirements and how to market the industry. Traditional economic and business criteria are now being used to define products and services to users. These users are responding with a rapidly increasing interest in and reliance on SPOT products and services. At this initial stage, SPOT's role in the next phase of commercialization is clear.

SPOT's Role

SPOT's primary purpose remains the reliable delivery of high-quality SPOT data products and services. The necessary financial and business resources have been committed to ensure the continuity of SPOT data to the user community. SPOT's global organization of receiving stations, processing centers, and commercial distribution outlets is now meeting users' requirements on a daily and routine basis. In addition, it is an essential responsibility to identify and open new markets for the use of SPOT data and related products and services. To this end, SPOT now also offers the necessary management services that are required to transform data into information.

This task is implemented through SPOT's project management

group that was formed to work with clients to design and deliver cost-effective remote sensing solutions to meet their information requirements. This "one-stop shopping" concept draws the entire range of capabilities within the industry, particularly those offered by value-added firms. The management of all component activities for the customer provides efficient and economical solutions to client problems without the client having to master the technologies of remote sensing or select and coordinate various subcontractors.

SPOT also provides increased access to industry-wide capabilities through several joint activities with value-added firms that help establish and build the commercial viability of the industry. For example, the SPOT Recommended^{TM} program provides users with a manageable selection of highly qualified value-added firms whose services and products can be confidently recommended.

The Industry Today

A necessary part of the transition from what remote sensing was to its current and future status is an understanding that changes in product definitions and pricing structures are inevitable. During the Landsat era, satellite data were provided from the perspective of government support to research and development studies. This dictated one set of product and pricing policies. The value of the data in particular situations, especially commercial applications, had not been established. The costs associated with acquiring and delivering the data were not subject to the business reality of a profit-and-loss statement. The primary objective of these programs was to get data into the hands of researchers to determine interest and applicability.

SPOT provides satellite data from a commercial perspective with business objectives. The future viability of the industry now rests on identifying commercial users whose information requirements can be met with SPOT and related remote sensing technology and, most important, on those same users determining the actual value they place on remote-sensing-derived products and services. The value established by the marketplace must generate sufficient revenue to cover acquisition and delivery costs if the industry is to maintain its commercial status.

The costs that must be recovered are not limited to SPOT. They include the value-added service firms that actually extract customer-requested information from SPOT data, and the image processing equipment manufacturers whose systems are the tools of the trade. Working together to recognize inter-dependencies and share the investments is essential to developing necessary markets. The experiences of the past clearly show that there is an identifiable value for deliverable products and services that will both cover costs and provide profits.

The development of the industry is on track, on plan, and on schedule for providing the necessary return on an investment that is fundamental to a successful commercial venture. Although it will be necessary for investments to continue, it is now recognized that remote sensing is a sound investment and a unique and exciting business opportunity.

SPOT's Products

SPOT's standard products are the data acquired by the SPOT satellites; SPOT's primary mission remains the collection and distribution of that unenhanced data. Users who require enhancement of SPOT data products will have those needs met by value-added firms at prices determined by a competitive marketplace. SPOT does not intend to offer value-added or enhanced products, but does maintain a flexibility in the types of products offered. Exercising this flexibility requires working closely with users and selected value-added firms to produce new solutions that meet users' requirements and support the necessary market research activity.

SPOT standard data products have become a simplified offering of digital and photographic products. These products serve as basic input for end users and value-added firms to begin the process of information extraction and to meet the input specifications of various applications and activities. SPOT's standard products represent a commitment to value-added firms and end users, a commitment that will allow them to develop their operations and activities based on a defined input from SPOT.

Digital products present SPOT data in its most powerful and flexible medium. Standardized formats allow all computerized image processing systems, from micros to mainframes, to manipulate SPOT data for maximum information extraction.

The film medium has the same amount of SPOT data as the digital product, but the medium isn't as flexible. Film "freezes" the image and the information and establishes the limits for photographic-based applications. These limitations explain why more and more applications are using digital media. Although SPOT cannot increase the flexibility of film media, we can maintain the integrity and clarity of the data. All film orders are produced by using digital enlargement techniques and producing film products directly from digital data.

SPOT's Pricing

The major component of our pricing policy for SPOT products is the data content. This reflects the successful completion of SPOT's primary mission, the SPOT satellites' acquisition of data as programmed, receiving those data on the ground, and processing them into standard products.

A secondary pricing component is the deliverable medium of the product. This component reflects both the costs incurred in producing the medium and the flexibility of information extraction available from that medium. Digital products offer the user the maximum amount of flexibility. Photographic products are the most expensive to produce. Both media include an equal amount of SPOT data. Over time, the market will establish prices for standard data products that show little difference based on medium.

Customers who task the satellite to acquire a new scene are now charged an acquisition fee. This fee is based on the significant and growing archive of available images, and the fact that meeting customer-specified acquisition requirements involves dedicating the SPOT system's most important resource, time on the satellite, thereby foregoing other acquisition opportunities. Approximately 40% of the satellite's acquisition capability is now reserved for fulfilling individual customer's requests.

Conclusion

SPOT's commercial approach establishes the standards for not only data suppliers, but also for other business components of the industry. Marketplace economics have replaced government policy as the singlemost significant component affecting the growth of the remote sensing industry. Flexibility, commercial responsiveness, and a business perspective are essential to ongoing industry development.

The combination of SPOT's advanced technology and commercial perspective is driving growth in all sectors of the industry. This has resulted in the accelerated development of new techniques and applications and increased accessibility to the entire technology. Commercialization has caused remote sensing to play an increasingly significant role in planning, management, and monitoring activities throughout the world.

Satellite Remote Sensing of Water Resources in the Yangtze and Yellow Rivers of China Based on Infrared Imagery of Cloud Distributions

R. J. Hung*
University of Alabama in Huntsville, Huntsville, Alabama
and
James C. Dodge†
NASA Headquarters, Washington, D.C.

Abstract

Although the two largest rivers in China originate in the same region separated only by the Bayanhar Mountains as a watershed, the Yangtze and Yellow Rivers behave in quite different ways. Most of the warm and humid air currents from the Arabian Sea and Bay of Bengal are blocked by the Bayanhar Mountains. As a result, the amount of water in the Yellow River is only 5 % of that in the Yangtze River. Based on the cloud coverage area and the cloud volumetric distributions, and also the thickness above 9.4 kms of the cumulus clouds located north and south of the Bayanhar Mountains from the goesynchronous satellite infrared imagery, the results suggest that a more detailed investigation is warranted in the hope that the proper modification of cumuli north of the Bayanhar Mountains would enhance the rainfall over the fountainhead of the Yellow River.

Introduction

The two largest rivers in China, the Yangtze and the Yellow, originate in the Bayanhar Mountains in Qinghai (Tsinghai) province that is a part of the Tibet Plateau, but they are separated by the watershed. The Tibet Plateau includes the areas of Xizang (Tibet), Qinghai (Tsinghai), and part of Sichuan province, China. The Tibet Plateau with an area of 1.9×10^6 km^2 and a population desity of 2.5 persons per km^2 has an average elevation of over 4.8 km, and in a large percentage of the area the altitude exceeds 7 km.

Copyright © 1990 by the American Institute of Aeronautics and Astronautics, Inc. No copyright is asserted in the United States under Title 117, U.S. Code. The United States Government has a royalty-free license to exercise all rights under the copyright claimed herein for Governmental purposes. All other rights are reserved by the copyright owner.

*Professor.
†Program Manager, Mesoscale Atmospheric Processes.

SATELLITE REMOTE SENSING OF WATER RESOURCES

Although these two rivers originate in the same region, they behave in quite different ways. The Yangtze River, which meanders its way through southwest, central, and east China, has plenty of water that benefits the basin it drains. On the other hand, the Yellow River, which flows through the northern part of China, is not only short of water but carries a tremendous amount of silt that causes frequent flooding. The problem was so serious in the past that the river came to be known as "China's sorrow."[1]

Because of the orographic effects of the Bayanhar Mountains, most of the precipitable water in the warm and humid air currents from the Arabian Sea and the Bay of Bengal is deposited on the southern slopes of the mountains so that the quantity of water in the Yellow River is only 5 % of that in the Yangtze River although both originate in the Bayanhar Mountains.[2]

To change this state of affairs, many proposals have been advanced to divert water and even the warm, humid air currents above the source of the Yangtze River to bring more rainfall to the fountainhead of the Yellow River. If 2 % of the water in the Yangtze River were shifted to the Yellow River, there would be a 40 % increase in the water in the Yellow River.

In the First GARP Global Experiment (FGGE), the American satellite GOES-1 was moved to the Indian Ocean at 58°E during the period of time from December 1, 1978 to November 30, 1979. The GOES-1 satellite was at the right location to obtain both infrared and visible imagery over the Bayanhar Mountains area of the Tibet Plateau.

Figure 1 shows a satellite visible image with a longitude-latitude overlay of the Tibet Plateau and surrounding areas at 0700 GMT, May 4, 1979. Figure 2 shows the geographical location of the Yangtze and Yellow Rivers and their watershed, the Bayanhar Mountains as viewed from a geosynchronous satellite located at 58°E.

By using infrared imagery from the GOES-1 satellite, Hung et al.[3,4] showed that heavy rainfall in the Tibet Plateau area is usually preceded by the high growth rate of convective clouds followed by a rapid collapse of the cloud top. They have shown the agreement between satellite observation and ground-based rainfall observation in the following ways: 1) good agreement between the collapse of the cloud as observed from the satellite and the beginning of the rainfall observed by the ground station; and 2) agreement between the dissipation of the cloud, observed from the satellite infrared imagery, and the ending of the rainfall observed by the ground stations. Recently, a comparison of the volumetric dissipation of clouds per unit area over a ground station location, based on satellite infrared imagery, with rainfall recorded at that station has been made for the Tibet Plateau area.[5] The result shows a

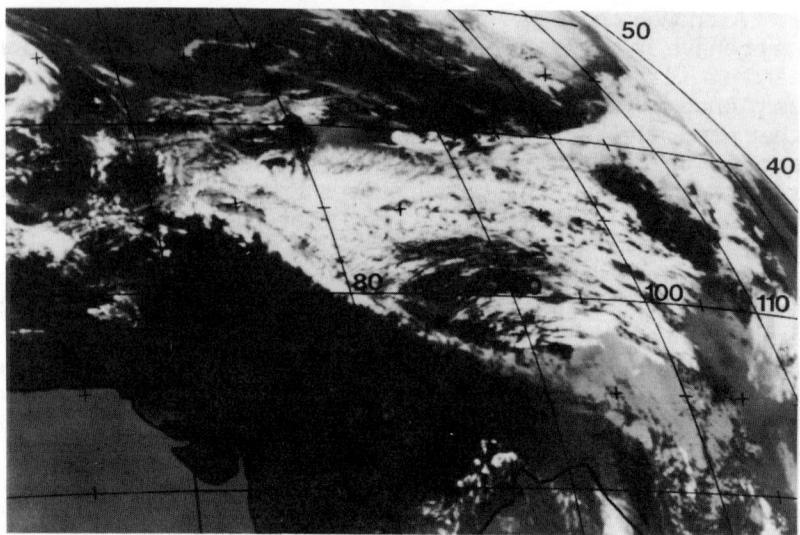

Fig. 1 Visible imager with longitude and latitude overlay covering the Tibet Plateau and its surrounding area at 0700 GMT, May 4, 1979, from GOES-1.

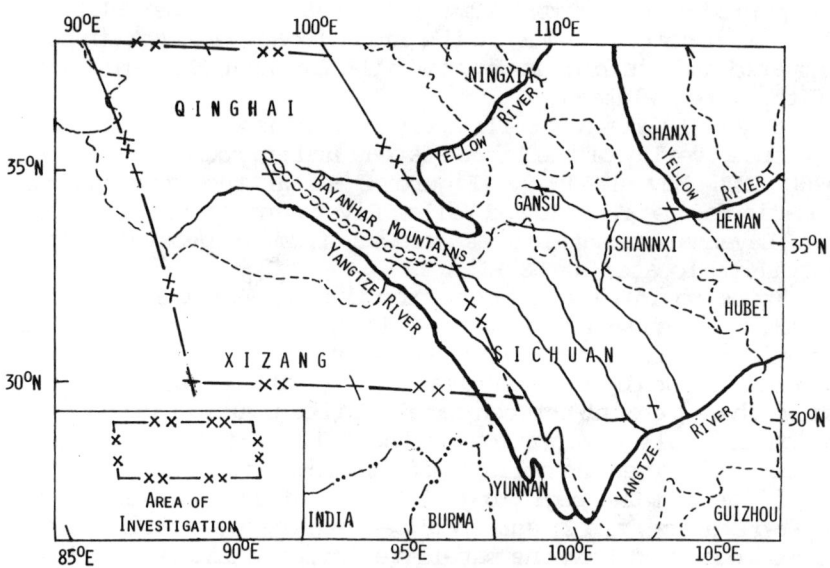

Fig. 2 Geographical locations of the Yangtze and Yellow Rivers and their watershed, the Bayanhar Mountains, as viewed from a geosynchronous satellite located at 58°E.

linear relationship between the volumetric dissipation of clouds per unit area and the amount of rainfall for amounts exceeding 8 mm. These studies indicate that the analysis of the volumetric dissipation of clouds per unit area using rapid-scan, if any, infrared imagery from a geosynchronous satellite can be valuable for a quantitative estimation of rainfall amounts.

Meterological Backgrounds

The Tibet Plateau significantly affects the initiation and development of heavy rainfall and severe storms in China, just as the Rocky Mountains influence local severe storms in the United States. Many rain-bearing synoptic systems in China originate over the plateau and then move eastward or northward, causing excessively heavy rains and outbreaks of severe storms in eastern and northern China. Often, synoptic systems passing over the plateau intensify rapidly on the lee side and eventually give rise to severe weather over a widespread area. The two most notable cases were the devastating flood of the Heihe River in North China in 1963 and the major flood of the

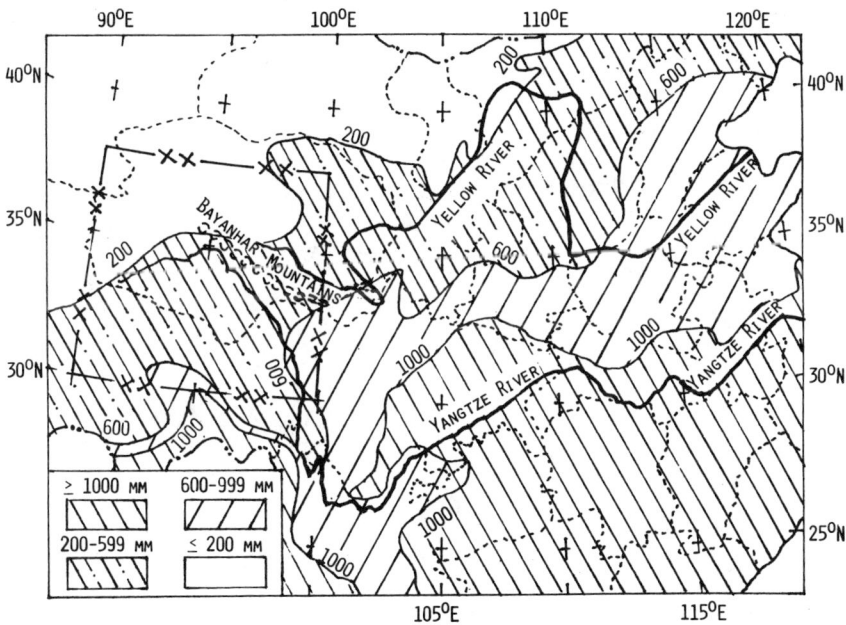

Fig. 3 Annual distribution of the quantity of rainfall at the Yangtze River and Yellow River basins and their surrounding areas (mm/yr) (China Meteorological Atlas, 1978).

Yangtze River in 1954.[6] The orographic effects of the Tibet Plateau also significantly affected the heavy rainfall over central China during the 1979 summer.[7]

The Tibet Plateau plays an important role in general atmospheric circulation as a source of both heat in the summer and cold in the winter. Particularly during the daytime in the summer, the heating effect from the intense solar radiation over plateau results in a high-temperature, low-pressure, high-humidity midtropospheric atmosphere with weak breezes. These are characteristics of a fairly unstable atmosphere with an abundance of convective clouds, thunderstorms, and hailstorms. The warm and humid air currents from the south that provide the major moisture sources in the Tibet Plateau area originate over the Arabian Sea and Bay of Bengal, propagating northward and reaching the western portion of the Qinghai province, the source region of the Yangtze River. Unfortunately, the major portion of these warm and humid air currents is blocked by the Bayanhar Mountains, resulting in a shortage of water in the Yellow River region.

Figure 3 shows the annual distribution of the quantity of rainfall in the Yangtze and Yellow River basins and their surrounding areas. This figure shows that most of the Yangtze River basin and its surrounding areas have an annual rainfall

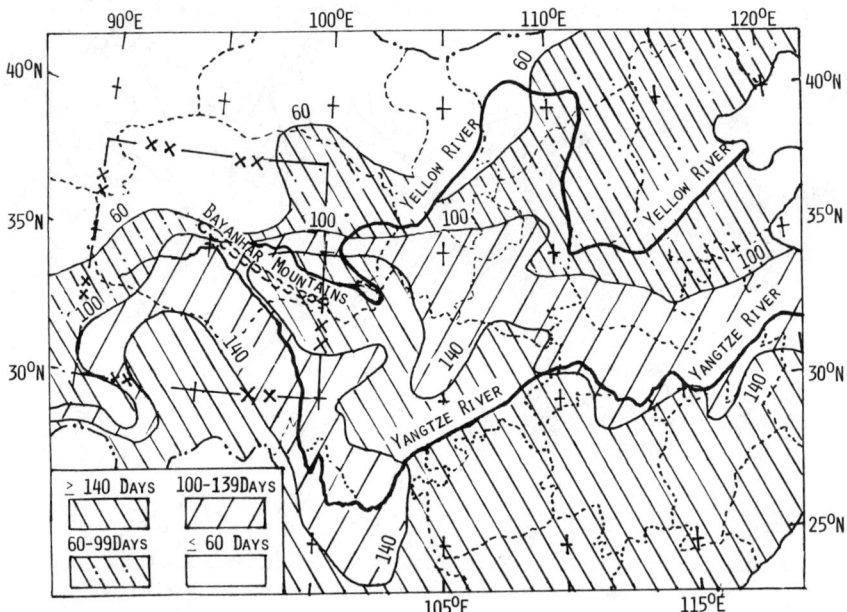

Fig. 4 Annual distribution of the rainfall days at the basins of the Yangtze and Yellow Rivers and thier surrounding areas (China Meteorological Atlas, 1978).

exceeding 1000 mm, whereas the rainfall of the Yellow River basin and its surrounding area is below 600 mm.[8] Figure 4 shows the annual distribution of rainfall days over the basins of the Yangtze and Yellow Rivers and their surrounding areas. Once again, this figure shows the uneven annual distribution of rainfall days in the corresponding regions of the Yangtze and Yellow Rivers, with more than 140 rainfall days in the region of the Yangtze River and less than 100 rainfall days in the region of the Yellow River.[8]

A comparison of Figures 3 and 4 shows that the Yangtze River is rich in water resources with a rather uniform rainfall rate, whereas the Yellow River is short on water or even semi-arid with an extremely uneven rainfall rate.

The question is whether or not there are enough systems of convective activity propagating northeastward from the Arabian Sea and Bay of Bengal across the Bayanhar Mountains, the watershed of the Yangtze and Yellow Rivers, to promise an increase in a precipitation over the Yellow River basin, if additional cloud modification could be applied. This investigation used the infrared remote sensing imagery of convective cloud distributions over the Bayanhar Mountains obtained from the American geosynchronous satellite over the Indian Ocean for a preliminary look at the problem.

Infrared Remote Sensing of a Cloud Distribution
over the Bayanhar Mountains

Previous Results

The convective clouds over the Tibet Plateau developed with high growth rates. During the daytime in the late spring and summer, convective clouds develop and thunderstorms and hailstorms are frequent, usually occurring several times daily.[6] Hung et al.[3-5] show that enhanced rainfall was associated with the collapse and dissipation of these tall convective clouds with cloud top temperatures below the temperature of the lower tropopause (double tropopause in the Tibet Plateau area), in agreement with the conclusions of a study of a tropical cyclone by Gentry et al.,[9] which showed that a high growth rate followed by a rapid collapse, in general, indicates a strong downdraft or precipitation. The probability of the development of convective clouds over the plateau is greatly enhanced due to the fact that the elevation of the plateau exceeds 4.8 km and the heat from the reflected and absorbed solar radiation is added directly to the middle and higher troposphere. Thus, the same amount of heat is apparently used more effectively over the plateau than over the adjacent low-level terrain.

Stout et al.[10] have estimated volumetric rainfall from convective clouds by using either visible or infrared image GOES geosynchronous satellites. Comparisons are based on the time-dependent measurements of the horizontal cloud area from visible imagery with digital gray level 172 and above, and also from infrared imagery with digital level 160 (equivalent to a blackbody temperature of $-26°C$) and above, with volumetric rainfall rates. The visible gray level is scaled in terms of brightness and the infrared digital level in terms of blackbody temperature. The higher gray value of the visible imagery implies higher brightness, whereas the higher values of the infrared digital level denote lower blackbody temperature. Stout et al.[10] show that convective precipitation climaxes typically 60 to 90 min before maximum cloud development. No study was made of the relationship between minimum cloud top temperature and the rainfall rate. It has been our experience that the horizontal cloud area at the tropopause temperature level keeps growing for a period of time after the collapse of the overshooting turret.[11-15] Hung et al.[3-5] conclude that enhanced rainfall is associated with the collapse and dissipation of tall convective clouds with a cloud top temperature below the tropopause temperature, whereas from Stout's study[10] of the horizontal cloud areas with temperatures $> -26°C$, one can only conclude that convective precipitation starts 60 to 90 min before maximum cloud development. Imagery from the systems of GOES geosynchronous satellites was used in the studies of Stout[10] and Hung et al.[4,14]

Present Investigation

The cloud distributions over the Bayanhar Mountains from May 1 to May 5, 1979 were studied. A typical result from 0930 to 2400 GMT, May 4, 1979, is illustrated in this paper. Local time is GMT plus 8 hr. The area from 90 to 100°E and 30 to 38°N with the Bayanhar Mountains near the center was selected for the cloud distribution analysis. The area of investigation is enclosed by double-cross lines as illustrated in Fig. 2 as seen from a geosynchronous satellite, and in Figs. 3 and 4 for geographical coordinates.

Figure 5 shows the time-dependent cloud top temperature distribution for the clouds south and north of the Bayanhar Mountains. It shows that the northside clouds were, on average, 2 to 9 K warmer than the southside clouds from 0930 to 1800 GMT, and then the northernside clouds remained, on average from 0 to 11 K warmer than the southernside clouds the rest of the day.

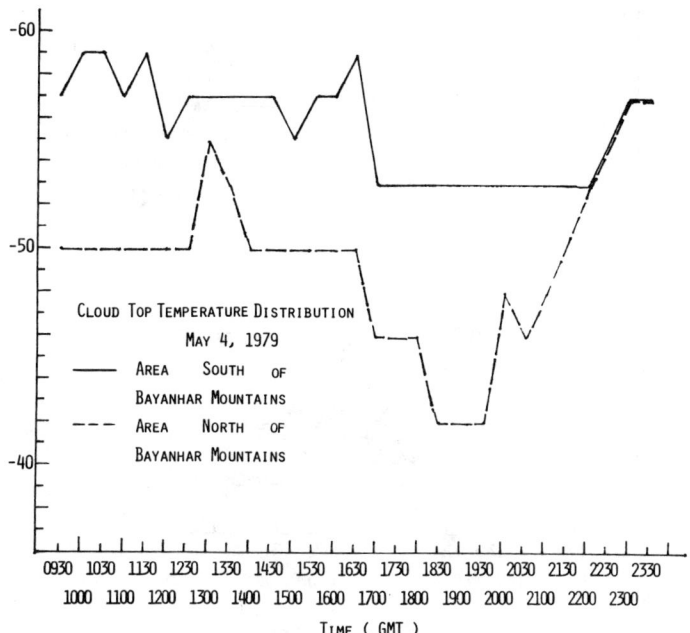

Fig. 5 Time-dependent cloud top temperature distribution for clouds south and north of the Bayanhar Mountains.

Figure 6 shows the time-dependent cloud coverage distributions north and south of the Bayanhar Mountains. The time average of cloud coverage of the nothern clouds before midnight local time (1600 hr GMT) was 6×10^4 km^2, whereas that of the southern clouds was 14×10^4 km^2 within the area of investigation. Figure 7 shows that the observed cloud coverage of the northern clouds before midnight local time was 25%, whereas that of the southern clouds was 75% within the area of investigation.

The quantitative rainfall rate has been estimated from the cloud volume dissipation observed from the satellite. The volume of each individual cloud at time t can be calculated as follows:

$$V(t) = \int P(T,t) \left(\frac{\partial T}{\partial z}\right) dz \quad \text{(Pizel - km)}$$

Here, V(t) denotes the volume of the cloud at time t; P(T, t) is the horizontal pixel number of the cloud top area with temperature T observed at time t; and z is the height of the cloud. The infrared imagery obtained from the satellite

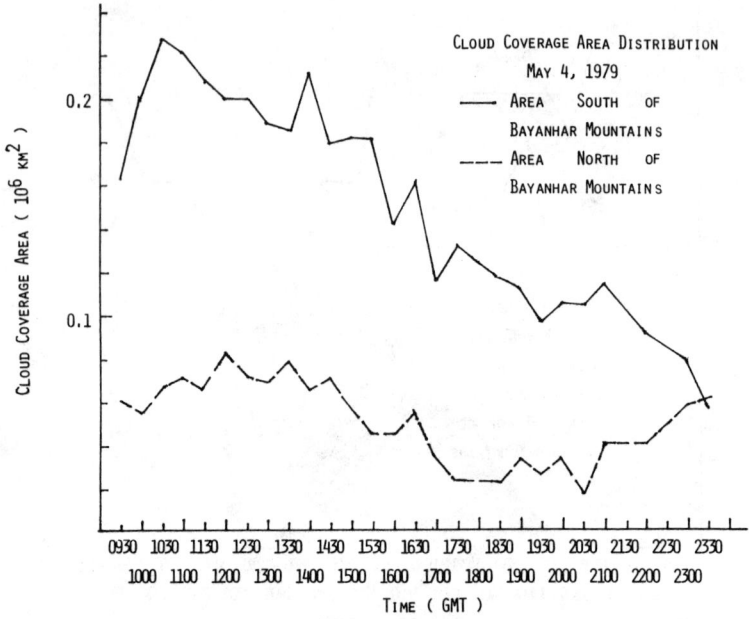

Fig. 6 Time-dependent cloud coverage north and south of the Bayanhar Mountains.

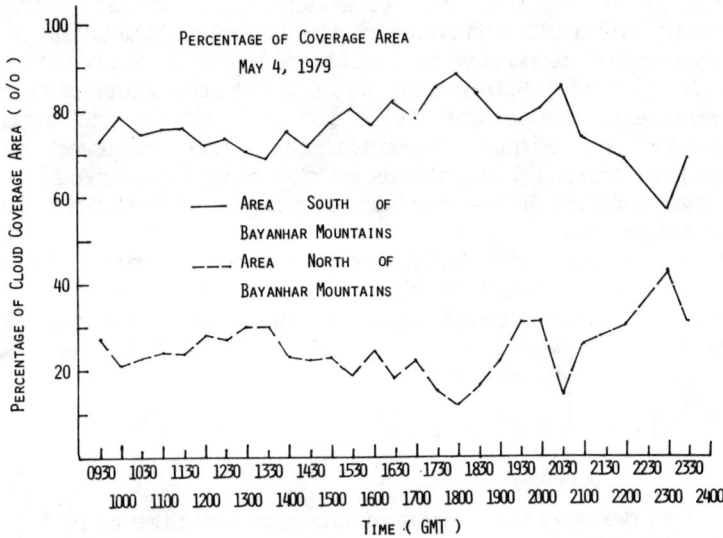

Fig. 7 Percentage of cloud coverage north and south of the Bayanhar Mountains.

SATELLITE REMOTE SENSING OF WATER RESOURCES   27

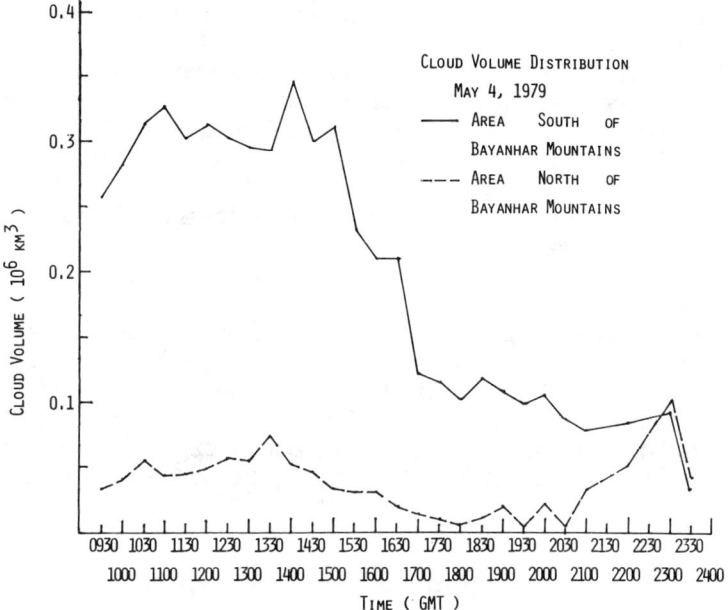

Fig. 8 Time-dependent cloud volume north and south of the Bayanhar Mountains.

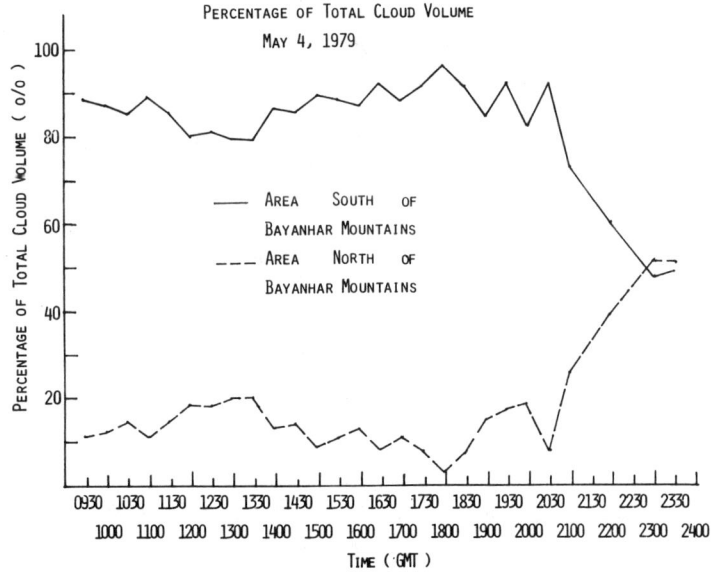

Fig. 9 Percentage of total cloud volume north and south of the Bayanhar Mountains.

proveides the temperature distribution of the cloud tops, with the number of pixels indicating the size. The temperature and height relationship, !T/!z, is available from the rawinsonde data near the location of the cloud.

In this study, the horizontal pixel number of the cloud top area with temperatue T observed at time t, P(T, t), was projected from the satellite coordinates shown in Fig. 2 to geographical coordinates shown in Fig. 3 to transform the pixel number in the area (km^2).

Figure 8 shows the time-dependent cloud volume distribution within the area under investigation north and south of the Bayanhar Mountains. The time average of the cloud volume of the northern clouds before midnight local time (1600 hr GMT) was 5×10^4 km^3, whereas that of the southern clouds was 24×10^4 km^3. Figure 9 shows that the average percentage of cloud volume within the area of investigation of the northern clouds before midnight local time was 10% to 20%, whereas that of the southern clouds was 80% to 90%.

Discussion and Conclusions

China is greatly affected by the monsoon circulation of Southeast Asia. The onset, advance, break, and withdrawal of

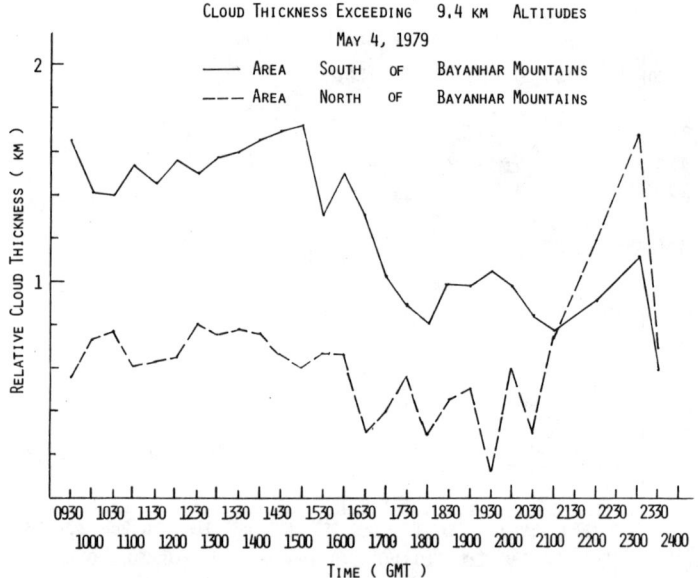

Fig. 10 Comparison of cloud thicknesses above 9.4 kms north and south of the Bayanhar Mountains.

the summer monsoons substantially determine the main characteristics and distribution of precipitation. Chinese meteorologists ascribe the activity of the summer monsoon in China to the thermal and dynamic effects of the Tibet Plateau.[6]

Cumulus clouds are of immense importance to man's livelihood and to his life. They produce more than three-fourths of the rain that waters our planet. Proper modification of cumulus clouds may greatly contribute to the enhancement of rainfall in the area.[16] Infrared remote sensing of cloud distributions over the Bayanhar Mountains over a six-day period shows that the percentage of average cloud coverage north (Yellow River basin) of the mountains was 25%, whereas that south (Yangtze River basin) of the mountains was 75%. Also, the result of infrared remote sensing of the percentage distribution of the average cloud volume of the northern clouds was 10 to 20%, whereas that of the southern clouds was 80 to 90%. This result suggests that proper modification of these clouds might enhance the rainfall over the area of the Yellow River basin.

Figure 10 shows the thickness of the clouds above 9.4 kms. The figure shows that the average of those clouds above 9.4 kms located north of the Bayanhar Mountains before midnight local time (1600 hr GMT) was in the range of 0.6 to 0.8 km, whereas that of the southern clouds was in the range of 1.4 to 1.7 km. For this period, these tall and huge cumulus clouds were potentially capable of producing heavy rainfall over the area. As a matter of fact, they produced excessively heavy rainfall over central China.[7] The reason that clouds located north of the Bayanhar Mountains did not contribute greatly to the rainfall over the Yellow River basin was due to the fact that they propagated eastward before precepitation started and brought heavy rainfall over the Sichuan and Hubei provinces, which cascaded to the Yangtze River (see Fig. 2).

Based on the area and volumetric distributions, and the height above the 9.4 km of the cumulus clouds located north and south of the Bayanhar Mountains, it seems as if the proper modification of the cumuli north of the mountains certainly might bring enhanced rainfall over the fountainhead of the Yellow River.

Acknowledgment

R. J. Hung would like to thank NASA Headquarters for research grant NAGW-1165 that supported the present research.

References

[1] Zhu, M. D., "New Idea to Produce More Rain in the North," China Daily, May 23, 1984, p. 5.

[2] Zhu, M. D., "How to Improve Arid Situation in Northwestern China," *Economic References* (Jingji Cankao), June 20, 1984, No. 715.

[3] Hung, R. J., Liu, J. M., Tsao, D. Y., and Smith, R. E., "Relationship Between Convective Clouds and Precipitation over the Qinghai-Xizang Plateau Area from Satellite Remote Sensing and Groundbased Observations," *International Journal of Remote Sensing*, Vol. 6, 1985, pp. 217-237.

[4] Hung, R. J., Liu, J. M., and Smith, R. E., "Infrared Remote Sensing of Convective Clouds and Amount of Rainfall over the Tibet Plateau Area," *Annales Geophysicae*, Vol. 3, 1985, pp. 767-776.

[5] Hung, R. J., Liu, J. M., Dodge, J. C., and Smith, R. E., "Remote Sensing of Cloud Distribution over the Bayanhar Mountains-Watershed of the Yangtze and Yellow Rivers," *International Journal of Remote Sensing*, Vol. 7, 1986, pp. 577-587.

[6] Tao, S. Y., and Ding, Y. H., Observational Evidence of the Influence of the Qinghai-Xizang Plateau on the Occurrence of Heavy Rain and Severe Convective Storms in China," *Bulletin of American Meteorological Society*, Vol. 62, 1981, pp. 23-29.

[7] Murakami, T., and Huang, W. G., "Orographic Effects of the Tibetan Plateau on the Rainfall Variations over Central China During the 1979 Summer," *Journal of Meteorological Society of Japan*, Vol. 62, 1984, pp. 895-903.

[8] "China Meteorological Atlas," China Atlas Publishers, Beijing, China, 1978, pp. 1-226.

[9] Gentry, R. C., Rodgers, E., Steranke, J., and Shenk, W. E., "Predicting Tropical Cyclone Intensity Using Satellite-Measured Equipment Black Body Temperatures of Cloud Tops," *Monthly Weather Review*, Vol. 108, 1980, pp. 445-455.

[10] Stout, J. E., Martin, D. W., and Sikdar, D. N., "Estimating GATE Rainfall with Geosynchronous Satellite Images," *Monthly Weather Review*, Vol. 107, 1979, pp. 585-593.

[11] Hung, R. J., and Smith R. E., "Remote Sensing of Tornadic Storms from Geosynchronous Satellite Infrared Digital Data," *International Journal of Remote Sensing*, Vol. 3, 1982, pp. 69-81.

[12] Hung, R. J., and Smith, R. E., "Remote Sensing of Arkansas Tornadoes on 11 April 1976 from a Satellite, a Balloon and an Ionospheric Sounder Array," *International Journal of Remote Sensing*, Vol. 4, 1983, pp. 617-630.

[13] Hung, R. J., Tsao, Y. D., and Smith R. E., "Case Study of Pampa, Texas, Multicell Storms," *Pure and Applied Geophysics*, Vol. 121, 1984, pp. 1019-1034.

[14] Hung, R. J., Phan, T., Lin, D. C., Smith, R. E., Jayroe, R. R., and West, G. S., "Gravity Waves and GOES IR Data Study of an Isolated Tornadic Storm on 29 May 1977," *Monthly Weather Review*, Vol. 108, 1980, pp. 456-464.

[15] Hung, R. J., Dodge, J. C., and Smith R. E., "The Life Cycle of a Tornadic Cloud as Seen from a Geosynchronous Satellite," *AIAA Journal*, Vol. 21, 1983, pp. 1217-1224.

[16] Simpson, J., and Dennis, A. S., "Cumulus Clouds and Their Modification," in *Weather and Climate Modification*, edited by W. N. Hess, Wiley, New York, 1974, pp. 229-259.

Earth-Orbiting Satellite Imageries for Geodetic Data: A Simulation Study

Sanjib K. Ghosh* and Zhengdong Shi†
Laval University, Quebec, Canada

Abstract

Modern technologies offer the possibility of using space imageries to determine the physical shape and size of the Earth by using photogrammetric procedures. A procedure is developed as a first step toward implementing this idea. The space imagery situation with regard to the NASA-Large Format Camera is simulated by using a rotating globe and a 35 mm amateur camera. Necessary programs are developed to calibrate the system for its use in obtaining mensural data. Subsequently, spatial triangulation is performed to complete the strip of photographs around the globe. This provides the possibility of obtaining mensural data on the surface of the Earth (the planetary body). The results obtained in this study are encouraging. Cost-effective geodetic information without the physical geodesy and reference-related constraints would be possible with this approach on a global basis as well as with regard to individual countries or regions.

Introduction

Most satellite imaging systems were developed to provide qualitative (interpretational) remote sensing data. However, the quantitative (metric) aspects cannot be ignored. Their study also involves our quest for knowing the Earth's shape and size more accurately. In this regard, geodesists have so far made Earth-based observations that suffer from two fundamental problems requiring, a priori, 1) the geophysical hypotheses concerning geomagnetic-gravi-

---
Copyright © 1990 by the American Institute of Aeronautics and Astronautics, Inc. All rights reserved.
*Professor of Photogrammetry.
†Research Associate in Photogrammetry.

tational fields that influence the observations; and 2) the fictitious bodies (ellipsoids) used for reference.

Satellite imageries as are currently available have opened up newer opportunities. The NASA-LFC (Large Format Camera), the French SPOT (Satellite Pour l'Observation de la Terre), etc. are systems in particular that provide stereoscopic (three-dimensional) coverage of the Earth's surface. Multiple profiles (Fig. 1) and ground clearance (height) information could provide data for determining the local shape and dimensions of the surface. Such information, when integrated in terms of multiple stereo-models and multiple passes of the satellite, would provide data to establish the physical shape and size of the Earth with regard to thousands of points on the surface. Such data in a geocentric universal system of three-dimensional coordinates (Fig. 2) would escape the two geodetic cartographic problems previously mentioned.

Furthermore, the fourth dimension (time) would help establish the differential changes and patterns. Also, the mutual correlations between the orbital parameters of the satellite sensor and the Earth are strong. They are, however, expected to be mutually complementary, and such knowledge would help other Earth science and environmental studies.

With the foregoing in mind, a simulation test was made on a rotating globe (of around 2 m diameter) of the Laval University Museum at Quebec, Canada. The globe has a rotat-

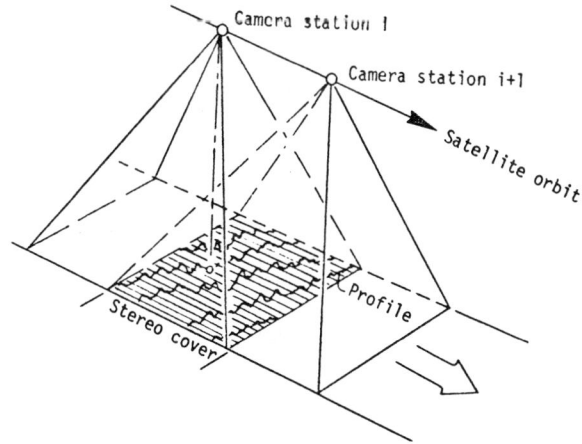

Fig. 1 Stereo-model and profiles with satellite imagery.

ing speed of 2.2 deg approximate. A system of satellite imagery was obtained by simulation in using a 35 mm non metric film camera, whereby 54 photos provided complete stereoscopic coverage along a "great circle." A "flying height" of around 42 cm in this case provides a near simulation of the geometry involved in the NASA-LFC system.

A complete procedure for the on-the-job calibration of the imaging system having been developed and performed, a system of analytical spatial triangulation around the globe has been developed. The resulting data would provide not only all necessary mensural information on the land surface of the globe, but also a double check on the orbital parameters. The latter set of information would be essential because in the real case one needs to overcome the problems caused by the lack of data over the oceans.

This simulation covers the following major aspects: 1) data-acquisition procedure; 2) sensor calibration to estabblish the "interior orientation" of the imaging system; 3) methodology study, two methods are tested for the purpose of recovering "terrain" models; 4) effect of image movement, i.e., to study the effect of the dynamism of the system; 5) misclosure analyses, i.e., misclosures involved in closing the strip around the globe, their analyses, and corrections for the propagated errors (dimensional and parametric).

The dimensional data may be directly implemented on geodetic and topographic mapping missions around the world.

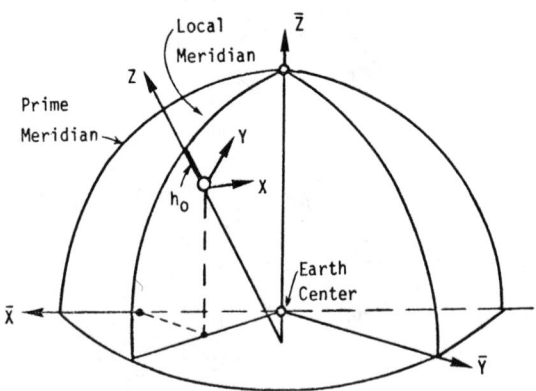

Fig. 2 Local coordinate system in relation to universal geocentric system.

Both the private and public sectors would benefit. Geophysical and space-related endeavors would be able to improve their mathematical models toward advancing their efforts.

The present study was made with only one strip (i.e. corresponding to one satellite pass). However, the procedure can be easily elaborated and several such strips can be integrated to cover larger areas or even the entire Earth's surface.

It is not known to us if any similar study has been made elsewhere. One can, however, mention the studies made by Schmid[4] with Earth-based photogrammetric observations of passive satellites, whereby 45 points around the Earth were established. In this study, Schmid successfully escaped the two fundamental geodetic-cartographic constraints. His methodology was, however, extremely costly and time-consuming and suffered from problems of logistics. His results, although limited to only 45 points, were internationally acclaimed as outstanding and unique at that time.

The objective of the present study is to develop a procedure that would be more thorough, more precise, more elaborate, and more economic than has been possible thus far.

Data Acquisition

A 35 mm camera (Konica FT-1) was used. It was stationary but detached from the globe. Fifty-four photographs were taken at the dynamic mode of the object (a rotating globe of about 2 m diameter). The forward overlap of the photos along a great circle (equatorial) was about 60%.

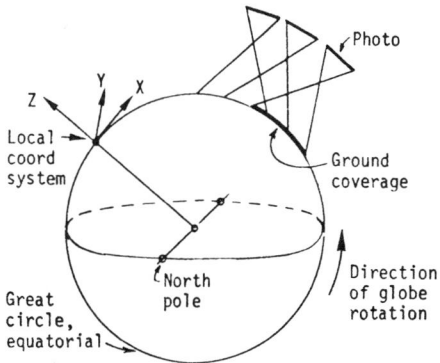

Fig. 3 Schematic of imaging process.

This simulates very closely an orbit of the NASA-LFC, in this case with a flying height of around 42 cm.

A local ground coordinate system (arbitrary and right-handed) is considered. The X axis passes through two selected natural points in the direction of the satellite orbit. One of the two points is on the great circle, the other being about 1 cm away from the circle. The Z axis lies on the plane of the great circle (Fig. 3).

With a view to obtaining the utmost precision possible, the original negatives were used for the studies. Three points common with the left and right adjacent photos were selected in the triple overlap area of each photo and were marked on this (middle) photo. One Wild PUG 3 point marking device was used for such marking whereby tiny holes of 60 μm diameter are drilled into the emulsion. The points are located in the corners of the stereo-models and at the nadir areas, as necessary for effective photo-triangulation. This constitutes the control network of 54 x 3 = 162 points. Some additional points were marked in two selected photos for the purpose of on-the-job self-calibration of the imaging system (the camera).

All points were measured twice to avoid reading blunders and to minimize the effects of instrumental and observational backlash. The observations were made in the stereo-mode with an analytical plotter (Wild BC1) having observational precision on the order of 1 μm.

Camera Calibration

The camera (imaging system) was calibrated by using the concept of self-calibration[2], which considers that the effects of influencing (unknown) parameters being contained in the observation data, they may be computationally derivable if appropriate mathematic models are used. The results would be verifiable with regard to their application in direct mensuration jobs. In the present case, the calibration procedure considers Eqs.(1)

$$V + A_1 \cdot \Delta_1 + A_2 \cdot \Delta_2 + A_3 \cdot \Delta_3 + W = 0 \tag{1a}$$

$$V_1 + \Delta_1 \qquad\qquad + W_1 = 0 \tag{1b}$$

$$V_2 + \Delta_2 \qquad\qquad + W_2 = 0 \tag{1c}$$

$$V_3 + \Delta_3 \qquad\qquad + W_3 = 0 \tag{1d}$$

where,
Eq.(1a) = linearized augmented collinearity condition equations, one for each of the photo coordinates (x,y)
V = vector of residuals of measured image-coordinates
W = discrepancy vector of parameters
A_1, A_2, A_3 = the coefficients to respective parameter groups
Δ_1 = unknown external orientation parameters $(X_o, Y_o, Z_o, \omega, \Phi, \kappa)$ defining the location and direction of the camera at each station
Δ_2 = unknown interior orientation parameters, $(x_o, y_o, f, k_1, k_2, k_3, p_1, p_2, p_3)$ defining interior geometry and distortions for the camera
Δ_3 = unknown object point coordinates (X, Y, Z) in three-dimensions
V_1, V_2, V_3 = residual vectors of the pseudo observations $(\Delta_1, \Delta_2, \Delta_3)$, respectively
W_1, W_2, W_3 = misclosure vectors in the equations.

Note: All symbols here are standard photogrammetric ones.[2]

The camera was calibrated twice in this project by using three photos in each of two areas on two opposite sides of the globe. In order to enhance calibration accuracy, more image points were selected in the triple overlap area than are normally used for control in the double overlap areas (Fig. 4).

Approximate ground coordinates of these image points (as necessary to initiate iterative solutions) were obtained by relative and absolute orientations of models in arbitrary local ground coordinate systems. The results of the two calibrations are shown in Table 1.

Table 1 indicates certain justifiable differences in the values of the parameters. In order to test the effect of these differences, one stereo-model was oriented twice by using values from the two calibrations. The results are presented in Table 2. Furthermore, at the ground point the maximum coordinate difference is less than 0.2 mm, which is about the same as their standard deviations. One, can

Table 1 Calibration results

Calibration	f σ_f mm	x_o σ_{x_o} mm	y_o σ_{y_o} mm	k_1	k_2	k_3	p_1	p_2
I	53.93 0.29	-0.01 0.36	0.13 0.37	0.126	0.000	0.000	-0.004	0.000
II	53.79 0.60	0.14 1.12	0.17 1.13	0.200	0.000	0.000	0.006	0.000

Note: f is the calibrated focal length.

x_o y_o are the principal point coordinates (fiducial reference).

σ are the respective standard deviations.

k_i p_i are the parameters involved in mathematical modeling the radial and tangential distortions, respectively.

infer, therefore, that the two sets of calibration parameters are within acceptable levels of difference. Their average values were used for subsequent photo-triangulation (bridging) along the orbit.

The other interesting facts obtained in the calibration results are that the radial lens distortion is always less than 2.5 µm and that the decentering (tangential) distortion is less than 1 µm, which thus can be simply ignored in subsequent work with this camera.

Spatial Triangulation

A spatial triangulation (bridging) was performed to recover the real "terrain" data from the 54 orbital photos. Because traditional analog instrumental bridging procedures are not adaptable for photos taken with a 35 mm camera, two different analytical procedures were used with the analytical plotter, Wild BC1. The strip, in each case, closes on the initial model.

1) *Independent model method.* Each model is independently relatively orientated. Coordinates (three-dimensional) of six model points plus the two perspective centers are obtained in each model. Then by means of simultaneous similarity ttransformation, all the models are transferred to the same ground coordinate system of the first model.[3]

Table 2 Test of the two calibrations

Photo no.	X_O / σ_{X_O} cm	Y_O / σ_{Y_O} cm	Z_O / σ_{Z_O} cm	ω / σ_ω grad.	ϕ / σ_ϕ grad.	κ / σ_κ grad.	Ref.
M511	180.42	92.54	517.68	6.146	32.723	-7.844	*
	0.01	0.01	0.01	0.000	0.000	0.000	
M501	193.62	91.48	507.79	7.755	42.168	-8.861	*
	0.06	0.08	0.06	0.134	0.103	0.083	
M511	180.42	92.54	517.68	6.146	32.723	-7.844	**
	0.01	0.01	0.01	0.000	0.000	0.000	
M501	193.63	91.47	507.78	7.778	42.137	-8.873	**
	0.07	0.08	0.06	0.135	0.103	0.084	

Note: Each case represents a dependent relative orientation where photo M511 was kept fixed.

* are values with parameters of calibration I.
** are values with parameters of calibration II.
X_O Y_O Z_O are coordinates of the camera perspective center.
ω ϕ κ are rotation angles of the camera axis.
σ are the respective standard deviations.

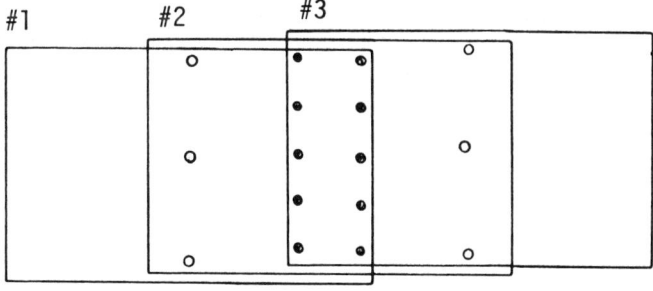

Fig. 4 Distribution of points for calibration.

2) *Analytical bridging method.* Analytical bridging (aeropolygon) is carried out with the instrument by using the built-in program PMO.[5] This method can be described as follows: 1) The first model (two photos) is absolutely oriented with three control points. Three transfer points (common with the next model) are measured on the first model after the absolute orientation. 2) The orientation parameters of the second photo of the first model and the ground coordinates of the three transfer points are input and subsequently, a dependent relative orientation of the strip's third photo and scale adjustment (i.e., coorientation and scale transfer) are performed. 3) By following steps 1 and 2, all subsequent models are connected and the "terrain" model of the area around the great circle is recovered.

In comparing these two methods, one sees that whereas the first solution is necessarily off-line, the second can be on-line. From the viewpoint of accuracy, one cannot say which would be better. Both methods were tested with regard to the data obtained with the BC1 analytical plotter. They gave almost identical results. The following observations are pertinent.

During the procedure with model coordinates (with a scale of approximately 1:12), the following are apparent: 1) Each model created about 1.2 deg Φ-crack error, i.e., the model tilt in the forward direction was larger than it should be. 2) These errors in Φ are purely systematic and are propagated from the first to last model, accumulating in the end to about 64 deg (Fig. 5), the direction of error propagation being the same as the direction of movement of the globe. 3) In terms of the misclosures of ground coordinates, whereas Y coordinates indicated very little misclosure, there were considerable misclosures with the X and

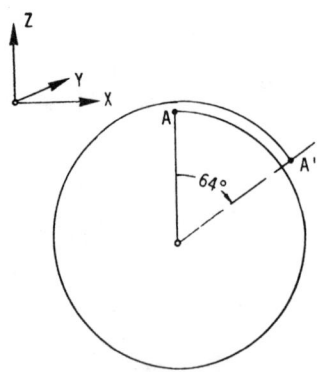

Fig. 5 Misclosure of strip.

Z coordinates. The longitudinal deformation of the strip yields a spiral form as against (ideally) a circle.

The foregoing seem justifiable in view of the dynamism involving the globe rotation and associated image smear[1] during and between the shutter openings. A simple calculation considering all aspects indicates a linear (on the surface) X movement of 0.59 mm for each model, thus yielding a total spatial displacement on the order of 0.59 x 54 = 31.86 mm. This corresponds to the actual misclosures encountered in the triangulation.

After applying correction for such systematic error propagation, the remaining residual misclosures are adjusted by using Eq.(2)

$$\bar{X} = \begin{bmatrix} S_x \\ S_y \\ S_z \end{bmatrix} \cdot M_{\Delta\alpha} \cdot (\bar{\bar{X}} - T) + T \qquad (2)$$

where

$$S_x = \frac{\bar{S} - dS_x}{\bar{S}}, \qquad S_y = \frac{\bar{S} - dS_y}{\bar{S}}, \qquad S_z = \frac{\bar{S} - dS_z}{\bar{S}}$$

dS_x, dS_y, dS_z are the scale corrections in X, Y and Z directions, respectively (Note: These are considered in view of possible scale-affinity in the entire strip);

$\bar{X} = (X, Y, Z)^T$ = corrected ground coordinates

$\bar{\bar{X}} = (X, Y, Z)^T$ = estimated ground coordinates before correction

$\bar{S}$ = estimated model scale before the correction

dS = scale correction

$T = (T_x, T_y, T_z)^T$ is the translation vector

$M_{\Delta\alpha}$ = the three-dimensional rotation matrix $M(\Delta\Omega, \Delta\Phi, \Delta K$ around the three axes of rotation)

Based on the foregoing, the remaining misclosures are distributed linearly to each model in the amount directly proportional to the triangulated number of models. This makes the 54 models fit without any misclosures in the local system of coordinates established for the first model. Subsequently, the three-dimensional coordinates of points along the great circle are used to obtain their averages, giving the mass center for the great circle. The idea is that similar great circles around the Earth may be integrated in a real case to obtain the coordinates of the mass center (i.e., the best fitting physical surface) of the Earth. A simple transformation of the whole data set can then be made to the geocentric universal system without any further possible error.

Subsequently, each mmodel can be separately oriented to generate terrain profiles in any direction, by integrating which one gets the information on the surface form as well as all relevant geodetic mensural data.

Conclusions and Recommendations

With regard to this simulation test, one can draw the following conclusions:

1) The camera used in the test was calibrated twice by using different sets of photos. Both calibrations gave slightly different interior orientation parameters of the camera, but the effect of the difference is negligible.

2) The dynamic effect of the rotating globe is very significant.

3) Double accumulation of image rotation can be discovered and corrected mathematically.

4) The whole procedure developed for the simulation test provides a system that can be used to calculate the physical parameters and size of the Earth from real satellite imageries.

The following recommendations are made:

1) To improve accuracy, each image point may be marked as necessary only on one photo. The corresponding points on other photo(s) can be measured in the stereomode.

2) Since the accuracy of self-calibration depends a great deal on the geometry of image points, including their distribution and quality, the sensor calibration should be repeated with different images from different locations.

3) A clear knowledge about the physical cause of the dynamism of the object is essential for the calculation of the effect of the movements. Precise analysis and calculation of the effect of the movements can improve final accuracy significantly.

4) In a round strip of stereo-models, the image rotation can cause distortions in the scale estimation. Obviously, these distortions are systematic. The exact mathematical models for these systematic errors are not fully developed yet. Their effect, however, can be corrected by considering Φ rotations and scale affinity as has been done in this study.

Acknowledgments

Partial support for this research was provided by the Natural Sciences and Engineering Research Council of Canada through Grant No. A1177. Laval University Museum Center (courtesy of Nicole Brindle) provided the object and facilities for the imaging. All mensural observations and data processing were done at the Laval University Photogrammetric Laboratories. The technical assistance of Paul Trottier, Jean-Paul Agnard, Jean Horvath, and Omari Cherkaoui and typing assistance of Danielle Guimond are acknowledged with deep appreciation.

References

[1] Ghosh, S.K., "Image Motion Compensation Through Augmented Collinearity Equations," Optical Engineering, Vol. 24, no.6, November 1985, pp. 1014-1017.

[2] Ghosh, S.K., Analytical Photogrammetry, 2nd ed., Pergamon Press, New York, 1988, pp. 201-208.

[3] Ghosh, S.K., Phototriangulation, Lexington Books, D.C. Heath and Co., Boston, Mass., 1975.

[4] Schmid, H.H., "Three-Dimensional Triangulation with Satellites," National Oceanic and Atmospheric Administration Professional Paper 7, 1974.

[5] Wild Heerbrugg, "Operational Manual of Wild BC-1 Analytical Plotter," Wild Heerbrugg, Switzerland, undated.

Remote Sensing Applications to Tectonism in West Tennessee

D. P. Argialas*
Louisiana State University, Baton Rouge, Louisiana
and
F. Shahrokhi†
The University of Tennessee Space Institute, Tullahoma, Tennessee

Abstract

It is difficult to recognize faults within the alluvial valley of West Tennessee because of the unconsolidated sediments. Therefore, Landsat satellite MSS images were evaluated for lineament identification using different spectral bands, seasons, enhancement, and image interpretation techniques. In addition, gravity trends were delineated on the gravity anomaly map of West Tennessee, and gravity lineaments were identified and mapped. The Landsat and gravity lineaments were quantitatively analyzed and compared by means of two-dimensional histograms and rose diagrams. The results indicated that the primary direction of both Landsat and gravity lineaments was NE. Weaker trends were found in the NW direction for both Landsant and gravity lineaments. The NE trend of the lineaments corresponded to faults and was in accordance with reactivation of the Reelfoot Rift near the Mississippi River. These results suggest that deeper features, maybe at the earthquake focal depth, probably extend to the land surface as Landsat-detectable lineaments.

Introduction

Lineaments have been defined as "mappable, simple or composite linear features of a surface whose parts are aligned in a rectilinear or slightly curvilinear relationship and which differ distinctly from the patterns of adjacent features and presumably reflect subsurface phenomena."[1] The importance of delineation and documentation of lineaments from satellite images has long been realized by the aerospace and remote sensing community.[2,3,4] Landsat images, in particular, show struc-

---
Copyright © 1990 by the American Institute of Aeronautics and Astronautics, Inc. All rights reserved.
*Assistant Professor of Civil Engineering and Remote Sensing and Image Processing Laboratory.
†Professor and Director, Remote Sensing Division.

tural features that cannot be detected by other means or that cannot be discerned on the surface due to their size. They also provide data on soil moisture and vegetation patterns that aid in the interpretation of lineaments.

In West Tennessee, the thickness of alluvium and loess prevents the traditional, direct evaluation and explanation of the structure and tectonic history of the area. Investigators have supported their judgments based on the analysis of nonsurface data such as gravity and aeromagnetic maps,[5] well and earthquake records,[6] and contours of older structures.[7] Lineations drawn from topographic maps and lineaments mapped on Landsat[8] and SLAR images[9] have previously been used for the purpose of defining fault zones in portions of the Mississippi embayment.

The present study provides an additional source of information on geologic structures through a detailed interpretation of surface expression as revealed from Landsat images. The present study adds to our knowledge of Landsat-detectable features for interpreting lineaments and tests the relationship between surface Landsat lineaments and deeper-seated gravity anomalies.

BACKGROUND

The study area is located between 35° and 37°N latitude and 88° and 90°W longitude. Geographically, it occupies portions of West Tennessee and southwest Kentucky, and it encompassed the eastern portion of the Mississippi embayment (Fig. 1).

Faults within the alluvial valley of the Mississippi River are difficult to recognize because of the unconsolidated sediments, and because of the generally small magnitude of fault displacement. The modifying influences of river meandering and alluviation are also apt to obscure evidence that may be available from a different perspective.[10] Therefore, the investigators had to support their judgments based on other evidence. A discussion follows providing some common fault evidence that has been of assistance in determining the geologic structure in the Mississippi embayment.

Fisk[11] was probably the first to indicate that there are lineaments in the lower Mississippi River Valley and to employ aerial photographs for their identification. Krinitzsky[10] summarized the surface expressions of fault systems in the same area and outlined additional features used for recognizing faulting in alluvial deposits. Hildenbrand et al.[5] and Stauder et al.[12] have shown that the epicenters of the earthquakes occurred in a zone that trends NNE, paralleling the Mississippi River in this area.

Ervin and McGinnis[13] suggested that the embayment is a reactivated rift structure, based on geophysical and geological information. Hildenbrand et al.[15] observed

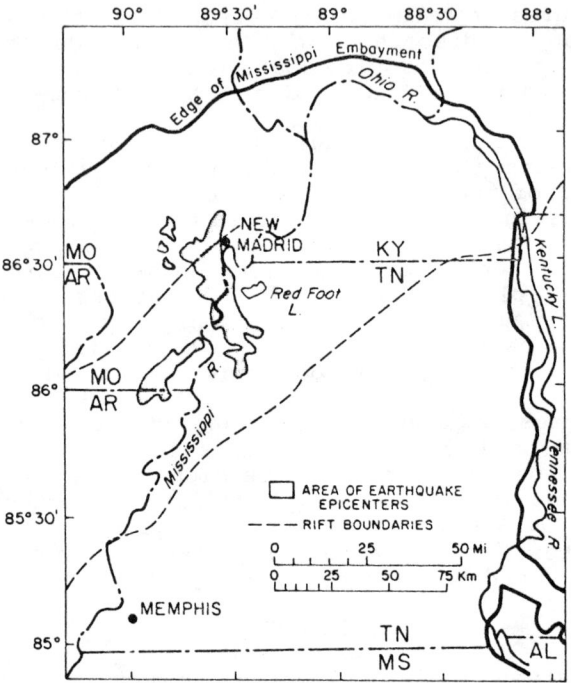

Fig. 1 Study area showing the Rift edges from Hildenbrad et al,[5] Earthquakes near the Mississippi River, from Stauder et al.,[12] are most frequent along the center of the Rift.

that the gravity and magnetic maps revealed several prominent anomalies that appeared to reflect the presence of major geologic or tectonic structures. These structures may be responsible for the generation of the seismic energy. O'Leary and Hildenbrand,[9] using side-looking airborne radar (SLAR) images and aeromagnetic data, suggested that the most prominent directional trend of the lineaments is the N40-45E, and that the alluvial part of the embayment is structurally controlled by a narrow half-graben trending N25E that has developed within the last two to three million years. O'Leary and Simpson [8] have found that the azimuth frequency of Landsat lineaments shows a good correlation with the magnetic anomaly trends in the northeastern region of the embayment.

Methodology

Data Acquisition, Enhancement, and Interpretation

Lineaments were derived from Landsat satellite images (Fig. 2), and gravity anomaly data. A number of

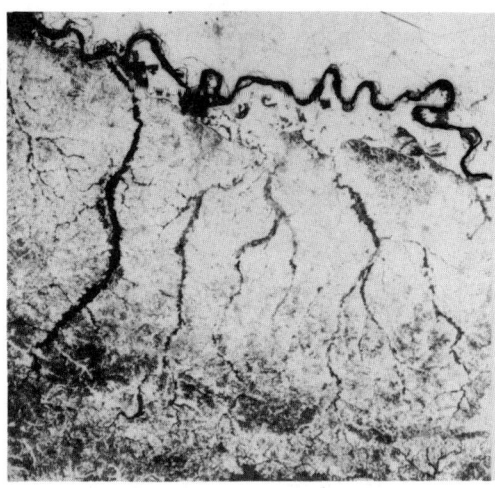

Fig. 2 Location of the study area shown on February 9, 1979, Landsat MSS band 7, contrast- and edge-enhanced image.

Landsat MSS images in two different scales, four different seasons, in both raw and enhanced form, and of the four spectral bands, band 4 (0.5 to 0.6 µ), band 5 (0.6 to o.7 µ), band 6 (0.7 to 0.8 µ), and band 7 (0.8 to 1.1 µ), were interpreted.[14,15]

Seasonal evaluations and contrast and edge enhancements were carried out. Seasonal enhancement was accomplished by choosing images for all seasons. Topographic lineaments were better interpreted in the winter scenes, where minimum vegetation growth, maximum exposure, and low sun angle prevailed. Vegetation growth that followed lineaments was best interpreted on fall and spring images when the difference between natural and cultivated plant growth was at a maximum. Total lineaments, due mainly to soil tones, were best expressed on spring images after most fields were bare from plowing and before crop growth effectively covered the soil.

Contrast enhancement, which brings out the brightness differences between adjacent image pixels, was performed through a linear contrast stretch by assigning each gray level encountered in the picture, pixel by pixel, to the entire brightness value range.[15] The effect of linear stretching was to enhance linear features that had been previously obscured.

Highpass filtering, a form of edge enhancement, was used to enhance lineaments by removing low-frequency components of the image that corresponded to uniform gray scale areas. Edge enhancement sharpened edges and therefore made lineaments more noticeable. Highpass filtering was accomplished by first finding an average value for every pixel in a 3-by-3 pixel area and then subtracting this value from the original pixel value.[15]

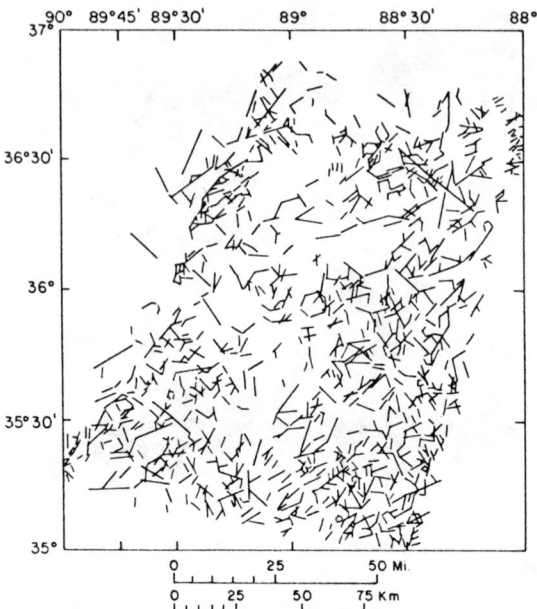

Fig. 3 Lineaments from Landsat imagery.

The contrast- and edge-enhanced images of spectral bands 5 (red reflectance) and 7 (near infrared reflectance) were determined to be the best. Both average image features and high-frequency details were more manifest than on any of the other image products. Band 7 images showed the sharpest detail because penetration of atmospheric haze was better in this part of the electromagnetic spectrum.

Lineaments shown by drainage and topographic alignments were relatively simple to identify and map. However, lineaments due to soil tone and subtle vegetation alignments were more difficult to interpret. The lineaments were identified as expressions of[14,15]: 1) straight segments of stream courses, or straight valleys; 2) edges of topographic highlands or aligned segments of them, such as bluffs, ridge crests, scarp edges; 3) dark or light vegetation zones, or vegetation borders; 4) linear soil tones, such as soil moisture zones and soil boundaries; and 5) lines parallel to lithological trends.

Linear features of cultural origin were eliminated by cross examination of color infrared aerial photography, at a scale of 1:120,000, and topographic maps of various scales. A zoom-transfer scope (ZTS) was used for correlating the Landsat scenes with the high-altitude color infrared photography or topographic maps. Figure 3 shows the interpreted lineaments in the study area.

REMOTE SENSING APPLICATIONS IN W. TENNESSEE

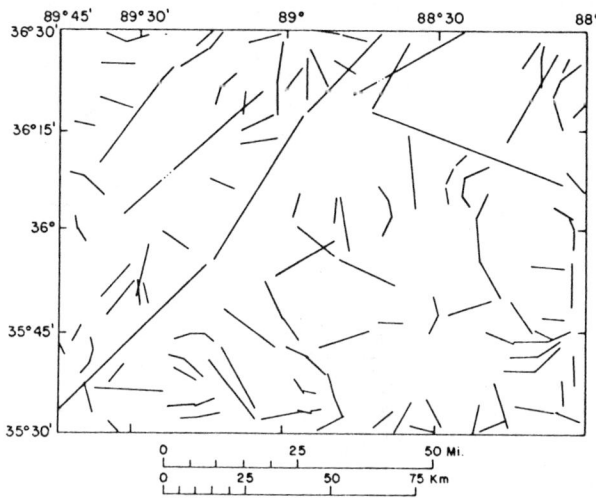

Fig. 4 Gravity lineaments for the study area.

In addition to the lineaments detected from Landsat images, gravity lineaments were mapped (Fig. 4) and analyzed from the gravity anomaly map corresponding to the study area.[15,16]

Comparison of Landsat and Gravity Lineaments

The area analyzed quantitatively was between 35°30' to 36°30'N, and 88°00' to 89°45'W. This was the same area for which the gravity map of West Tennessee was available. Histograms and rose diagrams of frequency data (total number of lineaments per area) and density data (total lineament length per area) were derived from the original data. The frequency distribution of lineaments was represented graphically using histograms.[15] The rose diagram provided a more convenient pictorial representation of the parameters of the histograms, i.e., azimuth class width and frequency distribution of lineaments as shown in Fig. 5. Five and nine degree azimuthal classes were employed for summarizing the lineaments.
Subsets of the original mapped area were also examined by partitioning the whole area into subregions, each approximately 2500 km^2, and then summarizing lineaments in each one of them separately. This partition resulted in splitting the original area into six subregions. Fig. 5a shows the Landsat lineament rose diagram for 20 azimuth classes. Fig. 6a shows the rose diagrams for the six subregions.
Quantitative summary of gravity lineaments was performed for the same areas and 20 azimuth classes as in the analysis of Landsat lineaments, so that the results

were comparable. The gravity lineament rose diagram for the study area is shown in Fig. 5b. Fig. 6b shows the rose diagrams of the gravity trends for the six subregions.

Results and Discussion

Examination of the histograms and rose diagrams for the Landsat and gravity lineaments yielded the following results[14]: 1) The main direction of both the Landsat and gravity lineaments is NE (Fig. 5a). 2) The secondary direction expressed in both of them is NW almost perpendicular to the first (Figs. 5a and 5b). 3) A close inspection of rose diagrams for the six subregions revealed a strong correspondence between the Landsat and gravity lineaments in the NE and NW trend (Fig. 6).

The foregoing results were compatible with the conclusions of other workers using different data, such as 1) large-scale black and white photography,[11] 2) shallow wells and local streams from topographic maps,[7] 3) magnetic, less detailed gravity maps and

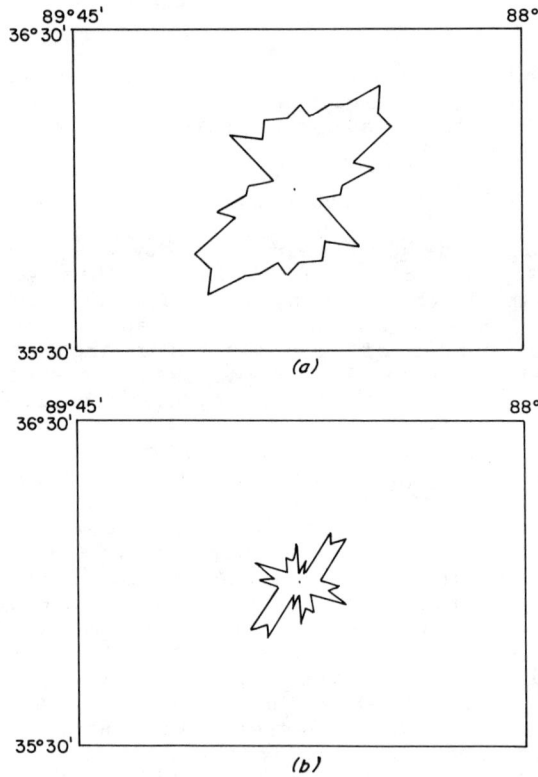

Fig. 5 Rose diagram plots in 20 azimuth classes for the area of study: a) Landsat lineaments, and b) gravity lineaments.

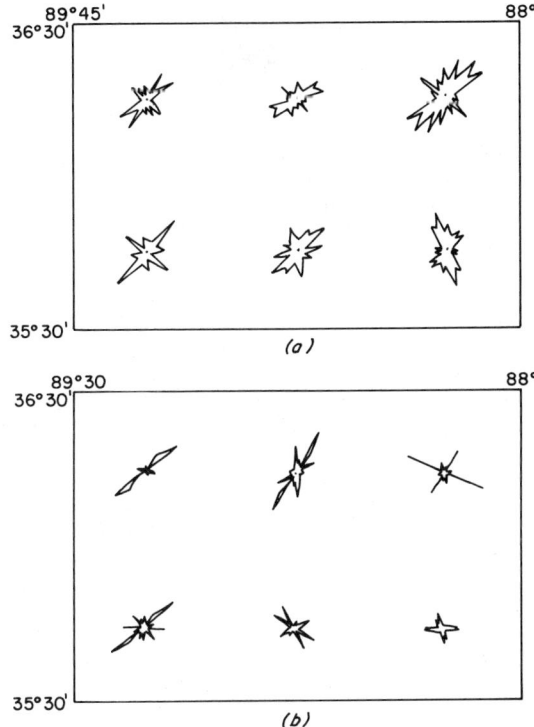

Fig. 6 Rose diagram plot in 20 azimuth classes for six subregions, each being approximately 2500 km^2: a) Landsat lineaments, and b) gravity lineaments.

earthquake epicenters,[5] and 4) magnetic, Landsat,[8] and SLAR imagery.[9]

The NE trend of the Landsat and gravity lineaments was parallel to the north-easterly trend of earthquake epicenters,[5] which coincides roughly with the axis of the basement trough, and the NE trends of hypothetical fault zones.[6] Thus, the embayment earthquakes could be genetically related to the lineament patterns. Presumably, the trends of the buried faults were expressed by the lineaments as well as the geophysical anomalies.

It was significant that the trends of lineaments were similar to fault trends interpreted by Stearns and Zurawski[7] and were roughly parallel to the axis of the embayment syncline. Stearns and Zurawski[7] used shallow drill holes together with trends of surface streams to map faults at the top Paleozic-base Cretaceous boundary and the shallow top of Porters Creek Clay (Paleocene). The present study supports their interpretation and was consistent with the notion that many, if not most, of the lineaments mapped here were faults extending downward through Paleocene and Cretaceous to focal depth. Parallelism of the lineaments (and faults) to the rift supports the rift model of Ervin and McGinnis[13] that was mapped by Hildenbrand et al.[5]

Conclusions

This study has demonstrated the utility of image interpretation of satellite data for geologic evaluations. It has been shown that important aspects of the geologic structure in West Tennessee could be inferred from the lineament analysis of Landsat images and gravity anomaly data.

Landsat imagery of all four seasons provided for the delineation of additional lineaments. Contrast- and edge-enhanced images have proven to be more effective, since more lineaments were identified on these than on the other image products. Landsat MSS bands 5 and 7 of the enhanced products were judged better than any other image set.

A detailed comparison was made between gravity anomalies and Landsat lineaments. Other comparisons were made between structure contour maps, faulting, and lineaments. Narrow zones of lineaments were observed to correspond (near the earthquake region) with independently drawn fault systems by others on the basis of drill holes and magnetic trends. The NE trend of the identified lineaments, which corresponded to these faults, was in accordance with reactivitation of the Reelfoot Rift near the Mississippi River. The NE lineaments could be concurrent to the development of the parallel embayment syncline.

The use of aerospace data has proven to be a valuable adjunct to traditional techniques. The synoptic view allowed detection of large area features, difficult to detect by ground-based measurements. The use of Landsat satellite data can potentially provide an additional source and different type of data for the detection of lineaments.

References

[1]O'Leary, D., Friedman, J., and Pohn, H., "Lineament, Linear, Lineation: Some Proposed New Standards for Old Terms," *Geological Society of America Bulletin*, Vol. 87, October 1976, pp. 1463-1469.

[2]Hodgson, R., Gay, S., and Benjamin, P., *Proceedings of the 1st International Conference on the New Basement Tectonics*, Utah Geological Association, Salt Lake, UT, 1974.

[3]Podwysocki, M., and Earle, J., *Proceedings of the 2nd International Conference on Basement Tectonics*, Basement Tectonics Committee, Denver, CO, 1979.

[4]O'Leary, D., and Earle, J., *Proceedings of the 3rd International Conference on Basement Tectonics*, Basement Tectonics Committee, Denver, CO. 1981.

[5] Hildenbrand, T., Kane, M., and Hendricks, J., "Magnetic Basement in the Upper Mississippi Enbayment Region: A Preliminary Report," Geological Survey Professional Paper 1236, 1982, pp. 39-53.

[6] Stearns, R., and Wilson, C., "Relationships of Earthquakes and Geology in West Tennessee and Adjacent Areas," Tennessee Valley Authority, Knoxville, TN, 1972.

[7] Stearns, R., and Zurawski, A., "Post Cretaceous Faulting in the Head of the Mississippi Embayment," *Southeastern Geology*, Vol. 17, June 1976, pp. 207-229.

[8] O'Leary, D., and Simpson, S., "Remote Sensor Applications to Tectonism and Seismicity in the Northern Part of the Mississippi Embayment," *Geophysics*, Vol. 42, April 1977, pp. 542-548.

[9] O'Leary, D., and Hildenbrand, T., *Structural Significance of Lineaments and Aeromagnetic Patterns in the Mississippi Embayment*, Publication No. 3, Basement Tectonics Committee, Denver, CO, 1981, pp. 305-313.

[10] Krinitzsky, E., "Geological Investigation of the Faulting of the Lower Mississippi Valley," Tech. Memo 3-311, U.S. Army Corps of Engineers, Waterways Experiment Station, Vicksburg, MS, 1950.

[11] Fisk, H., "Geological Investigation of the Alluvial Valley of the Lower Mississippi River," U.S. Army Corps of Engineers, Mississippi River Commission, Vicksburg, MS, 1944.

[12] Stauder, W., Herrmann, R., Chulick, J., Mascarenas, M., John, V., Leu, P., Shin, T., Yepes, H., and Finn, C., "Central Mississippi Valley Earthquake Bulletin," Quarterly Bulletin No. 39, Dept. of Earth and Atmospheric Sciences, St. Louis Univ, 1984.

[13] Ervin, C., and McGinnis, L., "Reelfoot Rift: Reactivated Precursor to the Mississippi Embayment," *Geological Survey of America Bulletin*, Vol. 86, 1975, pp. 1287-1295.

[14] Argialas, D., "Mapping and Comparison of Landsat Lineaments with Gravity Trends in West Tennessee," MS. Thesis, Univ. of Tennessee Space Institute, Aug. 1979.

[15] Argialas, D., Stearns, R., and Shahrokhi, F., "Mapping and Significant of Landsat and Gravity Lineaments in West Tennessee," *Journal of Aerospace Engineering*, American Society of Civil Engineers, Vol. 1, April 1988, pp. 174-187.

[16] Stearns, R., Keller, G., and Templeton, T., "Gravity Anomaly Map of Tennessee, West Sheet," Tennessee Division of Geology, 1980.

Satellite Technology in the African Center of Meteorological Applications for Development (ACMAD)

John A. Leese*
World Meteorological Organization (WMO), Geneva, Switzerland

Abstract

The African Centre of Meteorological Applications for Development (ACMAD), a cooperative located in Niamey, Niger, was designed to greatly enhance the capabilities of national meteorological services in Africa. The basic activities of ACMAD include the collection, processing, and analysis of meteorological data and the routine dissemination, to national meteorological services and other national and international organizations, of products applicable to a broad spectrum of human activities. Both meteorological and communications satellites will have significant roles in enabling ACMAD to collect input data from analysis and to transmit the output to users in the 50 countries of Africa. This paper outlines the history and structure of ACMAD and describes the essential role of satellite technology in the achievement of ACMAD's objectives.

Introduction

Mankind has always been influenced by the weather and climate, and this is especially the case in Africa. The ancient Egyptians were well aware of the seasonal flooding of the Nile, which was a large influence in their civilization. Throughout the centuries, man has constantly sought to acquire knowledge of these processes and to apply such knowledge to safeguard and improve his way of life. It is only relatively recently that he has been able to gain a deep enough understanding of atmospheric processes to enable him to forecast the weather in a skillful manner.

Meteorology is basically an observational science. Frequent and accurate measurements are made of various meteorological elements over large areas of the Earth's surface, and the rapid transmission of such data to meteorological centers and between centers of other countries is

---

Copyright © 1990 by the American Institute of Aeronautics and Astronautics, Inc. All rights reserved.
*Director, Institute for Naval Oceanography, Stennis Space Center, Mississippi.

needed. Progress in meteorology is only possible with international cooperation and standardization. This is one of the main trends of the inter-governmental World Meteorological Organization, a specialized agency of the United Nations, which in 1951 replaced the former nongovernmental body, the International Meteorological Organization.

The basic program of the WMO is the World Weather Watch (WWW) that provides for the constant exchange of weather data between all countries of the world, data on which all are dependent for the operation of their respective national meteorological services. One major development over the last 20 years has been vastly improved technology in the field of computers needed to process the enormous quantities of meteorological data to produce weather forecasts.

Another even more significant development has been in the field of outer space technology. Despite the widely acknowledged importance of observations, the global coverage of conventional surface-based meteorological observations has always been very uneven. The advent of meteorological satellites in 1960 provided a means for the first time of surveying the weather processes in the Earth's atmosphere on a broad scale.

Weather forecasting has therefore changed dramatically over the past 30 years, mainly because of developments in computers and satellites. At the start of the meteorological space age, we could provide useful forecasts one to two days ahead of time, but now with the wealth of high-quality satellite data, the arrival of supercomputers, and highly sophisticated numerical models, forecasts can be made with the same degree of accuracy up to five days in advance, and it looks probable that this can be extended up to two weeks by the turn of the century.

ACMAD as a Concept

As part of the World Weather Watch, the global data-processing system (GDPS) coordinates the data-processing activities of WMO members to make available all the processed information they require for both real-time and non-real-time applications in an effective and efficient manner. The GDPS is organized as a three-tier system, namely, world meteorological centers at the global level, specialized regional meteorological centers with geographical and/or activity specializations at the regional level, and national meteorological centers at the level of individual countries.

In recognition of the critical economic and social conditions in Africa and the important contribution of weather- and climate-related events to these conditions, the Economic Commission of Africa in April 1985 recommended the establishment of an African Centre of Meteorological Applications for Development (ACMAD) as a specialized/regional meteorological center within the WWW. It was anticipated that such a continental center would greatly enhance the capabilities of the respective national meteorological services, thereby making a valuable contribution to economic and social development. The basic activities of ACMAD were planned with this overall objective in mind. These activities include the collection, processing, and analysis of meteorological data and the routine dissemination to national meteorological services and other national and international organizations of output (information and the results of applied research) applicable to a broad spectrum of important weather- and climate-related human problems.

The specific objectives of ACMAD can be summarized as follows:

1) To produce more reliable and useful meteorological and climatological information (data and forecasts) through improved data networks and collection, processing, and analysis schemes.

2) To strengthen the capabilities of national meteorological services by developing various transferable methodologies for analyzing and modeling meteorological data and by contributing to the professional training of personnel.

3) To provide meteorological and climatological data and forecasts related to various specific weather- and climate-related events and, in particular, potential applications to both national and regional institutions.

4) To undertake applied meteorological research to enhance the understanding of relationships between weather and climate conditions and essential human activities, and to provide a sound basis for developing more useful products for application to important weather- and climate-related problems.

The establishment of ACMAD is expected to promote and strengthen technical cooperation among African countries, thereby enhancing their efforts to minimize the adverse effects of weather and climate events and to increase agricultural and industrial production. It will also provide a mechanism for improving the exchange of meteorological information among national meteorological services and for facilitating cooperative activities between the African continent and other regions of the world. ACMAD is designed to complement existing meteorological facilities and programs at the regional and national levels in Africa, including the AGRHYMET program and regional drought monitoring centers and regional area forecast centers. It will work in close cooperation with international agencies such as WMO, the Food and Agricultural Organization (FAO), the United Nations Environmental Program (UNEP), and the United Nations Education, Science and Cultural Organization (UNESCO).

ACMAD's services are expected to yield output of four principal types: 1) products associated with a meteorological/climatological watch that require routine dissemination; 2) output of technical reports derived from various applied meteorological/climatological studies; 3) techniques and methods from development work related to meteorological forecasting; and 4) output related to various training activities. A list of the anticipated output of each type is shown in Table 1.

The users of output from ACMAD will span a wide range of organizations. It was noted earlier that ACMAD is expected to greatly enhance the capabilities of the respective national meteorological services that will be the primary users of output from ACMAD. Another set of users will be the various international organizations and regional centers in Africa.

These national meteorological services and other organizations will pass on the information to a number of

Table 1--Preliminary list of ACMAD's products and services

Routine outputs: Meteorological/ climatological watch	1)	Daily Rain system bulletin Tropical cyclone watch Numerical diagnostic/ prognostic charts Contribution to pest watch
	2)	Weekly and/or 10-day periods Drought watch bulletin Precipitation accumlation charts Agrometeorological assessment bulletin
	3)	Monthly and seasonally Climatological charts Monthly outlooks Seasonal outlooks Environmental monitoring

(Table 1 continued on next page.)

Table 1--Preliminary list of ACMAD's products and services (Cont'd)

Meteorological applications output	1) Precipitation atlas (Africa) 2) Wind/radiation potential and evaporation studies 3) Seasonal climatic anomalies forecasting models 4) Drought characteristics 5) Meteorological input to crop yield forecasting models 6) Meteorological input to soil moisture estimation models 7) Agroclimatic atlas (Africa) 8) Meteorological/climatological data for specific applications 9) Meteorological/agrometeorological data bank 10) Experimental evaluation of use of ACMAD's products
Development output	1) Improved techniques for meteorological forecasting 2) Meteorological forecasts tailored to user needs 3) Depository library
Training output	1) Organization of training events and conferences 2) On-the-job training of senior staff and scientists 3) Support to regional training centers 4) Training materials (manuals, case studies, etc.) 5) Technical instruction manuals

governmental and nongovernmental organizations. Table 2 presents a list of the potential users of ACMAD products and services.

Role of Satellites in ACMAD

ACMAD will depend very heavily on satellites to provide input data for analysis and to transmit the output to users in the 50 countries of Africa. Meteorological satellites will play the major role in providing data and communication services. Communications satellites will also have a significant role in the collection and exchange of conventional observations plus the distribution of output from ACMAD.

Meteorological Satellites

<u>General Characteristics</u>. Meteorologists all over the world are now able to take advantage of a wealth of observational data and communications-related services flowing from specially equipped and highly sophisticated satellites. Several factors make meteorological satellite data unique compared with data from other sources, and it is worth noting a few of the most important:

1) Because of its high vantage point and broad field of view, a satellite can provide a regular supply of data from those areas of the globe yielding very few conventional observations.

2) The atmosphere is broadly scanned from satellite altitude and enables large-scale weather systems to be seen in a single view.

3) The ability of certain satellites to view a major portion of the atmosphere continually from space makes them particularly well-suited for the monitoring and warning of short-lived storms.

4) The advanced communications systems developed as an integral part of the satellite technology permit the rapid transmission of data from the satellite, or their relay from automatic stations on Earth and in the atmosphere, to operational users.

These factors are incorporated in the design of meteorological satellites to provide data and services through three major functions:

1) Remote sensing of spectral radiation that can be converted into meteorological measurements such as cloud cover, cloud motion vectors, surface temperature, vertical profiles of atmospheric temperature and humidity, snow and ice cover, ozone, and various radiation measurements.

2) Collection of data from in-situ sensors on remote field or mobile platforms located on the Earth's surface or in the atmosphere.

Table 2--Potential users of ACMAD's services

National and international organizations	National meteorological services in Africa
	Regional meteorological centers in Africa (e.g., AGRYHMET, drought (monitoring centers)
	Other regional centers and programs in Africa (e.g., agricultural research institutes)
	International organizations and agencies in Africa (e.g., WMO, FAO, UNEP, UNESCO)
	International organizations and centers outside of Africa (e.g., global and regional meteorological centers)
End-users (reached through national meteorological services and other organizations)	Governmental planners and policy makers
	Economic development planners
	Public health and safety officials
	Food relief organizations
	Agricultural extension agents
	Pest management agencies
	Farmers
	Fishing industry officials and fishermen
	Water and reservoir planners and managers
	Air traffic controllers
	Maritime shipping planners and navigators
	Energy supply-and-demand planners
	Construction contractors
	Industrial plant managers
	General public (commuting, recreation, etc.)

3) Direct broadcast to provide cloud cover images and other meteorological information to users through a direct readout station.

The operational satellite capability has evolved during the past 30 years. The 1960's witnessed the birth of the meteorological satellite as an unprecedented tool for observing broad-scale atmospheric phenomena. By the end of that decade, the meteorological satellite had grown to a highly sophisticated platform that could provide global coverage of cloud observations and was beginning to provide quantitative measurements of pertinent meteorological parameters. During the 1970's, a cooperative international network of meteorological satellites evolved. This effort culminated in the contribution to the Global Weather Experiment (FGGE) by a nearly complete global network of meteorological satellites in terms of sensor data and services. A more intensive effort now exists in the processing and applications of satellite data to increase the information obtained.

Meteorological Satellite Systems. The thrust of the current generation of meteorological satellites primarily is characterizing the kinematics and dynamics of the atmospheric circulation. The ability to achieve such objectives was demonstrated during the FGGE in 1979. This capability is now part of the global operations of the WWW.

The existing network of meteorological satellites, forming part of the Global Observing System of the WWW, regularly produces real-time weather information. This is acquired several times a day through direct broadcast from the meteorological satellites by receiving stations located in at least 125 countries. There are two major components to the current meteorological satellite network. One element is the various geostationary meteorological satellites, which operate in an equatorial belt and provide a continuous view of the weather from roughly $70^{\circ}N$ to $70^{\circ}S$. At present, there is a satellite at 0° longitude (operated by the European Meteorological Satellite Organization), a satellite at $74^{\circ}E$ (operated by India), a satellite at $140^{\circ}E$ (operated by Japan), and satellites at $135^{\circ}W$ and $75^{\circ}W$ (operated by the United States). The Soviet Union plans to operate a satellite at $76^{\circ}E$. All the present geostationary satellites collect data in the visible portion of the spectrum and in the infra-red "window" from 10.5 to 12.5 m. Several also collect data in other spectral intervals, especially in the water-vapor interval from 5.7 to 7.1 m. Spatial resolution varies among the different spacecraft, ranging from 1 to 5 km in the visible and from 5 to 11 km in the infra-red.

The second major element is the polar-orbiting satellites operated by the Soviet Union and the United States. The "Meteor-2" series has been operated by the Soviet Union since 1977. The polar satellite operated by the United States represents the evolutionary development of the TIROS satellite, first launched in April 1960. The NOAA series, based on the TIROS-N system, has been operated by the United States since 1978. These spacecraft provide global coverage and fly at altitudes of 850 to 900 km. Data acquired by the various sensors are handled in two ways by the on-board processors of the polar-orbiting satellites. Since the spacecraft is in view of ground stations no more than once per orbit (and sometimes not at all for several orbits), it is necessary to record the data on tape for later playback when the spacecraft comes in view of a ground station. At the same time that the data are being recorded, the spacecraft is broadcasting the data for the use of any properly equipped receiving station within range.

METEOSTAT. The METEOSAT satellite will undoubtedly be one of the most important satellite systems in providing input data and communication services for ACMAD. This satellite is the space segment of the Meteosat Operational Programme, executed by the European Space Agency (ESA) on behalf of the European Meteorological Satellite (EUMETSAT) Organization. METEOSTAT is a spin-stabilized satellite in geostationary orbit at an altitude of 35,800 km. The satellite is located over the Gulf of Guinea, at the crossing of the Equator and the Greenwich meridian ($0°N$, $0°E$). A series of spacecrafts are either in orbit or scheduled to be launched on a time scale to provide operational satellite continuity at least through 1995. EUMETSAT is now working with ESA to define the Meteosat second generation, which will continue the Meteosat Operational Programme after 1995.

The principal payload of the satellite is a multispectral radiometer. This instrument provides the basic data of the Meteosat system, namely, visible and infra-red radiances that are used to create images of the full Earth's disk as seen from geostationary orbit.

The three-channel radiometer includes 1) A visible channel comprising two identical visible detectors in the 0.5 to 0.9 m spectral region; 2) A thermal infra-red (window) channel in the 10.5 to 12.5 m region; and 3) An infra-red (water-vapor) channel in a spectral region characterized by water-vapor absorption (5.7 to 7.1 m).

Each infra-red image is composed of 2500 lines containing 2500 picture elements (pixels) with a spatial resolution of 5 km at the sub-satellite point (SSP). The visible image contains 5000 lines of 5000 pixels when both visible detectors operate simultaneously, or 2500 duplicated lines of 500 pixels when only one visible detector is in operations. Earth images are generated at half-hour intervals. The best area for the qualitative use of the images extends to about 65 deg great circle arc (GCA) from the SSP, but for quantitative use (extraction of meteorological products), this area is restricted to 55 deg GCA from the SSP.

One of the objectives of the Meteosat system is the extraction and distribution of meteorological parameters from the basic image data. The data products that are routinely distributed over GTS are 1) cloud motion vectors (CMV); 2) sea-surface temperatures (SST); 3) cloud analysis (CA); and 4) upper tropospheric humidity (UTH). A fifth product, cloud top height (CTH), is distributed via the satellite in the form of a WEFAX chart. These products, extracted by a fully automated set of software, undergo interactive quality control by meteorologists before the results are coded and distributed.

A major function of the Meteosat system is the dissemination of Earth images and other meteorological information by using the satellite as a relay. The transmitted data include processed Meteosat images in a variety of formats, as well as derived cloud top height maps and conventional meteorological charts received from the Deutsche Wetterdienst in Offenbach, Federal Republic of Germany. Images of the western Atlantic and the Americas, generated by the GOES satellite and reformatted by the Centre de la Meteorologie Spatiale (CMS) at Lannion in France, are also transmitted via Meteosat.

All these data are relayed via two dedicated dissemination channels to user stations located within about 75 deg of the SSP. At present, the channel operates in the following frequency bands:

Channel 1   1694.5 MHz   (WEFAX broadcast)
Channel 2   1691.0 MHz   (high-resolution digital)

Two forms of transmission are used: 1) Analogue (WEFAX), received by secondary data user stations (SDUS) on channels 1 and 2, and 2) High-resolution digital, received by primary data user stations (PDUS) on channel 2.

A SDUS in its simplest form consists of an antenna and front-end receiver and some form of recording device. This could be a recorder producing pictures by one of a number of processes including the use of electrolytic paper, electrostatic, photographic, or laser beam recording techniques. The possibility also exists for adding tape recorders to store images and for a certain amount of image processing by the addition of an analog-to-digital converter and a small computer system. This configuration, with the addition of bit-and-frame synchronizers and a microprocessor, would also enable the user to receive directly retransmitted data collection platform messages.

Low-resolution WEFAX transmissions are in a format basically compatible with those of other meteorological satellites, including the automatic picture transmission (APT) of the polar-orbiting satellites and transmissions from other geostationary satellites. The image data are transmitted in WEFAX formats after image processing in ESOC has been completed. Images are cut into convenient areas and have latitude and longitude grids and coastlines superimposed before transmission.

A PDUS is designed for the computer processing of high-resolution digital image data. A basic station comprises an antenna and front end (preamplifier and down converter), receiver, and frame synchronizer followed by a processing computer and display system.

A data collection system (DCS) is used to obtain data from in-situ platforms, on schedule or on command. Repeated environmental measurements are needed from locations that are difficult, hazardous, or expensive to access frequently. Measurements also are needed from locations not served by existing ground communication networks. The DCS service solves many of these problems by enabling rapid user access to measurements made by remotely located data collection platforms (DCP's).

Meteosat has a total of 66 telecommunication channels dedicated to the collection of environmental data from automatic or semiautomatic DCP's, which may be located at any point within the Meteosat coverage area. Half these channels are presently assigned for regional purposes, that is, for DCP's that operate exclusively with the Meteosat system, and the remaining channels are part of the international data collective system (IDCS) and are reserved for mobile DCP's likely to move through the coverage areas of any of the goeostationary meteorological satellites. Meteosat simply acts as a relay so that environmental data transmitted by the DCP are received in the central processing facility and then distributed to the users either by mail, in the form of telex messages, via the GTS, or by direct transmission via the satellite, in which case the messages are interleaved between WEFAX image formats.

DCP types fall into two distinct categories: 1) Self-timed DCP's, which transmit their reports automatically within preset time and frequency slots and are driven by a stable internal clock, and 2) Alert DCP's, which transmit a message only when a parameter value exceeds a preselected threshold.

Meteosat retransmits DCP messages, which enables a modified SDUS to receive selected DCP reports disseminated via the satellite as a 12.5 kbits/s data stream, during the pause that exists between two consecutive WEFAX broadcasts. The modulation selected permits the use of existing WEFAX receivers, which only have to be slightly modified by increasing the intermediate frequency bandwidth from 30 to 37 KHz. To complete the processing of the data, a bit synchronizer, a frame synchronizer, and a microprocessor-based processing system are added to the user's ground station equipment.

The meteorological data distribution (MDD) mission, which will be incorporated in the operational series of METEOSAT satellites, is primarily concerned with the transmission of meteorological information to Africa and the Middle East. The system for the MDD mission consists of: 1) Two uplink stations, each transmitting one channel of data in the 2106 MHz band, but each capable of uplinking the other channel; 2) Two downlink channels (A and B) from the satellite and operating in the 1695 MHz band; 3) The transmission rate of each channel will be 2400 bits/s; and 4) Data will be received by relatively low-cost ground stations with radio-frequency characteristics similar to those of a SDUS.

MDD channel A will be used primarily to disseminate digital facsimile data bulletins, and MDD channel B will be used for alphanumeric data bulletins (e.g., selected SYNOP's and TEMP's, SATOB's, SATEM's, ASDAR, ASAP, DCP messages, etc.). The basic format of message bulletins on both channels will be the same.

The use of standard message bulletins means that both types of data can be transmitted on either channel. Therefore, in the event of the breakdown of one of the uplink stations, a backup program comprised of interleaved channel A and B data can be provided by the other station.

Stations for the reception of MDD data will be designed to receive at least channel A and B data. Each station will require a parabolic antenna, low noise amplifier, down converter, and multichannel receiver. The processing of the incoming data streams will be performed using a microprocessor to decode, recreate messages or charts, select, display, and/or store data.

<u>Meteorological Satellites in Polar Orbit</u>. In the more than 25-year history of meteorological observations from space, the polar-orbiting satellites were the first spacecraft to provide data and services for weather analysis and forecasting. The present polar-orbiting meteorological satellites flying at relatively low altitudes (850 to 900 km), with inclinations from the equatorial plane of 80 to $100°$, provide two significant capabilities. First, due to its near-polar orbit, the spacecraft is able to acquire data from all parts of the globe in the course of a

series of successive revolutions. With its low altitude, on-board instruments of the polar orbiter can acquire data of higher spatial resolution than the high-altitude satellites in geostationary orbit.

The polar-orbiting satellites are principally used to obtain: 1) Daily global cloud cover [satellite pictures taken in visible (VIS) and infra-red (IR) portions of the spectrum]; and 2) Daily global sea-surface temperature (SST), vertical temperature, and water-vapor distributions in the atmosphere (quantitative satellite soundings).

Another capability of some spacecraft of the polar-orbiting meteorological satellite system is data collection and platform location. This DCS allows the spacecraft to receive data from stationary or moving platforms anywhere on the surface of the Earth or floating in the atmosphere. In addition, the position of moving platforms (e.g., drifting buoys in the oceans or balloons in the atmosphere) can be determined on the basis of a Doppler shift in the radio frequency received at the satellite. This data collection capability provides, from in-situ sensors, data that are not recoverable by satellite remote sensing such as surface wind and pressure, rainfall amount, river levels, sea salinity, subsurface oceanic temperatures, and many others. Changes in the platform location over time permit the determination of fluid motion such as ocean currents or atmospheric winds.

The data received from polar-orbiting satellites are forwarded to a central processing facility where a variety of products are produced. Most of these products are then made available to the world meteorological community via the global telecommunication system (GTS). The data acquired by the polar-orbiting satellites are also entered into one of the archives associated with the world centers of the WWW.

Automatic picture transmission (APT) services are provided by the polar-orbiting satellites. Data from two spectral channels are transmitted in a VHF band as the satellite orbits the Earth. The resolution of both visible and infra-red APT data is 4 km. The signals can be received on very low-cost equipment and displayed as an image or converted into digital products showing cloud, ocean, or land temperature values.

Like APT services, high-resolution picture transmission (HRPT) services are provided continuously from polar-orbiting satellites. Nevertheless, there are several differences between the services: HRPT provides five rather than two spectral channels with a resolution of visible images of 1.1 km instead of 4 km. The HRPT data must be sent on a broader bandwidth frequency, and consequently, much more expensive equipment is needed to receive and process the signal. This primary application is to provide more sophisticated observations of meteorological, oceanographic, and geophysical phenomena than can be obtained with APT readout. Some specific HRPT applications include the monitoring of pastoral areas for plant growth and development in Africa, the observation of surface currents in the world's oceans, and the study of desert locust breeding grounds in the Mediterranean.

Communication Satellites

Communication satellites used in WMO programs are of two types. Most widely used are those operated in common-carrier service for point-to-point communication, such as the INTELSAT's. There is a growing number of regional and national communication satellites as well in this category. Growing use is being made of these satellites in the GTS, resulting in both the increased reliability and rapidity of data exchange. The second class of satellites is those that provide specialized communication services, most notable of which is INMARSAT for communication to and from ships at sea.

Some ACMAD Benefits

It is readily apparent that ACMAD must make efficient and effective use of available space applications products and services in order to successfully achieve its objectives. ACMAD success should benefit all countries in Africa in a number of significant ways.

ACMAD will greatly enhance the use of meteorological data and information in member states, as well as stimulate the development of the regional network of meteorological telecommunications through the use of satellite telecommunications. This will not only be limited to meteorological applications but in all areas of satellite telecommunications that will contribute to the promotion of technical cooperation in the areas of production, distribution, and utilization of natural resources for the social and economic development of member states.

In the broad sense, ACMAD will help in the coordination of activities aimed at promoting regional intergration in Africa. It is envisaged that ACMAD will back up subregional and national meteorological applications capabilities by providing products that will assist national and subregional institutions in improving their climate and weather forecasting and other early-warning capabilities, and developing national manpower capabilities in activities that will ensure the effective application of meteorological data and information in production, particularly in the rural areas that are usually the first and worst hit by the vagaries of weather and climate.

ACMAD will also promote the exchange of information among farmers in similar production zones and the sharing of experiences on techniques and technologies in increasing agricultural production. In fact, agriculture accounts for a major share of the economy of Africa since 70% of the total population is rural. Of the entire land surface, only 16% lends itself readily to cultivation; most of the land can only be cultivated with difficulty and is particularly vulnerable to the vagaries of weather. The improved knowledge of weather phenomena and hence its forecasting and applications can and should help in making the best use of the land available.

It is important to remember the contribution of meteorology to the saving of human life and to obtaining optimal use of the investments that are required in order to achieve greater safety. ACMAD should make it possible to improve the quality of flood forecasts and the monitoring of tropical cyclones. A timely alert makes it possible to bring populations into safe shelter and thus avoid unnecessary loss of life. Statistical information on these phenomena (trajectories, threatened areas, intensity, etc.) would make it possible to adapt protective measures (building of dams and dykes, reinforcement of buildings and structures, etc.) to the need.

ACMAD, through its contribution to improving the assessment of water sources, will permit a more rational use of available water, leading to an increase in irrigated surfaces and more efficient management of farmland. It will also contribute to an increase in livestock production. Through improvements in forecasting and rapid warning, it will help to lay the groundwork for improved protection against natural disasters and will constitute an important factor in the reduction of food losses prior to, during, and after harvesting.

Through a better understanding of meteorological phenomena, better forecasting, and the application of such knowledge to agrometeorology, ACMAD will therefore be an invaluable tool in the improvement of agricultural technologies, making it possible for all nations to obtain hoped for self-sufficiency in food production by the end of the century.

The preceding discussion has restricted itself to furnishing a few examples to illustrate the advantages that can be derived from the continued and efficient operation of a center having the dual objectives of improving the effectiveness of the work of national meteorological services on the one hand, and furnishing them with increasingly more efficient means on the other, which they will then adapt to their particular needs.

It is, however, necessary to take into account the whole range of benefits to be gained from technologies developed by ACMAD and applied with its assistance, not only with respect to the attainment of self-sufficiency in food production but also with respect to the struggle against desertification and for a more rational use of water resources. This will lead, in turn, to an extension of the cultivated surfaces and to the development and use of new energy sources. All these activities contribute to environmental protection and to a more efficient use of national resources for the safety and well-being of present and future generations.

Acknowledgments

This paper was compiled from a number of WMO documents that either formed the basis for ACMAD as an institution for assisting all of Africa or provided recent information about the pertinent meteorological satellite systems that provide data and services over the geographical areas needed by ACMAD. The planning for ACMAD has involved many individuals in the WMO Secretariat and WMO members. The author gratefully acknowledges the work of all these people in making this paper possible.

Hydrologic Assessment of Critical Erosion Areas Using Satellite Data and a Geographic Information System

K. M. Morgan* and L. W. Newland†
Texas Christian University, Ft. Worth, Texas
and
S. A. Hayes‡
Water Utilities, City of Weatherford, Texas

Abstract

Sediment is the largest single source of pollution by volume in surface waters. Transported sediment increases the cost of water treatment, fills reservoirs, contributes to eutrophication, and erodes farm areas of valuable top soil. Proper watershed management practices provide a means for reducing the potential for erosion. However, areawide erosion is difficult to identify because sediment is removed from extensive source areas, intermittent in movement, related to climatic events, and affected by constantly changing land-use practices. Remote sensing along with other ancillary information can now be utilized in a geographic information system (GIS) to predict areawide soil losses and locate critical erosion sites within a watershed. This paper discusses the method involved so that developing countries might implement a low-cost program to monitor and protect their valuable soil resources.

Introduction

Areawide erosion is difficult to identify and monitor because of the diffuse nature of sediment pollution and the constantly changing land-use conditions in a watershed. Traditionally, field studies are used to provide detailed data concerning the erosion potential of specific areas. This method, although accurate, is costly and time-consuming.[1]

The development of the universal soil loss equation (USLE) provides a way to quantitatively predict long-term annual soil erosion from rill and sheet sources.[2,3] Although primarily developed for site-specific analysis of soil loss, the USLE does have applications in larger areas[4] and has

---

Copyright©1990 by the American Institute Aeronautics and Astronautics Inc. All rights reserved.
*Director, Center for Remote Sensing and Associate Professor of Geology.
†Director, Environmental Sciences Program, Professor of Biology and Geology.
‡Environmental Engineer.

proven to be a valuable tool in targeting existing and potential erosion problems.[5-6]

Remote sensing data, along with other ancillary information, can be used in conjunction with the USLE to predict area-wide soil losses and identify critical areas of erosion.[7] Remote sensing data provide a permanent record that can be used to detect and monitor land-use changes affecting the erosion potential of a watershed.[1] Ancillary data, such as climatic trends, soil maps, and topographic relief, can be combined with remote sensing data to form a valuable database for erosion predictions.[8] The development of large databases ultimately leads to the need for a geographic information system (GIS).

In recent years, there has been a dramatic growth in the interest and use of GIS's. Most agree that a computer-oriented GIS should have four functions[7] as diagrammed in Fig. 1:

1) data input/collection;
2) storage/retrieval;
3) manipulation/analysis;
4) and, display/output

Important features of a GIS include the ability to convert analog information into a digital format and to incorporate data from disparate sources such as imagery and soil maps. These data can then be digitized, stored, manipulated, and displayed as graphs, maps, or color images that meet a specific need. The computer provides an efficient method of

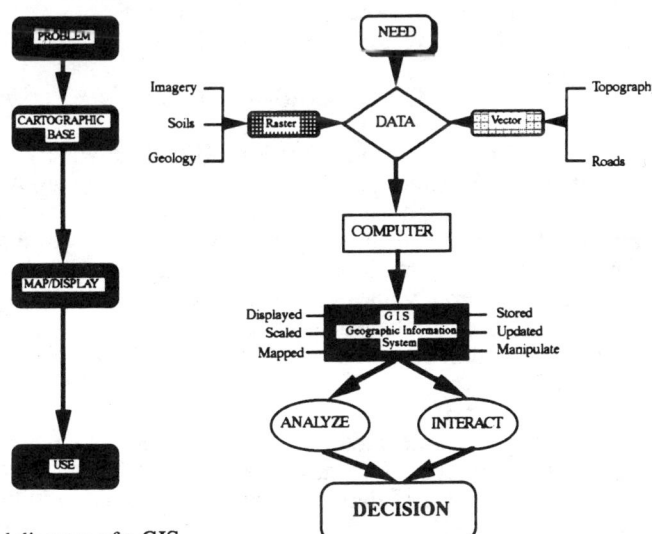

Fig. 1 Conceptual diagram of a GIS.

handling the voluminous data and analytical capabilities that are required.[9] Government-planning agencies have a need for the development of such systems to deal more efficiently with issues. Computer-oriented GIS models and satellite imagery provide a versatile method for keeping track of constantly changing conditions (i.e., land-use) in large areas such as watersheds.

Remote sensing is utilized by many disciplines for a variety of uses such as crop inventory, water quality work, vegetation mapping, and monitoring land-use changes.[10] The nature of the data enables the user to alleviate many of the problems normally associated with changes in land-use classification by providing up-to-date coverage of an entire watershed.[1] Landsat and SPOT data offer a means for the continual monitoring of urban and rural land-use changes, which is a notable advantage over analog map documents. Remotely sensed data also offer great promise in providing numerous modeling inputs, this due in part to the digital format of the satellite data.[11]

This paper describes how remote sensing data can be integrated into a GIS to calculate areawide soil losses and identify critical erosion sites within a watershed.

Setting Up the USLE/GIS Database

Areawide soil loss from sheet and rill (field) erosion for a watershed is usually calculated using the universal soil loss equation (USLE):

$$A = R \times K \times LS \times CP$$
$$A = \text{soil loss (tons/acre/year)}$$
$$R = \text{rainfall factor}$$
$$K = \text{soil erodibility factor}$$
$$LS = \text{slope factor}$$
$$CP = \text{land-use factor}$$

The universal soil loss equation (USLE) is widely used for targeting potential soil erosion problems and establishing allowable and actual sediment loading for a particular area.[11] In the equation, A represents the average tons of soil lost per year from each acre in a watershed. This equation involves numerous factors relating to the potential for erosion in an area. In general, values for the factors in the equation are taken from charts and tables published from field research. In the United States, the Soil Conservation Service has published much of the data needed to solve the USLE.

For a given watershed, the factors R, K, and LS essentially remain constant. However, CP may change from year to year as land-uses change. Landsat TM data have been used successfully to interpret the land-use/cover (CP) factor.[12] Remote sensing procedures can help alleviate the problem of monitoring land-use changes by providing up-to-date coverage of entire watersheds.[1]

A GIS can be utilized to store and analyze the USLE factor values. R, K, LS, and CP factor values are determined throughout the watershed from maps and imagery and entered into grid cells (raster format) as separate themes or files. The GIS is then used to calculate soil loss A on a cell-by-cell (pixel-by-pixel) basis [6] throughout the watershed. Critically high erosion areas can be identified numerically or by color coding.

Determining the Rainfall Factor R

In the USLE, the rainfall factor R represents the erosive force for average annual precipitation and can be determined for any area.[13] R is an index of the erosive potential of rainfall and represents the erosive force for the expected average yearly precipitation. It is a function of the kinetic energy of storm rainfall and rainfall intensity and is usually derived from 100 years of data. In the United States the R factor is obtained from isoerodent maps.

Determining the Soil Erodibility Factor K

The inherent susceptibility of a given soil to water erosion is quantitatively expressed by the soil erodibility factor. K values are based on actual measurements from selected natural sites over extended periods of time and are taken from published soil survey reports or calculated in the field.[14] Typical values range from 0.2 to 0.6. The higher the value, the more susceptible the soil is to erosion. Once the K values are extracted from published data tables or maps, the values are entered into the GIS database as a soil erodibility theme (or layer).

Determining the Slope Factor LS

Slope length L and percent slope S are important factors in sediment mobilization. These factors account for more variation in areawide erosion than any others in the USLE.[14] The longer and steeper the slope, the higher the potential for erosion. This factor is determined by

1) calculating soil lengths and gradients from topographic maps;
2) utilizing an LS conversion chart (94);
3) entering the LS factors into the pixel (raster) database.

Determining the land-use Factor CP

As mentioned earlier, the factors R, K, and LS essentially remain constant over time. However, land-use practices often change from year to year. Determining existing land-uses and then monitoring the changes can

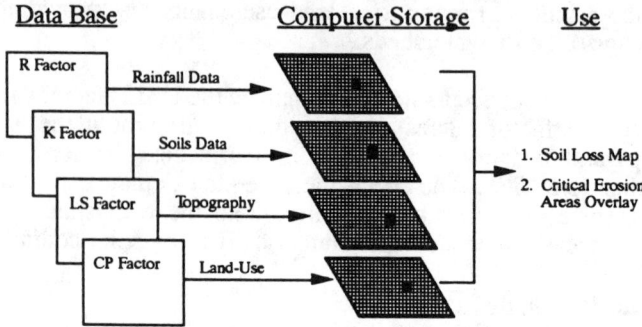

Fig. 2   Computer Integration of USLE into a GIS.

be accomplished by utilizing repetitive satellite data.[1] Designated land-use categories are classified and mapped from satellite imagery (or aerial photography). Actual CP factor values for each land-use category are obtained from published SCS tables and are then entered into the database. The steps necessary to construct the CP data file are the following:

1) acquire remote sensing data (MSS, TM, or SPOT);
2) designate "training sites" for various land-uses and conduct a classification;
3) produce a color-coded land-cover/CP factor map as a file in the GIS.

Estimating Areawide Soil Loss A

At this point, all the necessary data for soil loss calculations have been entered into the computer on a pixel-by-pixel basis. The pixel size is often determined by the satellite resolution cell, i.e., 80 m. for MSS, 30 m. for TM, and 20 m. for SPOT. The four primary data sets produced for the watershed are:

1) rainfall data (R);
2) soil erodibility data (K);
3) slope length/gradient data (LS); and
4) land-use/cover data (CP).

The watershed is now viewed as four layers of grid cell (raster) information (Fig. 2). Each pixel contains a discrete numerical value for rainfall R, erosion K, slope LS, and land-use/cover CP.

Average annual soil loss can be calculated using the USLE (A = R x K x LS x CP) and the GIS database to create a resulting layer called the watershed soil loss map. This new layer can be color-coded to correspond to soil loss estimates calculated for each pixel. Critically high erosion areas (pixels) can then be easily spotted by color (red is often used). The critical erosion sites can be extracted as an overlay and transferred onto the imagery so that exact locations within the watershed can be located.

Summary and Conclusions

The quality, quantity, and timing of watershed runoff are strongly dependent on land-use/cover. Satellite data integrated into a USLE/GIS can be extremely useful in handling the large volume of information deeded to calculate areawide soil losses. Traditional time-consuming, labor-intensive field work is replaced with satellite imagery, maps, and computer processing. Four data files constitute the land information base which include rainfall, soil erodibility, topography, and land-use. Over the last few years the authors have been able to conclude the following about the watershed USLE/GIS:

1. A USLE/GIS can be easily constructed to store, merge and calculate areawide soil losses.

2. Satellite data can be used to accurately map existing land-uses and monitor changes.

3. A USLE/GIS is a useful way to handle the large volume of data needed for watershed soil loss studies.and provides a permanent storage and retrieval system for erosion information.

4. Critical erosion areas can be easily located for immediate attention.

Watershed erosion monitoring provides an example of the evolution of environmental management programs utilizing a GIS. Monitoring and updating constantly changing land-use patterns is greatly simplified with the aid of remotely sensed data and a GIS. Once data are entered into a system, retrieval and manipulations are easily performed and new soil loss values immediately calculated. Various land-use management procedures may be applied to the individual cells, or pixels, to determine the most appropriate course of action. This allows the user to maximize benefits while minimizing cost.

References

[1] Morgan, K.M., and Nalepa, R., "Application of Aerial Photographic and Computer Analysis to the USLE for Areawide Erosion Studies," *Journal of Soil and Water Conservation*, Vol. 37, No. 6, Nov-Dec. 1982, pp. 347-350.

[2] USDA Soil Conservation Service, *Technical Guide Data for Estimating Soil Losses on Cropland*, Orono, ME, 1975.

[3] Wischmeier, W.H., and Smith, D.D., "Predicting Rainfall Erosion Losses, a Guide to Conservation Planning," *Agriculture Handbook. No. 537*, U.S.D.A., Washington, DC, 1978.

[4] Snell, E.A., "Regional Targeting of Potential Soil Erosion and Nonpoint-Source Sediment Loading," *Journal of Soil and Water Conservation*, Vol. 40, No. 6, Nov.-Dec. 1985, pp. 520-524.

[5] Pelletier, R.E., "Evaluating Nonpoint Pollution using Remotely Sensed Data in Soil Erosion Models," *Journal of Soil and Water Conservation*, Vol. 40, No. 4, July-Aug. 1985, pp. 332-335.

[6] Maas, R.P., Smolen, M.D., and Dresing S.A., "Selecting Critical Areas for Nonpoint-Source Pollution Control," *Journal of Soil and Water Conservation*, Vol. 40, No. 1, Jan.-Feb. 1985, pp. 68-71.

[7] Johannsen, C.J., and Barney, T.W., "Remote sensing applications for resource management," *Journal of Soil and Water Conservation*. Vol. 36, No. 3, May-June 1981, pp. 128-132.

[8] Walsh, S.J., "Geographic Information Systems for Natural Resources Management," *Journal of Soil and Water Conservation*, Vol. 40, No. 2, March-April 1985, pp. 202-205.

[9] Berry, J.K., "Computer-Assisted Map Analysis: Potentials and Pitfalls," *Photogrammetric Engineering and Remote Sensing*, Vol. 53, No. 10, Oct. 1987, pp. 1405-1410.

[10] Goodenough, D.G., "Thematic Mapper and SPOT Integration with a Geographic Information System," *Photogrammetric Engineering and Remote Sensing*, Vol. 54, No. 2, Feb. 1988, pp. 167-176.

[11] Phillips, K.M., Morgan, K.M., Newland, L.W., Koger, D.G., "Thematic Mapper data: A New Land Planning Tool," *Journal of Soil and Water Conservation*. Vol. 41, No. 5, Sept.-Oct. 1986, pp. 301-303.

[12] Langran, K.J., "Potential for Monitoring Soil Erosion Features and Soil Erosion Modeling from Remotely Sensed Data," Proceedings: *International Geoscience and Remote Sensing Symposium, IEEE*, New York, 1983, pp TPI-2.1-2.4.

[13] McGregor, K.C., and Mutchler, C.K., "Status of the R Factor in Northern Mississippi," *Soil Erosion: Prediction and Control*, Proceed-ings of National Conference on Soil Erosion, Purdue Univ., West Lafayette, IN, May 24-26, 1976, pp. 135-142.

[14] El-Swaify, R., and Dangler, E.W., "Erodibilities of Selected Tropical Soils in Relation to Structural and Hydrological Parameters," *Soil Erosion: Prediction and Control*, Proceedings of National Conference on Soil Erosion, Purdue Univ., West Lafayette, IN, May 24-26, 1976, pp. 105-114.

Applications of High-Resolution Remote Sensing Image Data

W. M. Strome,* D. Leckie,† J. Miller,‡ and R. Buxton§
PCI Inc., Richmond Hill, Ontario, Canada

Abstract

There are many situations in which the image resolution of satellite data is insufficient to provide the detail required for resource management and environmental monitoring. This paper will focus on applications of high-resolution (0.4 to 10 m) airborne multispectral and imaging spectrometer data acquired in Canada using the MEIS II multispectral line imager and the PMI imaging spectrometer. Applications discussed will include forestry, mapping, and geobotany.

Introduction

The availability of relatively high-resolution multispectral satellite data has revolutionized the way in which we manage our natural resources and monitor the Earth's environment. A single image covers large areas of land and water. The quality of the digital data, coupled with the rapid advances in computer technology, have made it economical to use digital image analysis techniques to extract resource information from the satellite images and to combine it with other geographically related data using geographic information systems (GIS).

Tremendous advances have occurred in our ability to gather and extract useful information from the data acquired by remote sensing satellites such as the U.S. Landsat and the French SPOT systems. Nevertheless, the spatial and/or spectral resolution of the sensors carried by these satellites is often insufficient to extract all the information required for some applications.

Copyright © 1990 by the American Institute of Aeronautics and Astronautics, Inc. All rights reserved.
*Chairman of the Board.
† Project Leader, Digital Remote Sensing.
‡ Associate Professor of Physics.
§ Marketing Representative.

Future satellites will carry sensors that will have much greater spectral resolution than the familiar Landsat MSS and TM or the SPOT HRV. The analysis methods that have proven useful in extracting resource information from these satellites will not, in general, be appropriate for the new sensors. High spectral and spatial resolution airborne data can be used to help develop the appropriate analysis tools for dealing with the future sensors to be carried aboard the polar platforms being designed in the United States, Europe, and Japan.

In Canada, scientists have been fortunate to have two airborne sensors available for conducting research into applications for the use of high-resolution multispectral data: the multispectral electrooptical imaging scanner (MEIS II, developed by MDA under contract to the Canada Centre for Remote Sensing [CCRS]) and the fluorescence line imager (FLI) (also called the programmable multispectral imager, or PMI), developed by Moniteq under contract to the Canadian Department of Fisheries and Oceans. Recently, a third sensor, the compact airborne spectrographic imager (CASI), which is similar to the PMI, has been developed in Canada by ITRES Research.

This paper will briefly describe the instruments involved, and will then describe several applications in which these systems have been used for solving resource management or environmental monitoring problems that could not be handled using existing satellite data alone.

High-Resolution Airborne Multispectral Sensors

The three Canadian sensors to be described are the MEIS, the FLI/PMI, and the CASI. There are several other airborne instruments that have similar characteristics, such as the Daedalus multispectral scanner, the airborne imaging spectrometer, and the airborne visible and infrared imaging spectrometer, but these will not be discussed here.

Multispectral Electrooptical Imaging Sensor (MEIS)

The MEIS II (see Refs. 1-3) is a solid-state, linear array, pushbroom multispectral line imager with up to eight channels in the visible and near infrared (VIS/NIR). (See Fig. 1.) It can also be configured to provide six nadir multispectral channels, plus two channels that provide fore and aft views for stereoscopic information. The stereo images can be used to obtain elevation data useful for the production of digital elevation models (DEM), topographic mapping, and tree height measurements.

The combination of fore, aft, and nadir views can also be used for research into the utility of multiple look ang-

HIGH-RESOLUTION REMOTE SENSING IMAGE DATA

les for the improved extraction of cartographic features. The swath width is 1024 pixels. The resolution of 0.7 mr yields a ground resolution ranging from 0.4 to 7 m with the Falcon aircraft in which it is flown, depending on the altitude above ground. Several different filter sets are available. The MEIS system provides digital image data with radiometric, spatial, and spectral performance which is superior to that obtainable from remote sensing satellites. The spectral bands can be as narrow as 3 nm while maintaining full radiometric resolution.

Programmable Multispectral Imager (PMI)

An airborne imaging spectrometer, sensitive in the visible and near-infrared regions, was developed by Moniteq

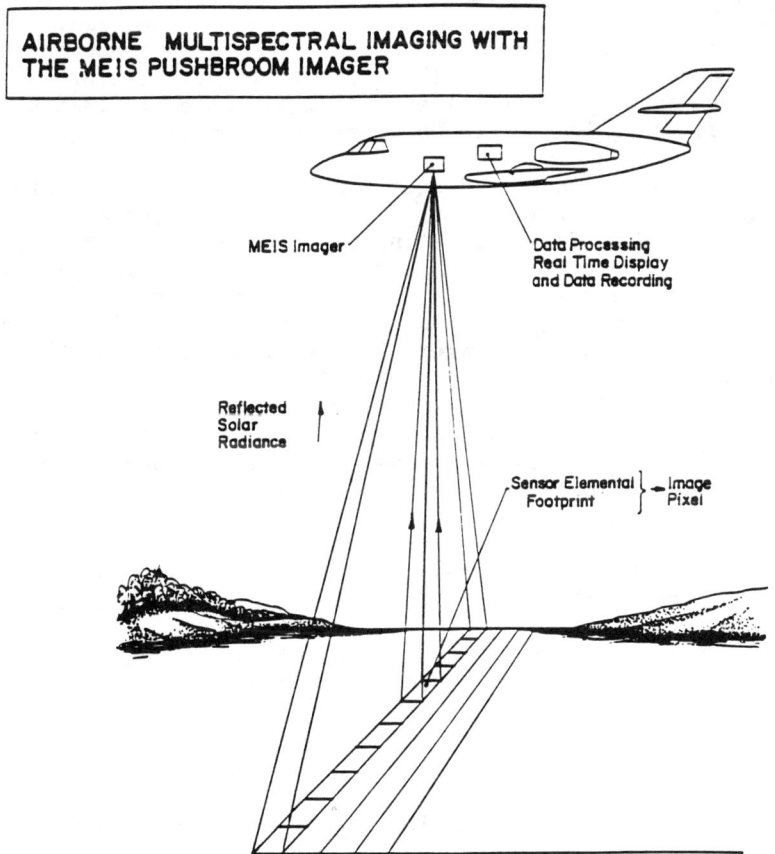

Fig. 1 Airborne multispectral imaging with the MEIS pushbroom imager.

for the Canadian Department of Fisheries and Oceans and has been used to acquire imagery over numerous vegetated and water targets in North America and Europe since its first flight more than five years ago. This system was first known as the fluorescence line imager (FLI), due to its original design objective as a proof-of-concept sensor to image the solar-stimulated fluorescence of chlorophyll in plankton in water from space. It is also called the programmable multispectral imager (PMI) due to its unique ability to reconfigure its spectral bands. For most sites, imagery has been acquired in both the multispectral mode (using a bandset appropriate for the application) and in the spectral imaging mode (as an imaging spectrometer) to provide a comprehensive data set for analysis.

The FLI/PMI is a pushbroom imager that incorporates a two-dimensional CCD (charge-coupled device) array at the focal plane of a spectrometer to provide 288 spectral samples over the 430 to 805 nm range in each of 385 spatial samples over a 15 deg field of view. (See Ref. 4.) (See Fig. 2.) The resulting spectral resolution is 2.6 nm sampled at 1.3 nm increments. Five of these spectrometer-detector modules are used to increase the spatial dimension to 1925 pixels over 70 deg (only four modules are currently functional).

As the electronics and recording facilities cannot handle all the data available from the arrays, the data volume and rate are limited by operating the sensor in either of two modes: In the full spatial mode, all the spatial pixels are recorded and the spectrum is sampled into eight selectable bands, causing the sensor to perform as an eight-band multispectral pushbroom imager. The ability to define band start and end wavelengths to within 1.3 nm provides unprecedented precision in band characteristics. In the full spectral mode, all the color detectors are recorded while the dimension across the swath is sampled in eight noncontiguous look angles per module, causing the sensor to perform as an array of spectrometers forming a coarse ground-resolution image. In this mode, a full 188 bands of spectral data are provided for 40 along-track lines of pixels separated by **gaps of four pixels, yielding a "rake-" like pattern of** coverage.

The selected data from the detector arrays are digitized to 12 bits and summed into the spatial or spectral groups to 16 bits.

A radiometric calibration is provided for each pixel in real time that allows for responsivity and dark signal correction. The necessary dark current data are recorded during each flight and stored in memory. The nonuniformities of the response of pixels in the CCD arrays are removed by applying uniformity correction factors, which are derived

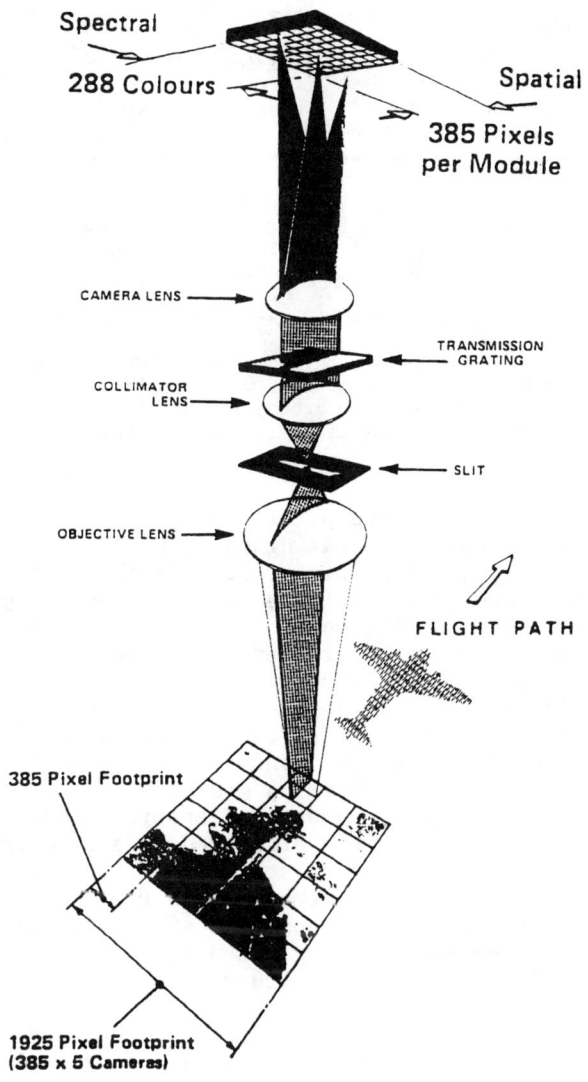

Fig. 2 Spectral imaging in the PMI with area arrays.

from a laboratory-based calibration activity; an integrating sphere is used as a reference source. The sphere itself is periodically calibrated spectrally, and a correction is provided for its spatial nonuniformity. The precision of the sphere calibration is approximately 0.6%.

The data from the PMI are recorded onto a high-density digital tape (HDDT) in the aircraft, from which computer-compatible tapes (CCTs) are produced postflight.

TABLE 1

Typical pixel footprint

Altitude AGL ft	Mode	Crosstrack m	Alongtrack m
2000	Spatial	0.4	7.4
	Spectral	4.0	12.1
5000	Spatial	1.0	8.2
	Spectral	10.0	12.8
10,000	Spatial	2.0	9.4
	Spectral	20.0	14.0

Note: The line-to-line separation along track is given by the ground speed and integration time, which is independent of altitude. Also, the gap between ground footprints in the spectral mode is approximately four times the cross-track pixel size.

The dimensions of the pixel footprint of the FLI/PMI depend on the sensor's operating mode and the flight parameters. Typical pixel footprints for several altitudes above ground (AGL) are given in Table 1. The spectral characteristics also depend on the mode. Several bandsets for the spatial mode have been created that take into account the application, atmospheric absorption, detector blemishes, and the anticipated signal levels. A single spectral mode is currently in use that forms a "rake" view of the ground. The data format of the CCT conforms to the LGSOWG (Landsat ground station operators working group) standard. For the spatial mode, this corresponds to 8 spectral bands of 16-bit data for 1925 spatial pixels organized as a band interleaved by line (BIL). A radiance scale factor for each of the 8 bands converts the digital numbers into radiance units. For the spectral mode, it corresponds to 40 spatial pixels, each with 288 spectral elements organized as a band interleaved by pixel (BIP). There are 288 radiance scale factors for this mode.

Image processing software has been developed to handle both the 16-bit and high spectral dimensionality of the data. This entails moving the data between devices (CTT, disk, and display), as well as formatting the data display. The spatial mode data are handled much like that from other multispectral imagers. The spectral mode data require a different treatment. It has been especially useful to show a three-color image on the left of the screen, and a plot of the spectrum of any chosen point within the scene on the right.

More effective use of the spectral mode data can be achieved in two ways: 1) the use of models (physical or mathematical) to extract parameters of interest, or 2) the development of appropriate bandsets that can be used on subsequent flights to provide data for modeling or effective classification. There is a need for research in both these areas in many application disciplines.

Compact Airborne Spectrographic Imager (CASI)

The CASI is similar in concept to a single module of the five-module FLI/PMI. The nominal field of view across the direction of motion is 30 deg corresponding to 385 spatial pixels of the detector. The system has up to 288 user-selectable spectral bands in the range of 450 to 900 nm, with a minimum width of 1.6 nm. The unit can be programmed to scan at rates from 10 to 100 lines per second. The user can define several spectral bands for full image data with precise spectral boundaries. In addition, the full spectrum for any single along-track line may be gathered.

Applications

Land Applications

Given the VIS/NIR spectral range of the instruments and their deployment to date over targets in temperate climates, the primary benefit over land has been in the investigation of vegetation, notably trees. There have been several areas of interest: 1) forest inventory; 2) forest damage assessment; 3) mineral exploration; 4) hydrocarbon exploration; 5) landfill site monitoring; 6) agriculture; and, 7) atmosphere. The objective in most cases was to determine the extent to which the increased spatial and/or spectral resolution could be used to detect subtle effects attributable to various known causes of vegetation stress.

Forest Inventory. Forest inventory, and insect and disease damage assessment, are applications in which high-resolution imagery is important. Lower-resolution satellite data have shown good capabilities for mapping broad forest types and areas of extensive severe forest damage. Although useful, they do not provide sufficient information for most forest management decisions.

Forest planning in most countries is based on forest management inventories and stand maps that provide specific information on species composition, age, height, density, and often site type of forest stands. The source of this information is generally 1:10,000 to 1:20,000 scale aerial photographs supplemented by varying degrees of field visits and sample plots. In order to interpret the forest parameters just listed, images are required on which each overstorey crown is discernible. The spatial resolution required will depend on forest type, but will generally be less than 2 m. In many areas, resolutions of less than 1 m are required. Sometimes, detail within a single tree crown is important for interpretation, leading to a requirement for very high resolution. A large number of combinations of species and density are possible in a natural forest. Resolutions that result in a large percentage of mixed field components will be inadequate for determining the forest parameters necessary for management inventories.

Very high-resolution linear array imager data have shown capabilities for interpreting tree species. Leckie and Dombrowski (see Ref. 7) reported good discrimination of five softwood species using principal component enhancements of 1.2 m resolution MEIS II data. MEIS II imagery at 0.4 to 1.5 m resolution from various locations across Canada has shown good capabilities for discerning individual trees, determining density, and delineating stand boundaries. The visual interpretation of high-resolution-enhanced linear array imagery therefore looks promising for the production of management inventory maps. The automated classification of species on a single-tree basis using very high-resolution imagery has met with mixed success (see Refs. 8,9) and requires much further research. Such capabilities would be an important advancement in forest inventory methods.

Height is an important requirement of management inventories. A stereo-viewing capability is therefore necessary. Linear array imagers is designed so that stereo imagery is produced. This requires that accurate geometric correction of the imagery be conducted. (See Ref. 10.) A preliminary analysis of visual interpretations of 0.4 to 1.5 m resolution stereo MEIS II imagery shows that it is likely feasible to estimate stand height to the accuracy currently obtained from aerial photographic interpretation for management inventories.

Another key feature of digital imagery is that it permits accurate and efficient geometric correction to cartographic coordinates. (See Ref. 5.) Thus, for the production of forest inventory maps, forest stands derived from digital imagery can be related to existing forest inventory information residing on a GIS. Furthermore, geometrically multispectral imagery derived from an airborne sensor has also been used to produce the inventory map directly on the GIS. (See Ref. 6.)

High-resolution imagery is necessary for management inventories. Low-resolution data are appropriate for reconnaissance inventories, regional inventories, or specific-purpose inventories such as forest fuel-type mapping. The resolution will depend on the specific-purpose and required forest types of each survey. Current satellite data can also be very useful in giving broad-scale patterns such as old burned areas and ecological stratifications that are often very helpful for management inventory mapping.

Forest Damage Assessment. Remote sensing plays an important role in forest damage assessment. Damage information is used to plan salvage logging, determine long-term harvest scheduling, manage insect control programs, update forest inventories, and monitor forest environmental conditions. The appropriate spatial resolution is dependent on the purpose of the damage survey. Sometimes, only a general indication of the location and level of damage is needed. However, detailed stand-by-stand damage assessments are becoming increasingly important to forest management. When **extensive areas are damaged, it is often impractical to** survey the damage in detail; low-resolution imagery might be used, although the information may be suboptimum for the intended purpose. In other cases, detailed damage appraisal down to the single-tree level is required. A mix of broadscale mapping and stratification, followed by detailed sampling, is often the best methodology.

For damage surveys that require only general damage mapping, satellite data or airborne data at resolutions larger than 5 m have shown some capabilities. Moderate and sometimes inconsistent results have been obtained from satellite imagery, but it has been effective for various operational damage surveys. Airborne imagery at resolutions between 5 and 10 m also has capabilities for broad-scale damage assessment. (See Refs. 11-14.) As with inventory mapping, a problem arises due to mixed pixels, (see Ref. 15) especially mixtures of different species or densities. For example, there were difficulties in supervised classifications of spruce budworm defoliation of balsam fir in mixed wood stands due to the presence of different proportions of hardwoods masking the presence and level of defoliation. (See Refs. 11,14.) Change detection, use of a priori knowledge of forest type, optimum spectral bands, or visual interpretation of enhanced imagery may help to alleviate this problem. Similarly, the PMI has been used for several surveys over forests in North America and has provided valuable spectral reflectance information related to species and environmental conditions. When detailed information was acquired from field samples as well, the red-edge characteristics of the ground and airborne data showed good agreement. (See Ref. 16.) Similar work has been conducted in Europe in the Black Forest and Austria. (See Ref. 17.)

High-spatial-resolution data for interpretation on a single-tree basis are appropriate for several applications. They may supplement damage surveys, which generally incorporate ground plot surveys. For example, cumulative spruce budworm defoliation (loss of foliage) was easily detected on 0.4 m resolution MEIS II imagery on Cape Breton Island, Nova Scotia, Canada. Patterns of defoliation within a tree crown were also discernible. Automated procedures for defoliation estimation on a single-tree basis would prove to be an important sample plot survey tool. Similar procedures would also be effective for sampling sites as part of a multistage or multiphase survey that uses low-resolution data for stratifying damage zones. For some forest damage such as mountain pine beetle, it is often desirable to locate and cut single trees suffering damage in order to prevent the spread of the damage. Preliminary analysis of MEIS II imagery by P. Murtha (see Ref. 18) shows good capabilities of 0.7 m resolution data for detecting single trees with symptoms (red coloration) of mountain pine beetle attack. Work with digitized large-scale photographs indicated the benefits of very high-resolution imagery, giving information within a tree crown. (See Ref. 19.) Forest decline damage in Germany was detected on a single-tree basis by Koch et al. (see Ref. 20) using 0.75 m multispectral scanner data.

High spectral resolution can also be important for damage assessment. Damage symptoms are often subtle and highlighted in specific spectral bands. Leckie et al., (see Ref. 21) using spectral reflectance measurements of individual trees, reported the benefits of using narrow spectral bands for detecting red coloration of trees caused by the feeding of the spruce budworm. The damage symptoms of foliage loss due to the spruce budworm (cumulative defoliation), however, were broadband features. (See Ref. 22.) The analysis of needle reflectance of trees attacked by the mountain pine beetle showed that selected narrow spectral bands may be useful for assessing early indications of stress caused by the pine beetle. (See Ref. 23.) Fine spectral features associated with forest decline in Germany and the northeastern United States have been reported and imaging spectrometer data from the FLI/PMI have proven useful for detecting these differences. (See Refs. 24-26.)

Much of the attention has been focused on the chlorophyll red edge, which is known to exhibit variations in position and shape as a possible response to abnormal environmental conditions. These conditions can be numerous, including atmospheric pollution, precipitation pH, species, temperature and precipitation history, soil type, metal and hydrocarbon concentrations in the soil, season, and toxic substances in the soil. As a result, data interpretation has tended to be site-specific when one of these conditions is considered dominant: If anomalies in the imagery correlate with known conditions, similar anomalies elsewhere in the imagery can therefore be attributed to the same cause.

The principal analytical tool used to date is the inverted Gaussian model for the red edge. (See Refs. 26-28.) This curve is fitted to the pseudoreflectance derived from the spectrum of each pixel (either mode) to yield four parameters: 1) the infrared shoulder reflectance; 2) the reflectance minimum corresponding to the chlorophyll absorption maximum; 3) the wavelength at the reflectance minimum;

and, 4) the red-edge slope inflection wavelength.

The quality of the fit, and therefore the accuracy and precision of the four derived parameters, are far superior using the PMI spectral mode rather than the PMI spatial mode or the MEIS II, due to the vastly greater number of spectral samples obtained in the same wavelength interval (approximately 100 vs 5 or 6). This improved ability to fit the spectrum must be balanced against the homogeneity of the larger ground footprint of each pixel.

An RGB (red/green/blue) image is then produced from these parameters and is referred to as a "stress" image, with a "blue shift" of the edge and reflectance shoulder variations indicative of stress. Such an image may be further manipulated by conventional image processing techniques.

As useful as this procedure has been, it is believed that other information present in the red-edge shape requires further research. In addition, the information in the blue and green regions has yet to be utilized in conjunction with the red-edge parameters. Consequently, other effective ways of determining parameters of the site from the airborne spectrum remain to be found.

Blue-shift analyses of high-spectral-resolution data have been applied to detecting forest decline. In the Camel's Hump Mountain in Vermont, red spruce have been in decline for some time. (See Refs. 24,26.) One can display the spectral mode PMI image as a color composite of red-edge parameters such that the region of greatest damage appears red. The spectra, which are usually shown to the right of the image, can be used to interpret the result for specific cursor positions. The image for this test showed a high correlation between the red regions in the scene (i.e., stressed) with known high damage sites, thereby demonstrating the effectiveness of this technique for determining perturbed areas.

Similar analysis of FLI/PMI's spatial mode is under way for two maple wood lots in southern Ontario, in which ground investigations have shown detrimental effects attributed to acid rain. Preliminary analysis shows wood lots with a higher level of damage as having a greater concentration of red (more blue-shift pixels). Such stress imagery is not detectable by other mapping techniques.

Geobotany. Many stress images and associated spectra have been generated for forested mineralized sites in the Canadian shield in Ontario. These images were derived from PMI imagery by applying the inverted Gaussian model to spectral data. The stressed areas revealed by this technique for one site exhibit differences that correlate with known geochemical anomalies, which is confirmed by the two sets of spectra from the two regions in the scene. (See Ref. 29.)

Similar results have been presented for other sites. (See Ref. 30.)

Miller et al. (see Ref. 31) have conducted field/airborne remote sensing experiments to investigate the botanical and spectral reflectance response in leaves of selected trees (balsam fir, black spruce, aspen and white birch) that were innoculated with solutions of copper nitrate, nickel nitrate, and sodium arsenite. The concentrations used ranged from 250 to 1000 ppm. This was followed by frequent optical/botanical sampling during a ten-day period. MEIS II data were then collected with spectral channels at 480, 548, 675, 698, 710, 734, 746 and 776 nm with spectral bandwidths varying from 6.3 to 17.8 nm and with a spatial resolution of 38 cm. The data were analyzed to evaluate the potential to detect metal-induced stress from the resulting reflectance changes.

Laboratory experiments involving the treatment of plants with high concentrations of metal ions have provided evidence of the blue shift of the reflectance red edge as an expected plant response to metal-induced stress. (See Refs. 32-34.) Attempts to observe a similar metal-induced response in natural settings with trees living over mineralized deposits have been relatively unsuccessful. Thus, an experiment was developed whereby mature trees at the Petawawa National Forestry Institute (PNFI) in Ontario, Canada, were injected with metal ions. The experiment was conducted between August 1 and 15, 1986. Both natural and plantation vegetation were used.

Details on the test plots and methods of injection have been provided elsewhere. (See Ref. 31.) Detailed laboratory spectral reflectance measurements were made on excised leaves at a portable laboratory on the same day that samples were collected. A J-Y grating monochromator between 500 and 900 nm at 4 nm resolution was used for the measurements. Leaf stacks of both first- and second-year needle clumps were measured separately. Field spectral reflectance measurements were made with a ten-channel radiometer in situ from the top of a ladder on August 14, 1986. The MEIS II data were collected over the balsam fir site on August 9 and 29, 1986.

By using field data for calibration, the MEIS II data were compared with the laboratory and field spectra of individual trees. The primary analysis was the extraction of red reflectance edge parameters (the inflection point and the chlorophyll-well position).

Of the treated balsam fir, only injection with sodium arsenite produced visible symptoms on first- and second-year needles. Symptoms were increasing defoliation rather than chlorosis. Analysis of MEIS data showed a consistent blue shift in the red inflection point of 7 nm for two of the individual trees treated with sodium arsenite. The red-edge analysis was found to be insensitive to canopy radiometric texture, in agreement with previous work, (see Ref. 35) provided that closed canopy conditions were met.

Agriculture. The MEIS II has been used for several agricultural studies in southwestern Ontario. The FLI/PMI has been used to investigate agricultural applications in Ontario and Switzerland.

Water Applications

These instruments have been used for many water application activities. The examples given here are primarily for the FLI/PMI.

Bathymetry. Multispectral scanning technology, coupled with the ability to geocode the imagery, can be used for shallow water bathymetric measurements. (See Ref. 36.) In shallow environs, where vessel captains have the most concern, conventional surveys with launches may be difficult. This technique of producing hydrographic data is very effective in monitoring changes in dynamic features such as sand bars.

Chlorophyll Mapping. Chlorophyll concentration in Lake of the Woods, Ontario, was derived from spatial mode data by a band ratio method. The results compare very favorably to in situ data that were obtained to verify the data processing methodology.

FLI/PMI also provides another, and unique, method of determining chlorophyll concentration in water by measuring the chlorophyll's fluorescence amplitude caused by stimulation by sunlight. This is a weak signal with a narrow spectral bandwidth at 685 nm, which must be measured in the presence of the much larger water reflectance; FLI/PMI has successfully imaged this signal in lakes and in the ocean.
The measurement of chlorophyll concentration allows algal blooms in lakes, rivers, and the ocean (see Ref. 37) to be mapped for commercial fishing and recreation environment assessment purposes.

Near-Shore Water Quality. A simulated natural color image of the outflow of the Humber River into the Toronto, Ontario harbor was derived from FLI/PMI spatial mode data. A pseudocolor enhancement of this data reveals (in red) the treated sewage from the subsurface discharge and its interaction with the silt plume (in buff) from the river. The capability to detect and distinguish such materials and to map their distribution can therefore be used to monitor water quality in industrialized areas.

Future Availability of Data

In order to continue the research areas outlined here as well as enable multitemporal projects, Moniteq made the

FLI/PMI available for data acquisition in North America, Europe and elsewhere in the world in 1989 and offers PMI- and MEIS-type units. Moniteq and Rem/Sense Mapping Technologies offer geometric correction systems and services for airborne imaging systems. Similarly, ITRES Research makes the CASI available worldwide and is offering units for sale. The MEIS II is on loan from CCRS to Innotech Aviation, which had the system installed in a Falcon jet that is capable of flying missions.

Conclusions

The applications described require much higher spatial and/or spectral resolution than is currently obtainable from satellites. In order to plan for the new sensors being developed, instruments such as the MEIS II, FLI/PMI, and CASI can provide the data sets that will help us to develop the processing and analysis methods required.

Acknowledgments

The authors appreciate the contributions of many, including Dr. A. Hollinger and L. Gray of Moniteq, Prof. B. Rock of the University of New Hampshire, V. Singhroy of the Ontario Centre for Remote Sensing, Drs. K. Itten and K. Staenz of the University of Zurich, D. McLaughlin of the Ontario Ministry of the Environment, Dr. C. Banninger of the Graz Research Institute, J. Freemantle and Dr. D. Jayasinghe of the Institute for Space and Terrestrial Sciences, York University, and Hilda van Walraven of PCI Inc. We also acknowledge the support of the Canadian Department of Fisheries and Oceans, the Canada Centre for Remote Sensing, The Canadian Forestry Service, Supply and Services Canada, The Ontario Ministry of the Environment, The Ontario Ministry of Natural Resources, and the Jet Propulsion Laboratory.

References

1) Till, S.M., Neville, R.A., Leckie, D.G., and Strome, W.M., "Advanced Airborne Electro-Optical Imager," Proceedings of the 21st International Symposium of Environment, Oct. 26-30, 1987.

2) McColl, W.D., Till, S.M., and Neville, R.A., "Enhanced CCRS Data Acquisition Capabilities," Proceedings of the 9th Canadian Symposium on Remote Sensing, Aug. 14-17, 1984, pp. 617-626.

3) Neville, R.A., McColl, W.D., and Till, S.M., "Development and Evaluation of the MEIS II Multidetector Electro-Optical Imaging Sensor," Proceedings of Society of Photo-Optical Instrumentation Engineers, Advanced Infrared Sensor Technology, Vol. 395, 1983, pp. 101-108.

4) Hollinger, A.B., Gray, L.H., Gower, J.F.R., and Edel, H.R., "The Fluorescence Line Imager: An Imaging Spectrometer for Ocean and Land Remote Sensing," *Proceedings of Society of Photo-Optical Instrumentation Engineers, Imaging Spectroscopy II Symposium*, Vol. 834, 1987.

5) Gibson, J.R., Park, W.M., Hollinger, A.B., Dunlop, J.D., and O'Neil, N.T., "Geometric Correction of Line Scanner Data," *Proceedings of the Canadian Symposium on Remote Sensing*, 1987.

6) Reid, N., and Fisher, J., "Creation of Forest Inventory Maps Using Airborne Digital Imagery and Geographic Information Systems," *Proceedings of GIS '89 Symposium*, 1989, p. 165.

7) Leckie, D.G., and Dombrowski, A., "Enhancement of High Resolution MEIS II Data for Softwood Species Discrimination," *Proceedings of the 9th Canadian Symposium on Remote Sensing*, Aug. 14-17, 1984, pp. 617-626.

8) Hughes, J.S., Evans, D.L., Burns, P.Y., and Hill, J.M., "Identification of Two Southern Pine Species in High Resolution Aerial MSS Data," *Journal of Photogrammetry*, Vol. 52, Aug. 1986, pp. 1175-1180.

9) Gougeon, F.A., and Moore, T., "Classification Individuelle des Arbres a Partier d'Images a Haute Resolution Spatiale," *Proc. 6e congres de l'association quebecoise de teledetection*, May 4-6, 1988.

10) Gibson, J.R., "Processing Stereo Imagery from Line Imagers," *Proceedings of the 9th Canadian Symposium on Remote Sensing*, Aug. 14-17, 1984, pp. 471-487.

11) Leckie, D.G., and Gougeon, F.A., "Assessment of Spruce Budworm Defoliation Using Digital Airborne MSS Data," *Proceedings of the 7th Canadian Symposium on Remote Sensing, Winnipeg, Manitoba, Sept. 1981, pp. 190-196*.

12) Koch, B., and Kritikos, G., "Integrative Investigation on Forest Damage Detection on Air-Borne Multispectral Scanner Data," *Proceedings Integrated Approaches in Remote Sensing*, April 1984, pp. 109-113.

13) Ahern, F.J., Bennett, W.F., and Kettela, F.G., "Evaluation of Two Digital Airborne Images for Surveying Spruce Budworm Defoliation," *Journal of Photogrammetric Engineering and Remote Sensing*, Vol. 52, 1986, pp. 1647-1654.

14) Leckie, D.G., and Ostaff, D.P., "Classification of Airborne Multispectral Scanner Data for Mapping Current Defoliation Caused by the Spruce Budworm," *Journal of Forest Science*, Vol. 34, Feb. 1988, pp. 259-275.

15) Leckie, D.G., "Factors Affecting the Assessment of Forest Damage Using Airborne Multispectral Scanner Data," *Journal of Photogrammetry*, Vol. 53, Dec. 1987, pp. 1665-1674.

16) Miller, J.R., Hare, E.W., Hollinger, A.B., and Sturgeon, D.R., "Imaging Spectrometry as a Tool for Botanical Mapping,"

Proceedings of Society of Photo-Optical Instrumentation Engineers, Imaging Spectroscopy II Symposium, Vol. 834, Aug. 1987, pp. 108-113.

17) Banninger, C., "Spectral Response Characteristics of a Metal-Stressed Coniferous Forest as Measured by the FLI Airborne Imaging Spectrometer," Proceedings of International Geoscience and Remote Sensing Symposium, 1988, p. 1331.

18) Murtha, P.A. (personal communication).

19) Murtha, P.A., and Wiart, R.J., "Cluster Analysis of Pine Crown Foliage Patterns Aid Identification of Mountain Pine Beetle Current Attack," Journal of Photogrammetric Engineering and Remote Sensing, Vol. 55, pp. 83-86.

20) Koch, B., Ammer, U., Kritikos, G., and Kubler, D., "Untersuchungen zu Beurteilung der Vitalitat von Fichten Anhand Multispektraler Scannerdaten," Fortstwissenschafteliches Centralblatt, Vol. 103, 1984, pp. 214-231.

21) Leckie, D.G., Teillet, P.M., Ostaff, D.P., and Fedosejevs, G., "Sensor Band Selection for Detecting Current Defoliation Caused by the Spruce Budworm," Journal of Remote Sensing of Environment, Vol. 26, 1988, pp. 31-52.

22) Leckie, D.G., Teillet, P.M., Fedosejevs, G., and Ostaff, D.P., "Reflectance Characteristics of Cumulative Defoliation of Balsam Fir," Canadian Journal of Forestry Research, Vol. 18, 1988, pp. 1008-1016.

23) Ahern, F.J., "The Effects of Bark Beetle Stress on the Foliar Spectral Reflectance of Lodgepole Pine," International Journal of Remote Sensing, Vol. 9, 1988, pp. 1451-1468.

24) Rock, B.N., Hoshizaki, T., and Miller, J.R., "Comparison of in Situ and Airborne Spectral Measurements of the Blue Shift Associated with Forest Decline," Journal of Remote Sensing of Environment, Vol. 24, 1988, pp. 109-127.

25) Herrman, Rock, B.N., Ammer, U., and Paley, H.N., "Preliminary Assessment of Airborne Imaging Spectrometer and Airborne Thematic Mapper Data Acquired for Forest Decline Areas in the Federal Republic of Germany," Journal of Remote Sensing of Environment, Vol. 24, 1988, pp. 129-149.

26) Hoshiza, T., Rock, B.N., and Wong, S.K.S., "Pigment Analysis and Spectral Assessment of Spruce Trees Undergoing Forest Decline in the North East United States and Germany," Geojournal, 17, 1988, pp. 173-178.

27) Miller, J.R., Hare, E.W., and Wu, J., "Quantitative Characterization of the Vegetation Red Edge Reflectance: An Inverted-Gaussian Reflectance Model," submitted to International Journal of Remote Sensing, Jan. 1989.

28) Miller, J.R., Hare, E.W., Neville, R.A., Gauthier, R.P., McColl, W.D., and Till, S.M., "Correlation of Metal Concentrations with Anomalies in Narrow Band Multispectral Imagery of the

Vegetation Red Reflectance Edge," Proceedings of the 4th Thematic Conference on Remote Sensing for Exploration Geology, April 1-4, 1985, pp. 143-153.

29) Singhroy V., and Fortescue J., "Report to the Geoscience Working Group of the Canadian Advisory Committee for Remote Sensing," 1987.

30) Singhroy, V., Stanton-Gray, R., and Springer, J., "Spectral Geobotanical Investigation of Mineralized Till Sites," Proceedings of the 5th Thematic Conference on Remote Sensing for Exploration Geology, 1986, pp. 523-543.

31) Miller, J.R., Boyer, M.G., Wu, J., Gauthier, R.P., Hare, E.W., and Belanger, M., "Detection of Spectral Effects in Individual Tree Crowns of Metal-Injected Trees Using High Resolution Pushbroom Imagery," Proceedings of the 16th Congress of the International Society for Photogrammetry and Remote Sensing, July 1-10, 1988, Part B, pp. 847-856.

32) Horler, D.N.H., Barber, J.P., Darch, D.C., and Barringer, A.R., "Approaches to Detection of Geochemical Stress in Vegetation," Advanced Space Research, Vol. 3, 1983, pp. 175-179.

33) Horler, D.N.H., Dockray, M., Barber, J., and Barringer, A.R., "Red Edge Measurements for Remotely Sensing Plant Chlorophyll Content," Advanced Space Research, Vol. 3, 1983, pp. 273-277.

34) Horler, D.N.H., Dockray, M., and Barber, J., "The Red Edge of Plant Leaf Reflectance," International Journal of Remote Sensing, Vol. 4, 1983, pp. 273-288.

35) Rencz, A.N., Bonham-Carter, G.F., van der Greint, C., Miller, J.R., and Hare, E.W., "Preliminary Results from Modelling Vegetation Spectra Derived from MEIS Data," Proceedings of the 10th Canadian Symposium on Remote Sensing, pp. 909-917.

36) O'Neill, N.T., Kalinauskas, A.R., Dunlop, J.D., and Hollinger, A.B., "Bathymetric Analysis of Geometrically Corrected Imagery Collected Using a Two-Dimensional Imager," Proceedings of Society of Photo-Optical Instrumentation Engineers, Ocean Optics VIII Symposium, Vol. 637, 1986.

37) Gower, J.F.R., and Borstad, G.A., "Mapping Phytoplankton Fluorescence with the FLI," Proceedings of International Geoscience and Remote Sensing Symposium, 1988, p. 1391.

Remote Sensing Applications to Earth Resources Survey in Pakistan

Salim Mehmud*
Pakistan Space and Upper Atmosphere Research Commission, Karachi, Pakistan

Pakistan's national space agency, the Pakistan Space and Upper Atmosphere Research Commission (SUPARCO), has given high priority to satellite remote sensing applications in view of the significant role remote sensing can play in Earth resources surveying and management. SUPARCO initiated its remote sensing applications program as early as 1973, followed in 1978 by the establishment of a full-fledged Remote Sensing Applications Centre (RESACENT) equipped with extensive visual and digital interpretation facilities. More than 60 research studies covering diverse fields such as agriculture, forestry, water resources, geology, environment, and land use have been undertaken at RESACENT using remote sensing data.

As the national coordinator for remote sensing, SUPARCO acquires, archives, and disseminates remote sensing data of Pakistan, provides technical support to interested national user agencies in the application of the data, organizes training programs and seminars for their benefit, and participates in various activities at the international level. Remote sensing activities received further impetus when SUPARCO's ground station for acquisition of Landsat Multi Spectral Scanner (MSS)/Thematic Mapper (TM), System Probatoire d'Observation de le Terre (SPOT) Haute Resolution Visible (HRV), and National Oceanic And Atmospheric Administration (NOAA) Advanced Very High Resolution Radiometer (AVHRR) data was commissioned at the end of 1988. Further plans include significant expansion in aerial remote sensing capabilities, research and development work (R&D) on sensors, development of Geographic Information System (GIS) and establishment of an ARGOS network in Pakistan.

Copyright © 1990 by the American Institute of Aeronautics and Astronautics, Inc. All rights reserved.

*Chairman, Pakistan Space & Upper Atmosphere Research Commission; currently with Defence Science and Technology Organization.

REMOTE SENSING APPLICATIONS IN PAKISTAN

I. Introduction

Pakistan's national space agency, SUPARCO, has always recognized the potential of remote sensing to contribute to the national economic development process through its ability to provide useful information on natural resources and environmental processes. The repetitive and synoptic coverage of the Earth's surface, possible through remote sensing satellites, can significantly aid in the optimum exploration and management of natural resources and in the monitoring of various environmental processes and problems. For the developing countries in particular, where typically resource information is scanty or out-dated, satellite remote sensing can play a highly effective role in mapping, resource management, and socioeconomic development. Even in the advanced countries, remote sensing can effectively supplement other conventional methods used for the collection of resource and environmental information.

In view of the potentialities offered by remote sensing, SUPARCO has always accorded high priority and importance to remote sensing applications in its research programs and overall scheme of things. In fact, even before the launch of the first United States (US) Landsat in 1972, SUPARCO had convened a conference of potential users drawn from various departments/agencies in Pakistan in 1971 to acquaint them with this new technology and its potential applications. SUPARCO was quick to respond to the opportunity offered by the launch of Landsat-1 and within a year of its launch i.e. in 1973, formally initiated a broad-based remote sensing applications program with the formation of a research group for remote sensing.

Over the next five years there was appreciable expansion in remote sensing activities, and in 1978 SUPARCO established a full-fledged Remote Sensing Applications Center, called RESACENT, at Karachi under a newly constituted Remote Sensing Applications Division. This Centre plays a pivotal role in remote sensing activities in Pakistan as the country's national remote sensing center.

RESACENT has built up a sizeable archive of satellite remote sensing data of the country, acquired since 1972, and is well-equipped with laboratory facilities for visual and digital analysis of remote senisng data. It is adequately staffed with remote sensing professionals and earth scientists who carry out research studies related to the application of remote sensing data to a variety of resource and environmental problems.

In addition to being a major user of remote sensing data, SUPARCO as the national space agency responsible for all space-related programs is also the national coordinator for remote sensing activities in the country. In the capacity of national coordinator, SUPARCO acquires and archives remote sensing data of Pakistan, disseminates these data to other national user agencies on request, provides technical support and guidance to these agencies in the use of remote sensing data and techniques through joint collaborative projects and otherwise, and organizes in-house short training courses on remote sensing on a regular basis as well as seminars and workshops on specific thematic applications for the benefit of national user scientists.

In addition to the main Centre (RESACENT) at Karachi, (capital of Sind Province), SUPARCO has recently set up regional centers at Islamabad (the federal capital) and Peshawar (the capital of North-West Frontier Province). It will soon be establishing a third center at Lahore (the capital of Punjab Province) to be followed by a fourth at Quetta (the capital of Baluchistan Province). The regional centers are being equipped with basic data interpretation equipment also, thereby providing users in different parts of the country easy access to remote sensing data, information, and basic interpretation facilities.

II. SUPARCO's Data Archives

RESACENT's data archives contain nearly 2500 Landsat Multi Spectral Scanner (MSS) scenes, each in three or four bands, in the form of transparencies, as well as more than 225 MSS Computer Compatible Tapes (CCTs). These data represent, on an average, approximately 35-40 times repetitive coverage of the entire country from 1972 onwards. In additon, a limited number of Landsat Return Beam Vidicon (RBV) and TM, NOAA AVHRR, Metric Camera, Nimbus 7 Coastal Zone Color Scanner (CZCS), and Shuttle Imaging Radar SIR-A images are also available at the Centre. Over the last two years, a sizeable amount of SPOT HRV data have also been acquired. The data archives are being constantly expanded.

III. Data Processings and Interpretation Facilities

RESACENT has well-equipped laboratory facilities for visual and digital interpretation of remotely sensed data. Some of the major electrooptical equipment used for visual interpretation include a density slicer, color additive viewer, zoom transferscope, twin Reflectance/transmission densitometer and zoom stereoscope. More electrooptical equipment is being procured for RESACENT as well as for the regional centers established in other cities.

A full-fledged photographic laboratory is functioning at RESACENT. It has, among other usual facilities, enlargers, contact Printer, diazo printer and developer. The photograph laboratory produces black and white prints at various scales as well as color composites using the color substractive Diazo Process. Other equipment includes a Hasselblad 500 centimeter Polaroid, and other sophisticated cameras, plain paper photo copier, and slide and overhead projectors.

Increasing emphasis is now being placed on computer-based digital processing and interpretation techniques. SUPARCO's digital processing facilities include an IBM 4331 mainframe computer used for generation of gray-level maps and a dedicated, interactive, microprocessor-based digital image processing system. The image processor is hooked to the expanded model of an IBM PC AT with various peripherals such as a 1600 Bands Per Inch (bpi) tape unit, color copier, digitizer, and a color display monitor of 512 x 512 resolution. Two sophisticated image analysis software packages are being used for image processing work with this system. Another IBM PC AT and an IBM PS/2 have also been acquired to augment digital processing and computing facilities at the Centre.

A hand-held ratioing radiometer, with filters corresponding to Landsat MSS, TM, and SPOT HRV spectral bands, has also been procured. The equipment will be used to measure radiances of different features and surface classes in different spectral regions corresponding to the above-mentioned spectral bands thereby facilitating correlation between ground-basd and satellite-based radiance values.

IV. Satellite Ground Station

From the operational viewpoint, the success of a multi-disciplinary remote sensing applications program is critically dependent upon the availability of real-time/near real-time satellite data on a regular basis. SUPARCO is accordingly setting up a satellite ground receiving station near the federal capital, Islamabad, for the acquisition of satellite remote sensing data. The station will be capable of directly acquiring and preprocessing Landsat MSS and TM data, SPOT HRV data in both multispectral and panchromatic modes, and NOAA AVHRR data. The SUPARCO station will be one of the most sophisticated ground stations for receiving satellite remote sensing data in the Asia-Pacific region. The station was expected to become operational by the end of 1988.

The station will have three main subsystems, viz.: 1) the data acquisition subsystem; 2) the data processing subsystem; and 3) the photograph processing subsystem.

The data acquisition subsystem will have the capability to automatically track Landsat, SPOT, and NOAA satellites and acquire image and ancillary data from these satellites. The subsystem will basically comprise:

1) A 10-m Cassegrain dual feed antenna for simultaneous X-band and S-band reception with associated tracking and receiving equipment for Landsat and SPOT satellites.

2) 2.5-m antenna with parabolic reflector and five elements prime focus tracking feed for L-band reception with associated tracking and receiving equipment for NOAA satellites.

Although primarily designed to cater to Landsat, SPOT, NOAA and other satellites orbiting 600-950 km above the Earth, the subsystem will have the flexibility to adapt to other future satellite systems with minimum hardware charges.

The data processing subsystem will have two high-density digital tape recorders; two VAX 11/780 computers, each with four Mbytes of main memory (MOS memory with error correcting code) and associated peripherals; Floating Point Systems' type 5210 array processors; I^2S image terminals; hard copy units; high-precision laser beam film recorders; and a geometric work station. The subsystem will receive ancillary and raw sensor data from the data acquisition subsystem and store these on the high-density digital tape recorders. The raw sensor data will be radiometrically and geometrically corrected to various levels of precision as required. Facilities for production of quick-look imagery will also be available. All data processing functions will be controlled by the VAX computers. The subsystem will generate standard or special CCTs for users/customers, besides performing various other functions such as calculating antenna pointer angles, etc.

Radiometric corrections will take into account sensor, sun angle, and atmospheric effects. Geometric corrections will be performed at two levels, leading to generation of system-corrected and precision-corrected data, respectively. System corrections will correct all errors due to irregularities in satellite and sensor movements. Precision corrections will require, in addition to the basic system corrections, accurate ground control points data. Further, depending on the level of precision required and/or the particular type of application being considered, digital terrain elevation data may also be used. System-and precision-corrected data will be available in the form of CCTs with densities of 1600 bpi and 6250 bpi in BIL or BSQ formats. The subsystem will, therefore, be able to accept CCTs from such foreign ground receiving stations using compatible formats for further processing, if so required.

The photograph processing subsystem will have processors, enlargers, and related miscellaneous equipment for production of satellite imagery in photographic form for users/customers. The final products will be high-quality black and white films or transparencies and paper prints as well as high-quality multispectral color composites in film or paper print form.

V. Application of Remote Sensing Data

Satellite remote sensing data acquired by SUPARCO are used both within and outside SUPARCO for a large variety of natural resources and environmental studies. About 55 user organisations in Pakistan are making or have made use of these data in one way or the other. Some of these user agencies have also conducted joint collaborative research projects/studies with SUPARCO. So far these studies have been mainly based on the use of MSS data, but SPOT HRV data are now being increasingly applied. Limited use has been made of other types of data. Some of the more important studies conducted in SUPARCO and other user agencies in the country are briefly described below.

A. Studies Conducted in SUPARCO

1. Agricultural Studies

A number of research studies have been carried out at RESACENT to examine the utility of Landsat MSS data for the identification and estimation of acreage for major field crops in Pakistan such as wheat, rice, and cotton. These studies involved the use of Landsat data of selected test sites for different parts of the year and different stages in the crop growing season. Visual and digital interpretation of the multitemporal and multispectral data was followed by ground truth surveying. It was found that MSS spatial resolution of 80-m although useful for delineation of vegetal cover in general, is inadequate for reliable crop detection in Pakistan because of the small sizes of agricultural fields and the prevalent mixed cropping patterns. It is expected that high resolution TM and SPOT data would appreciably improve the situation regarding identification of crops in Pakistan.

The other important studies have been related to monitoring of water-logging and salinity, detection of pest infestation in the cotton crop, identification and area estimation of mangrove vegetation in the Indus deltaic region, identification and estimation of fallow land, study of riverine forests, and mapping of forests in hilly areas.

Some of these studies were carried out in collaboration with other national user agencies. The study of pest infestation in the cotton crop was a collaborative effort between

SUPARCO and Pakistan Central Cotton Committee. SUPARCO is cooperating with the Sind Forest Department in the study of riverine forest. SUPARCO is also involved in a collaborative study with the Pakistan Agricultural Research Council regrding measurement of radiance values of wheat and maize using both ground-based radiometers and satellite data.

2. Water Resources Studies

The first application of Landsat MSS data in Pakistan was the estimation of the area affected by the 1973 floods in the Indus Basin. Landsat imagery was able to provide accurate figures for flood inundation quickly, conveniently, and inexpensively. The Indus basin floods of 1976 and 1978 were similarly studied. Recently, NOAA Automatic Picture Transmission(APT) images, routinely acquired at SUPARCO, have been used to obtain approximate estimates of flood inundation whenever higher resolution Landsat or SPOT data were not available.

Landsat MSS data have also been used to compute the areal extent of snow cover and river runoff for eventual use in river runoff prediction. Some empirical relations between maximum snow cover extant at the end of winter season and corresponding total river runoff during the subsequent spring/summer melting season have been derived for different subbasins. More satellite data are being analyzed and correlated with runoff data for deriving more reliable relationships suitable for use in river runoff forecastings.

In a recent study, Landsat data have been used to map glaciers over the inaccessible northern mountainous range of Pakistan. The approximate sizes of some of the glaciers have also been computed.

Landsat MSS data have also been used to study turbidity and siltation processes in the reservoirs of the Tarbela and Mangla Dams. The study has shown that the extent and position of suspended sediments can be monitored with the help of MSS data. Also, the data can qualitatively differentiate between different turbidity levels. Mathematical equations relating reservoir water level and water surface area for each dam have also been derived.

3. Geological Studies

Several studies based on photogeological interpretation of Landsat MSS imagery have been undertaken. One study involved the delineation of different type of geological features in some parts of the Makran coastal range in the Baluchistan Province. Various types of geological structures such as probable faults, folds, and other lineaments were identified and delineated. Hydrological processes in the coastal region were also studied.

A study based on Landsat MSS and RBV data was undertaken to assess the nature of sediment dynamics, morphological changes, and characteristics of lagoonal and estuarine water, tidal flats, environmental conditions, turbidity, and circulation patterns in various lagoons along the Pakistan coast. The complex nature of sediment deposition accompanying morphological changes in one of the lagoons over several decades has been studied through comparison among satellite, aerial, and historical ground data.

Another study, a hydrogeological investigation of the Bela Plain of Baluchistan Province, was carried out. Dominant rock-types, geological structure, drinage patterns, sedimentary patterns, soil moisture concentration, and vegetation were studied as surface indicators of subsurface hydrological conditions.

4. Environmental Studies

The morphology of the Indus Delta and the adjoining coastal belt of Pakistan was studied using Landsat data and a limited amount of aerial remote sensing data. Seasonal/diurnal changes in the high-water line, salt water intrusion, wave direction, sedimentation distribution, coastline changes, vegetation, and land use patterns were studied.

A study to monitor desertification processes in the arid areas of the country is currently underway in collaboration with the Pakistan Council of Research in Water Resources.

Both Landsat MSS and SPOT HRV data have been used for several land use/land cover studies both on a regional scale and in the urban areas of Karachi and Lahore.

Landsat MSS data have also been used to prepare a wetlands map of Pakistan showing 17 different types of wetlands.

SUPARCO and the University of Bristol, U.K. are collaborating in an R&D program for operational use of NOAA AVHRR data in several environmental applications. The first stage of collaborative program will be concerned with the development of software enabling operational application of NOAA AVHRR data to cloud cover monitoring, rainfall estimation, vegetation indexing, and snow cover mapping.

B. Studies Conducted by Other User Agencies

Satellite remote sensing data are being used by about 55 other user organisations in Pakistan for a very large variety of applications in the fields of agriculture, water resources, land use, forestry, geology, geography, oceanography, cartography, soil science etc. The Water and Power Development Authority (WAPDA) has applied Landsat data for studying water logging and salinity, dam monitoring, snow surveying, hydrogeological investigations, and other water resources studies. The irrigation departments of the provinces of Punjab and Sind have used Landsat MSS data for flood plain mapping, planning of flood control measures, and routine river surveying and mapping. Satellite data are also being used for a variety of geological applications such as basinal geological evaluation, regional geological mapping, lithological studies and structural and tectonic studies, especially with reference to their potential in oil/gas/mineral exploration. The Geological Survey of Pakistan, the Pakistan Atomic Energy Commission, the Oil and Gas Development Corporation, and some oil exploration companies are the major users of satellite data for geological applications. Landsat data have also been used by the Survey of Pakistan and the Soil Survey of Pakistan for surveying and updating of existing maps. Some universities in Pakistan have acquired Landsat data from SUPARCO for their research studies.

NOAA images acquired through APT stations at Karachi, Lahore, and Peshawar are also being routinely used by the Pakistan Meteorological Department in conjunction with ground-based meteorological and hydrological data for weather and flood forecasting.

VI. Aerial Remote Sensing

Aerial remote sensing can provide a useful input in studies requiring detailed observations or in situations where satellite data are unavailable. SUPARCO is in the process of expanding the scope of its aerial remote sensing activities. In additon to the multispectral camera, which has been flown on several aerial missions, an assembly of four Hasselblad ELX 500 cameras complete with necessary accessories has recently been acquired. These sophisticated cameras will be used for precision aerial photography. Other sophisticated cameras, scanners, and radar imaging devices are also being procured.

SUPARCO is contemplating the acquisition of a light aircraft of its own, to be fitted with different types of sensors, procured or developed in-house, for aerial remote sensing as a supplement to the satellite remote sensing program.

VII. Training Activities

SUPARCO has played a pioneering role in introducing and promoting remote sensing in Pakistan. As the national coordinator for remote sensing activities, SUPARCO regularly organizes short training courses twice a year for the benefit of scientists from other national user agencies. Twelve such courses have been held at the Remote Sensing Applications Centre at Karachi, in which a total of more than 200 scientists from diverse disciplines and professions and belonging to various federal and provincial government departments, autonomous bodies, and research and academic institutions have been provided training in remote sensing applications. In addition, SUPARCO has helped to organize, in collaboration with other user organizations, training seminars on special thematic applications in various cities of the country.

In 1977 SUPARCO and the United Nations (U.N.) jointly organized a Pak-U.N. Regional Seminar on Remote Sensing Applications at RESACENT for the benefit of member countries of the Economic And Social Commission For Asia And the Pacific (ESCAP) and Economic Commission for West Africa (ECWA) regions. The seminar was of two weeks duration and was attended by 44 pariticipants from 13 ESCAP and ECWA region countries, including Pakistan. In 1984 SUPARCO and Food And Agriculture Organization (FAO) jointly organized a Pak-FAO National Training Workshop on Computer Assisted Remote Sensing Techniques for Land use Management and Agricultural Crop Assessment at Karachi. The workshop was also of two weeks duration and was attended by 24 participants. In July 1989 SUPARCO hosted, in cooperation with the U.N. Outer Space Affairs Division, a workshop on remote sensing applications. SUPARCO and ESCAP also organized a remote sensing symposium under the ESCAP's Regional Remote Sensing Programme (RRSP)

SUPARCO has been assisting the Department of Physics, University of Karachi since 1973 in teaching an optional paper on space physics at the Master of Science level. Remote sensing now forms a part of this optional paper and is taught by SUPARCO scientists on a regular basis. Also, lectures on remote sensing are delivered to Master of Science (geography) students of the Karachi and Peshawar universities. Remote sensing has also been included in the Bachelor of Science and Master of Science space science syllabus of the Institute of Space Science, University of the Punjab, Lahore. SUPARCO is also providing teaching support to this Institute in various subjects of space science, including remote sensing.

Over the last 15 years, SUPARCO has also been sending its own scientists and engineers abroad to participate in various international/regional training courses, seminars, and workshops on different aspects of remote sensing technology and applications, which have been hosted or sponsored by foreign governments/space agencies and international/regional bodies such as the U.N., FAO, European Space Agency (ESA), ESCAP, Committee on Space Research (COSPAR) of the International Council of Scientific Unions (ICSU), etc.

VIII SUPARCO's Participation in International Activities

As the national space agency and coordinator for remote sensing, SUPARCO serves as Pakistan's focal point for all international and regional space-related activities and programs. In this capacity, SUPARCO has been liaising with ESCAP in connection with the RRSP and has participated in its various aspects. Senior SUPARCO scientists also participate in the meetings of the directors of the national remote sensing centers in the ESCAP region, the associated meetings of the Intergovernmental Consultative Committee (ICC), and the Asian Conference on Remote Sensing.

Pakistan is also a member of the U.N. Committee on the Peaceful Uses of Outer Space and COSPAR. SUPARCO represents the country on both these bodies and takes an active part in the deliberations on space-related activities, which include remote sensing. Recently SUPARCO has become a member of the International Union for Conservation of Nature and Natural Resources (IUCN) in order to collaborate in the use of remote sensing for monitoring changing ecological conditions, the impact of long-term climatic changes on environment, etc.

IX Future Activities

With the commissioning of SUPARCO's ground receiving station near Islamabad, user scientists in Pakistan will have access to Landsat MSS, TM, SPOT HRV, and NOAA AVHRR data in near real-time on a regular basis. The timely availability of satellite data will enhance their usefulness, particularly for the study of dynamic processes and phenomena. The station will hopefully provide a tremendous impetus to the remote sensing program in Pakistan.

Other future programs include further development and expansion of data analysis/interpretation facilities, especially in the area of digital image processing; use of data from different types of satellites/sensors for various resource and environmental applications; significant expansion of aerial remote sensing facilities and activities; development of GIS; R&D work on different types of remote sensors; and establishment of ARGOS facilities in Pakistan for collection of meteorological, hydrological and oceanographic data from unmanned platforms.

These developments should bring SUPARCO closer to its ultimate goal of launching an operational remote sensing program.

Use of the Spectroradiometer LI-1800 to Solve Problems of Preservation of the Environment

Miloslav Krizek*
Remote Sensing Center, Prague, Czechoslovakia

Abstract

In the Remote Sensing Center we are solving problems within the framework of the state plan of technical development, "Aerospace research of the environment of Czechoslovakia." For measurements in situ we are using spectroradiometer LI-1800. This device enables to measure spectral reflectance in the range of wavelengths from 300 nm to 1100 nm. It means that visible and near infrared (NIR) regions of spectra are covered. LI-1800 can be used to measure under laboratory conditions or in situ. Under laboratory conditions it has been used to measure more than 100 samples of typical soils. We correlate our measurements with main pedologic factors. In cooperation with the Slovak Academy of Sciences we measure in situ spectral reflectance of the different soil types simultaneously with overflight of the Landsat Thematic Mapper.

For purposes of observation of negative effects of open-cast coal mining on the surrounding environment, we have measured spectral reflectance of the forests. There are great differences between healthy and damaged trees, especially at the NIR spectral region. We are looking for dumps with rock-oil waste and its influence in water. In cases of ecological breakdown caused by leaks of rock-oil and its products, we measure reflectance of the vegetation. We use our spectroradiometer onboard a helicopter, too. In this case we can simultaneously take photographs and measure reflectance. We derived a relationship between optical density of photographs on different spectral regions and spectral reflectance.

Copyright © 1990 by Miloslav Krizek. Published by the American Institute of Aeronautics and Astronautics, Inc. with permission.
*Research Engineer, Geodetic and Cartographic Enterprise.

I. Introduction

For purposes of studying some spectral properties of the reflectance of different samples we are in our Remote Sensing Center using spectroradiometer LI-1800. This device enables measurement of spectral reflectance in the range of wavelengths from 300 nm to 1100 nm with measuring steps from 1 nm to 10 nm.

With this instrument it is possible to measure under laboratory conditions or in situ. The spectroradiometer has been used onboard a helicopter, too. In this case we can simultaneously take photographs of the Earth's surface and measure spectral reflectance.

We are looking for those methods that help preserve the environment.

II. Measurement of Soil Reflectance

Under laboratory conditions all main soil types, as much as 100 samples, have been investigated.[1] Each sample was measured in two physical states: air-dry and field water capacity wet. The set of typical soils was described by the following features: locality, soil type according to classification of the Food and Agricultural Organization, substratum, quantity and quality of humus, color, and granularity.

The relationship of the quantity of humus to the average value of reflectance was described by a regression curve with a third-degree polynomial. The value of the correlation coefficient was 0.58. When only one soil category was considered, e.g., loess, the relationship became resemblance. In this case the coefficient of correlation rose up to 0.72.

Humus horizons have increasing reflectance in both of the physical states in the measured range of wavelengths from 400 nm to 1100 nm.

III. Relationship between Reflectance and Film Density

During aerial photography from onboard the helicopter we measured spectral reflectance with our spectroradiometer. With help of the densitometer Theimer DDM 4, we measured the density of film used in the R, G, and B components. We than obtained the set of typical density coefficients listed in Table 1 for six different samples of the Earth's surface.

For this experiment three different films were used. One of them was color reversal film Orwochrom UT18 (50 ASA).

Reflectance ς was measured in the range of wavelengths from 400 nm to 1100 nm. In Table 1 there are reflectances corresponding to wavelengths as follows:

$$\lambda_R = 700 \text{ nm}$$
$$\lambda_G = 546 \text{ nm}$$
$$\lambda_B = 439 \text{ nm}$$

The altitude of taking photographs and measuring reflectance from onboard of helicopter was 200 m.

A relationship between optical density of exposed film and spectral reflectance was found:

X_i /i = 1...6/ are values of densities of different samples.
Y_i^1 /i = 1...6/ are values of spectral reflectances at one spectral band.

We found vector B as follows:

$$B = (X^T \cdot X)^{-1} \cdot X^T \cdot Y$$

After matrix processing, in the case of color reversal film Orwochrom UT18 we can obtain three resulting equations:

$$\varsigma_R = -0.1942\, D_R + 0.2110\, D_G - 0.0577\, D_B + 0.1888$$

$$\varsigma_G = -0.0446\, D_R - 0.0509\, D_G + 0.0274\, D_B + 0.1844$$

$$\varsigma_B = -0.0157\, D_R + 0.0823\, D_G - 0.0713\, D_B + 0.0597$$

Using these equations we can determine spectral reflectances of other samples of the Earth's surface, given knowledge of their density components.

Table 1 Density and reflectance of samples

	Alfalfa	Asphalt	Lake water	Mixed forest	Town dump	Meadow
D_R [1]	1.45	1.13	1.80	2.53	0.85	2.06
ς_R [%]	12.7	13.9	6.1	3.6	6.1	8.5
D_G [1]	1.17	1.03	1.44	2.35	0.74	1.85
ς_G [%]	10.1	11.2	7.0	2.7	4.8	6.1
D_B [1]	1.49	1.11	1.43	2.74	1.01	1.94
ς_B [%]	2.7	7.3	4.8	1.8	1.8	2.4

IV. Reflectance of the Healthy and Damaged Coniferous Trees

During observation of the surrounding environment of open-cast coal mining in northern Bohemia, we measured spectral reflectance of coniferous trees in situ. There were three types of spruces: silver spruce (Picea Pungens), Norwegian spruce (Picea Abies), and Canadian spruce (Picea Glauca).

In the case of Picea Glauca, there were minimal differences between spectral reflectances of healthy and damaged trees at both spectral bands (visible and NIR). In the case of Picea Pungens, there were no great differences at the visible spectral bands, but at NIR it was about 10 %. In the case of Picea Abies, by analogy the differences at NIR were about 15 %. The greatest damage to these coniferous trees is caused by emissions of sulfur oxides.

Very good discrimination for this purpose is possible with the help of color infrared film, e.g., Kodak infrared Aerochrom 2443.[3]

V. Summary

Spectroradiometer LI-1800 helps us to know which film materials are better to use for different types of samples. With its help we can also better understand the relationship between the density of used film and spectral reflectance. Measurements in situ and from onboard a helicopter help us to know more about spectral properties of our Earth's surface.

Acknowledgment

I want to thank to my colleague Milan Parik for his help with measurements on the densitometer, and spectroradiometer and in taking photographs.

References

[1] Krizek, M., Muricky, E., and Sefrna, L., "Spectral Reflectance of Soils for Aims of Remote Sensing," Fifth Symposium of the Working Group Remote Sensing, International Soil Science Society, Budapest, Hungary, Apr. 1988, pp. 186-189.

[2] Escadafal, R., Girard, M. C., and Courault, D., "Modeling the Relationship between Munsell Soil Color Spectral Properties," Fifth Symposium of the Working Group Remote Sensing, International Soil Science Society, Budapest, Hungary, Apr.1988, pp. 190-201.

[3] Kodak Data for Aerial Photography, Kodak publication M-29, Rochester, NY, 1982, pp. 74-77.

Chinese Very Small Aperture Terminal System for Ministries

Sen Dan*
China Broadcasting Satellite Corporation, Beijing, China

Abstract

The objective and technological approach of the Chinese very small aperture terminal (VSAT) system of data communications is described in this paper. The system is primarily designed for the management business of many governmental ministries and administrations. It consists of a centralized processing and switching facility and a number of groups of remote terminals. The network is constructed in a star configuration because of simplicity and the inherent nature of the management business. Either Intelsat or Chinese domestic communications satellite can be used for the space segment. The system performance has been verified by field trials. Some results of system analysis can be used for traffic design.

Nomenclature

BER	= bit error rate
BPSK	= binary phase-shift keying
b/s	= bit/second
CES	= central earth station
C/Io	= carrier power/interference density
C/IMo	= carrier power/intermodulation density
C/No	= carrier power/noise density
C/T	= carrier power/noise temperature
dB	= decibel
dBi	= decibel relative to an isotropic radiator
EIRP	= effective isotropic radiated power
E/No	= energy ratio
G/T	= figure of merit
Hz	= hertz
K	= Kelvin
PI	= primary interface

Copyright © 1990 by Sen Dan. Published by the American Institute of Aeronautics and Astronautics, Inc. with permission.
*Vice president and senior scientist.

SCPC   = single channel per carrier
SI   = secondary interface
TDM   = time-division multiplexing
W   = watt

1. Introduction

Many Chinese governmental ministries and administrations are lacking efficient communication links, which are of vital importance for management and information collection. The existing terrestrial facilities are limited in both quantity and quality. The VSAT system enables office-to-office communication with the advantages of flexibility, expedient installation, and low initial investment from users.

The China Broadcasting Satellite Corporation has implemented a satellite data communication system for users, who require private channels but not an independent system. Many governmental departments, including the State Seismological Bureau, Civil Aviation Administration, Ministry of Railway, Ministry of Energy, State Meteorological Bureau, and State General Administration of Customs, have joined this project. These users usually need reliable and continuous low-speed data links.

The data processing facility in the CES (hub station) and a number of VSATs were set up in July 1988. The performance demonstrations and field trials have shown satisfactory results. The system capacity will be greatly expanded to meet increasing demands in the next few years.

2. System Overview

The system is designed to provide a cost-effective solution for data communication in a very wide coverage, including some rugged areas. The system consists of a hub station located at Yungang, southwest of Beijing, and a number of remote terminals, as shown in Fig. 1. For each administration there is a central communication office connected to the hub station by terrestrial lines. By the nature of management the administrations have their own groups of terminals. Since the major capacity is used for traffic between Beijing centers and remote VSATs, star network configuration is a reasonable choice.

The outbound signal from Beijing to VSATs is a time-division multiplexed data stream of 57.6 kb/s packeted in central processor. All VSATs receive the TDM signal and transmit inbound signals with BPSK/SCPC modulation at a bit rate of 2400, 4800, or 9600 b/s. The data stream is

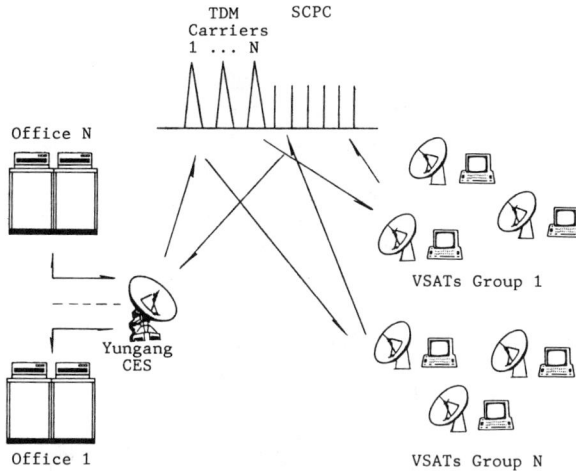

Fig. 1 Satellite data communication system.

differentially encoded, convolutionally encoded using 1/2 code rate, and then spread by encoding it with a pseudorandom signal to reduce power spectral density in accordance with the limitation set by the Consultative Committee of International Radio (CCIR).

The CES is equiped with a 13 m antenna and redundent low noise amplifiers, which provide G/T of 32.6 dB/K. Users' communication centers are connected to CES by multipoint-to-point digital microwave links.

A number of protocols have been developed for data processing and message switching in point-to-point and multipoint applications. Data to and from the user's host computer is interfaced to the satellite transmission facility through microprocessor-controlled data interface equipment. Each port on the host computer is served by a dedicated PI unit. At the VSAT end a SI unit is used between channel decoder and data terminals, such as a microcomputer, facsimile, and sometimes a vocoder or still-picture equipment. The central and remote interface modules establish and maintain contact using supervisory packet transmission. A microcomputer-based management system controls and monitors the network.

The VSAT station is a very compatible set with an antenna of 1.8-2.5 m diam in C-band, 2 W solid-state power amplifier, and low noise block of 70 deg K. Outdoor assembly consisting of antenna and mount, orthogonal mode transducer feed, low noise block, soild-state power amplifier, and upconverter can be easily installed on the roof of buildings. The indoor unit is housed in a table top enclosure. RS232C interfaces are available for connecting with user's data terminals. Fifty VSATs have been installed in different regions of China since July 1988. The objective in the next two yrs is 300 VSATs in operation.

Field trials have shown satisfactory BER performance. For uncoded BPSK, the theoretical value of E/N_o is 11.4 dB to achieve BER of 10E-7. Taking into consideration a coding gain of 5 dB with 1/2 forward error correction (FEC) and hardware implementation degradation of 2 dB, the practical E/N_o is 8.4 dB for a BER of 10E-7.

3. Information Transmission

There are two modes of data transmission. If the host computer is configured to operate in a point-to-point mode, the PI at CES works with and emulates one specific VSAT data terminal; whereas in multidrop configuration it serves as the interface to all remote VSATs and their attached communications terminals sequentially.

PI and SI have stored parameters that determine call sign and address, terminal equipment protocol support, crystal frequencies, and receiver tuning. They also participate in link management, diagnostics, and error recovery functions. The configuration of each PI and SI is stored at the system management microcomputer in the configuration file and can be downloaded to the electrically erasable programmable read-only memory (EEPROM) of PI and SI.

Since the system handles a data stream, whatever its content, any information data having the appropriate bit rate and level can be transmitted through the network. Although this VSAT system is not designed for telephone service, voice communication is found to be a need in many cases.

A linear predictive voice coder has been developed in the Beijing Institute of Satellite Telecommunications. It works in a 2400 b/s channel but actually at a bit rate of 2152 b/s, because some time slots in a packeted frame are allocated for data packeting parameters. Two-way transmission is necessary for both voice and synchronizing signals in response to a poll from VSAT or CES. The major problem encountered in such an application is time delay, which may exceed 1 s because of queuing, processing, polling, and space delays.

Another application is still-picture transmission. Television pictures taken by a television camera are picked and stored in frame memory, which is then driven by clock from VSAT. The sight pictures can be transmitted through the system frame by frame. The picture transmission is very useful in case of emergency, where pictures from remote sites need to be reported on a timely basis and terrestrial lines are usually inefficient. An experiment for monitoring floods along the Yellow River was successfully conducted in August 1988. An airborne television camera took flood pictures, which were transmitted to a shore VSAT via UHF channel. After frame memory the pictures were further transmitted to the Ministry of Waterconservancy.

4. Link Budget

The Chinese domestic communications satellite "CHINASAT-1" was launched in March 1988. The satellite has shown to be efficient for VSAT system operation. Link calculations are listed below:

Parameters
  Satellite
    Satellite location   87.5 deg K
    Saturated flux density   -82 dBW/sq.m
    EIRP (saturated)   34 dBW
    G/T   -4.5 dB/K
    Transponder saturated gain 116 dBW/(dBW/sq.m)
  CES
    Antenna diameter   13 m
    Transmit gain   56.4 dBi
    Receive gain   53.4 dBi
    G/T (clear sky)   32.6 dB/K
    G/T (faded)   31.6 dB/K
  VSAT
    Antenna diameter   1.8 m
    Transmit gain   39.2 dBi
    Receive gain   35.4 dBi
    G/T (clear sky)   14.1 dB/K
    G/T (faded)   13.2 dB/K
Inbound link
  Uplink
    EIRP   40.0 dBW
    Transmitted power (at feed) 1.2 W
    Spreading loss   -163.1 dB/sq.m
    Rain fade and other losses -4.0 dB
    power flux density at satellite -127.1 dBW/sq.m
    Gain of 1 sq.m antenna   -37.0 dB/sq.m
    Satellite G/T   -4.5 dB/K
    Uplink C/T   -168.6 dBW/K

Uplink C/No	60 dB-Hz

Downlink
Transponder linear gain with 4 dB output back-off	120 dBW/(dBW/sq.m)
EIRP per carrier	-7.1 dBW
Downlink free space loss	-196.5 dB
Rain fade and other losses	-1 dB
G/T of CES	31.6 dB
Downlink C/No	55.6 dB-Hz

Performance
Noise increment due to intermodulation C/IMo	70 deg K
C/IMo	58.9 dB-Hz
Noise increment due to interference and crosspole	300 deg K
C/Io	52.6 dB-Hz
Total C/No	49.8 dB
E/No	9.9 dB

Outbound link
Downlink
Data rate	57.6 kb/s
Required downlink C/No	59.0 dB-Hz
G/T of VSAT	-13.2 dB/K
Rain fade and other losses	1 dB
Downlink free space loss	196.5 dB
Required EIRP/carrier	14.7 dB

Uplink
Power flux density at satellite	-105.3 dBW/sq.m
Spreading loss	163.1 dB/sq.m
Rain fade and other losses	4 dB
EIRP/carrier	61.8 dBW
Gain of 1 sq.m antenna	-37 dB/sq.m
Satellite G/T	-4.5 dB/K
Uplink C/No	81.8 dB-Hz

Performance
Noise increment due to intermodulation	20 deg K
C/IMo	69.2 dB-Hz
Noise increment due to interference and crosspole	65 deg K
C/Io	63.2 dB-Hz
Total C/No	57.2 dB-Hz
E/No	9.6 dB

Since the minimum E/No with FEC decoding is 8.4 dB for a BER of 10E-7, results of the preceding calculations are valid for further analysis in traffic design.

In summary, a VSAT with a 1.8 m antenna and 1.2 W radiated power provides data transmission of 9.6 kb/s at a

BER of better than 10E-7. EIRP per SCPC is -7.1 dBW. In the hub station, 4 W power per carrier is required with a 13 m antenna to support data of 57.6 kb/s. Satellite EIRP in this case is 14.7 dBW. The large difference in EIRP for outbound and inbound has significant influence on channel arrangement of the system.

5. Traffic Evaluation

The objective of traffic design is to allocate outbound and inbound channels and hence to determine the number of VSATs incorporated, so that maximum use of a transponder's capability can be achieved. The total number of VSATs and outbounds is subject to power and bandwidth constraints of a satellite transponder. Based on the results of link calculations, a procedure of power and bandwidth tradeoff is given below for traffic consideration.

The channel parameters are defined as follows:
Pso = satellite operating EIRP
Pob = outbound EIRP per carrier
Pib = inbound EIRP of SCPC
Bt = total traffic bandwidth
DFo = outbound channel spacing
DFi = inbound channel spacing
Nob = number of outbound channels
Nvs = number of VSATs

The power and bandwidth constraints are approximately expressed by the following equations:

$$Pso = Nob*Pob + Nvs*Pib \qquad (1)$$

$$Bt = Nob*DFo + Nvs*DFi \qquad (2)$$

Bt should be equal or less than the bandwith of a transponder, which is assumed to be 36 MHz.

Fig. 2 illustrates these relationships with EIRPs of 26, 28, and 30 dBW, respectively. Channel parameters are adopted from link budget and existing hardware. The channel spacings are taken at 230.4 and 43.2 kHz for outbound and inbound, respectively. Although a specific group of parameters has been chosen for analysis, the results still have general meaning for traffic evaluation of the VSAT system.

It is interesting that the number of VSATs is around 700-800, though satellite EIRP varies from 26 to 30 dBW. Increased EIRP is used mainly by outbounds. This can be explained by the fact that the number of VSATs is bandwidth

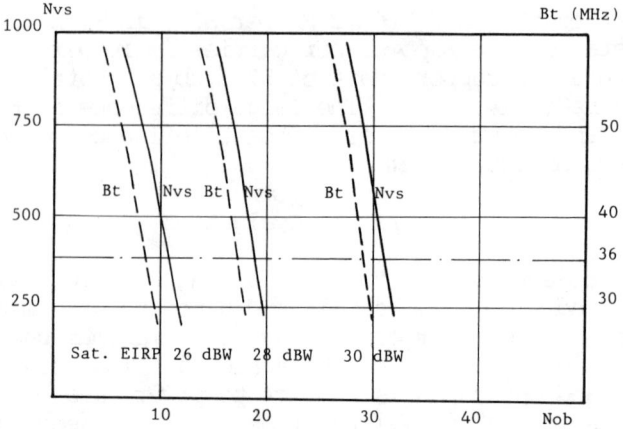

Fig. 2 Relationships between quantities of VSATs and outbounds.

limited, whereas outbounds are power limited. The desired average number of VSATs communicating with an outbound depends on the user's network scale, the nature of transmitted data, and priority requirements in the data packeting process.

6. Conclusion

The VSAT data communication system is a useful tool of national telecommunication with the capability of continuous data collection, distribution, and exchange.

Use of Satellite Communication for Technology Development and Transfer in Developing Countries

W. T. Weerakoon*
Center for Space Science and Technology, Kelaniya, Sri Lanka

Abstract

This paper examines the general architechture required for satellite communication systems as they become a powerful tool in accelerating the process of technology development in developing countries.

Formulating a satellite communication system's architecture involves three basic considerations : space, Earth and policy. The various problems and issues presently experienced in regard to these building blocks are discussed in order that developing countries might arrive at a suitable architecture for commercializing the space program of their respective nations. Since space-related hardware is always expensive, satellite systems already in operation, such as the existing INMARSAT and mobile SATCOM system in the Molniya orbit, are identified as suitable for this purpose.[1] The architecture proposed here is therefore designed around these two systems. It is anticipated that this type of project will later demonstrate the strength and weaknesses, cost-effectiveness, and simplicity of the architecture and then pave the way for creating a truly commercialized global network that can be put into operation within the space sector of developing countries.

INTRODUCTION

Since the launching of the Early Bird in 1964, based on Arthur C. Clarke's concept of global communication, space technology and its applications have demonstrated the potential for using satellite systems as a powerful tool in

---
Copyright © 1990 by the American Institute of Aeronautics and Astronautics, Inc. All rights reserved.
*Director-General.

socio-economic development. During the past few years, such systems have played a prominent role in the fields of telecommunication, meteorology, remote sensing, data transfer, and mapping of Earth resources. Pilot projects have also shown that such systems are cheaper, more effective, and more efficient in the long run.[2] Past experience in countries like Indonesia, the Soviet Union, Canada, United States, South America, Europe and Australia has demonstrated the capability, power, and usefulness of these systems as a tool for development.[3] In 1986, INTELSAT launched the "Project Share" experiment primarily to provide limited facilities to developing countries for acquiring the basic professional skills required to operate and man such systems for future development strategies. This project was funded by INTELSAT, and it is said to have been successful in acquiring preliminary data and demonstrating the usefulness of satellite communication systems for developing countries[4].

The rural satellite program launched by United States of America International Development Agency in 1980 to assist developing countries in exploring the use of satellite communications for rural development also has proved its credibility and paved the way for the use of such systems permanently in South America and Australia. Japan, China and Indonesia are now experimenting with the same idea. All these pilot programs at first were designed mainly to enable both developed and developing countries to apply satellite technology to areas of development and to find the most effective and appropriate means of using such technology as a development tool. These programs also had another dimension: to encourage national governments, international agencies, development banks and industries to utilize satellite communication systems as a tool for accelerating their development strategies.

Harnessing science and technology for national development is a must, and each country should therefore plan its strategy with a combination of technologies and techniques appropriate and suitable to its conditions. Since the gap between the existing national level of science and technology and the state of the art internationally is usually high, it becomes necessary to adopt techniques, equipment, technologies, and methods developed elsewhere. This can be done easily by cooperating with developed countries and sharing their technologies through agreed technology – transfer schemes. Cooperation in this field implies the use of large and versatile scientific and technological knowledge, powerful and advanced industrial

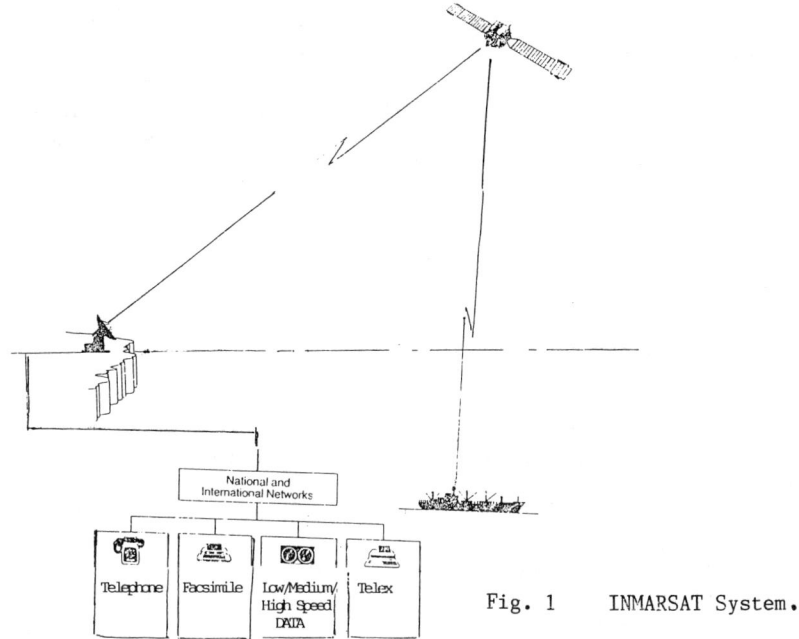

Fig. 1   INMARSAT System.

capacity, substantial financial resources, worldwide telemetry, and other supporting services. Therefore, bilateral cooperation in space activities becomes useful.

It was also clear in the recent COSPAR 25th plenary session in 1984 that although decisions regarding development activities rest with the scientists and policy makers of that country, external resource's can play a major role in amplifying, accelerating, and desseminating such benefits to developing countries.[3]

It appears that we have a long way to go to fully realize the fruits of the current space research program, which is now expanding rapidly. In developing countries a challenge also awaits scientists, engineers, policy makers, and entrepreneurs to utilize such technology, developed by others and now freely available to them to accelerate their speed of development and to keep abreast with the New World.

Technology Transfer via Satellites

The existing telecommunication networks of developing countries, including both hardware and software, need to be critically examined and updated in light of changes that have taken place during the past few years and are also envisaged for the future. This is considered the prerequisite that will determine the rate of acceleraton of the commercialization programs of developing countries in the future. It is also not possible to immediately expect radical changes in technology development from any efficient system without giving consideration to those who operate them. The level of excellence one can expect from their own staff, overseas and locally trained with limited practical experience is less than that encountered in industrialized societies.

It must also be noted that the current high level of expertise in more developed countries has been acquired through years of experiment, professional interaction, and experience. Therefore, it becomes important at first to develop simple systems and gradually expand, at the same time concentrating on training the manpower required to operate such systems in the shortest possible time. The use of satellites for training manpower to meet this challenge is no more a dream than a reality. Taking into consideration the foregoing facts, a pilot project directed toward technology transfer should be organised to:

1) Provide an intergrated communication system for selected technology areas.
2) Undertake technology transfer when necessary to organise projects that will generate products for both export and local markets using imported technology.
3) Undertake the transfer of data and information considered to be vital for the socioeconomic development of developing countries.

It is also felt that to understand and outline the actual requirements for a general architecture necessary to propagate satellites communications, the design of a simple system using the existing resources in the space sector is extremely useful at the initial stages. A pilot project of this nature is also capable of providing an opportunity for local scientists and professionals to interact with their counterparts and participate in futuristic development strategies in a realistic manner. On the other hand, with developing countries under the guidance of recognized institutions in developed nations in initiating the development of satellite communication technology, money lending institutions like the World Bank can be convinced to lend its money to upgrade the existing telecommunication infrastructures considered vital to the introduction and success of satellite communication technology. In this manner, developing countries can become equal trading partners and contribute their share to the future in terms of a global development strategy.

It has now been accepted that all developing countries should acquire within a short period the technological capabilities to meet future challenges.[5] Without this, their very survival will be at a risk. It is in this context that satellite communication will become useful for developing countries in accelerating their technology development activities as they enter the 21st century. These countries can acquire a technological capability either by developing modern technologies in their own nations or by working with developed countries through electronic media, thereby acquiring the necessary skills through technology – transfer processes to continue their development process. As pointed out earlier, the former is a difficult task, involving time and resources, skilled and experienced scientists and engineers, and many other inputs from the world communications infrastructure. The latter is much cheaper, easier, quicker, and reliable in the long run. It has also been pointed out that until developing countries acquire technological skills, the forces linked to technological growth and cost reduction in services and equipment that enhance productivity and strengthen a national economy will always generate from developing countries who presently control the technology.[5]

After taking these factors into consideration, it seems much wiser for developing countries to make the second choice rather than the first. The information required to undertake any kind of new technology is now available in developed countries and is easily accessible through well-designed satellite communication systems and technology-transfer agree-ments. It is also important to consider undertaking specific technology development projects by collaborative methods using satellite technology to develop useful in-house technology capabilities that will then spill over into an existing system making it richer and more meaningful in terms of development.

Basic Issues in Telecommunications

Developing countries are distributed all around the globe and are at various development stages and of various sizes. Each has its own culture, traditions, political system, people, languages, social and religious practices. Any network designed to accelerate technology development must take into consideration these factors as well as some of the present issues in telecommunications. A telecommunication's network acts as the nervous system of a country, and only with a good telecommunications system can any country look forward to acquiring benefits from the recent advanced space research and development programs.

Although space technology has triggered the gift of many benefits to mankind, such rewards have yet to come to developing countries. International telephones, point-to-point data communications, the facsimile, dissemination of television and radio programs, the transmission of signals by terrestrial transmitters and the distribution of TV programs by a direct broadcasting system using satellites are a few such benefits.

For more than 25 years, the role of telecommunications in overall socioeconomic development has been discussed and analyzed, and various proposals have been offered by many experts and international agencies. Developing countries, however, still not been able to solve their problems. This may be due to the fact that the five components of telecommunications development, finance, capital, maintenance, and quality service, were not considered in trying to solve the present problems experienced in providing efficient service and expanding the tele - communication facilities within developing countries. In mid-1960, the United Nations and many other international agencies addressed this question, but with the emergence of satellite communication, such organisations began to look at ways in which satellites could be utilized for accelerating development. Due to this diversification of international strategies together wth factors to be outlined next, developing countries were left behind.

One of the unique features of rapidly expanding economies is the possibility of the transfer of information from one point to another in an increasingly efficient manner. According to Goldschmidt,[6] this is not possible yet in developing countries because of the following conditions :

1) Available facilities are overloaded and overused, leading to poor overall quality of service.
2) Major economies of scale inherent in telecommuni-

cations are lost and incur tremendous cost to the existing network when expanding its facilities.
3) The private and public sector investments have not yet made a meaningful impact, mainly because investors give to areas where communication is already available, thus limiting economic activities in urban areas.
4) Inability to plan and replace existing telecommunication systems in a systematic and methodical manner.
5) Telecommunication systems are operated in an inefficient and unprofitable manner, becoming an added financial burden to consumers, thus limiting system expansion.
6) Preventive maintenance systems are very poor, with breakdowns occuring very frequently, thus making the system unsuitable and inefficient.
7) The work force is not properly trained, as such, it contributes to frequent breakdowns, additional costs, loss in confidence, and loss of all the innovations brought to the system.

Another important aspect to consider when developing telecommunication system is the nature of investments. Since development is capital-intensive and many international lending institutions are reluctant to provide large capital to or invest their money in developing countries at a time when third-world debts are very high, faster development is not possible. Because of these circumstances, some of the basic issues to be identified are as follows:

1) state of the art of telecommunications in the developing country;
2) high cost of product and services related to telecommunications;
3) poor quality maintenance;
4) lack of venture capital;
5) objective manpower development policies;
6) absence of sound technology policies.

Among these issues, venture capital and poor maintenance require the highest priority. For years to come, capital formation will remain a serious challenge and problem for developing countries as they attempt to adapt and build up their existing systems to develop an information-based economy that is badly needed for technological development. Many international development agencies now accept telecommunications as an essential growth factor, and their contributions are increasing, but because of the heavy debt incurred by developing countries, this policy is now being reexamined. Therefore, developing countries need to embark on new strategies for raising funds. In industrial societies, the growth of hightech industry has been greatly stimulated and funded by venture capitalists. A venture capitalist can be found anywhere in societies where a liberalized economy exists. The concept of involving the venture capitalist in development is somewhat new to developing countries.

However, attempts must be made once again both by international as well as national institutions to restructure their economy in order to allow foreign participation in local development. This will necessitate the formation of some new economic and political principles and strategies based on the new international order and a North-South dialogue which was initiated by United Nations in Geneva. However, many techniques are already available for financing development projects initiated by developing countries. Venture participating, piggy backing on commercial invest= ments, sharing cost regionally, and changing development strategies to suit existing conditions are a few such techniques that can now be used for the establishment of telecommunications in developing countries.

Poor-quality maintenance is another issue that needs closer examination. This can be attributed to the lack of objective manpower development policies. With the modernization of both software and hardware, training and education become crucial to maintaining quality services in telecommunications. Not enough emphasis has been placed on this area. It will be some time before developing countries can fine tune their creative and innovative abilities to provide the necessary expertise to harness science and technology and offer quality service in telecommunications. Until such time,

1) It can be arranged to transfer high-tech products as well as technology from industrially developed countries.
2) Developing countries can engage in partial production and final assemblies of electronic products for domestic markets based on imported technologies.
3) Developing countries can undertake research and development to improve their indigineous technologies using innovative techniques that can be transferred from more developed countries.

In this manner, developing countries can keep abreast of global progress through an electronic network of the type proposed in this paper using the hardware of existing space sectors.

General Architecture

Having discussed the problems and issues related to the space, Earth and policy sectors, we will now attempt to outline a general architecture for the purpose of SATCOM network for the Asian region. The basic facilities required by the space sector are satellites; controlling stations and exchange facilities to accept and transfer data and information rapidly, effectively, and economically in the desired manner and not only in voice; pictures and radio text, also in color and fascimile form in both a compressed and uncompressed manner, coded or otherwise. By scanning existing facilities in a space sector, it can be concluded whether or not such facilities are adequate to provide the required services and can be easily utilized without much modification and additional cost.

It is therefore proposed that both the international maritime satellite communication system (INMARSAT) and the mobile satellite communication system (mobile SATCOM) be used for such an architecture. However, if mobile SATCOM systems are found more suitable, after a pilot proejct a decision can be made to use only such systems. Since a decision of this nature will depend very much on a cost-benefit analysis based on a time parameter, it is not possible at this stage to say one is better than the other. Earth stations are always costly to develop and need trained manpower for their management. INMARSAT is developing an Earth station cheaper than existing ones, and in time, by testing it in the context of this proposed network, will evaluate its feasibility. Mobile SATCOMS have a number of advantages as compared to geo-stationary satellite systems :

1) The cost of mobile SATCOM Earth stations is very much less than that of geo-stationary satellite systems.
2) Geo-stationary satellites provide good global coverage, but suffer from severe propagation problems.
3) Mobile SATCOMS have 100 times more antenn-to-antenna power availability from base stations than geo-stationary satellites.

4) For all satellite operating in the MOLNIYA orbit, although it uses a three-satellite constellation for 24-hour coverage, the launching cost is half that of the geo-stationary satellites.

5) The capital cost of a mobile SATCOM system tends to be directly related to the amount of radio frequency power needed for the link. This means that for any configuration that reduces the power needed per voice channel, the system becomes more commercially attractive.

The INMARSAT system has three major components; the satellite leased by the organization, the coast Earth stations, and the ship Earth stations as shown in Fig. 1.[7]

It is a geo-stationary satellite system parked in orbit 36,000 Km. above the equator. It is connected by direct leased lines to satellite control stations, from which the

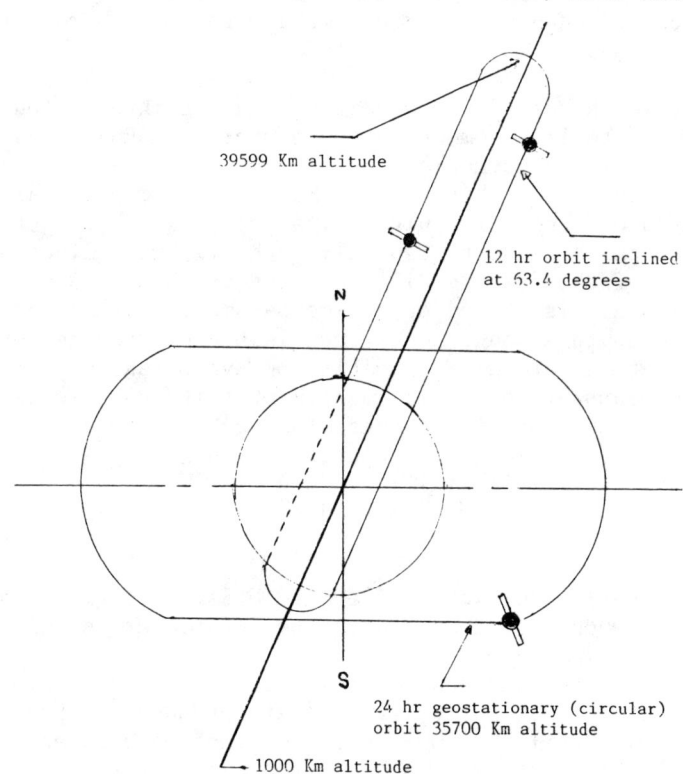

Fig. 2   Molniya Satellite Orbit.

SATELLITE COMMUNICATION APPLICATIONS

satellite is linked by its own ship stations to the Indian Ocean and Atlantic satellites and to all coast stations around the world.

The mobile SATCOM proposed is a 12-hour capital Molniya orbit satellite system that is, at present, used extensively by the Sovitet Union. See Fig.2. For 24-hour coverage, three satellites in three elliptical orbital planes are needed. A system of this nature is less costly to operate and is within the reach of developing countries.

The Earth sector is the one that needs to be carefully modified and redesigned to accomodate the changes taking place in worldwide telecommunication networks. All local networks msut be redesigned so that they can be easily connected to terminals processing artificial intelligence and also to microprocessors, and microprocessor-controlled transmitters and receivers. They must also be capable of collecting only selected information from data banks, processing this information if necessary, and transferring it to the required destination in an efficient and economical manner.

One of the most important criteria to be satisfied in regard to the Earth sector is the compatibility of the local network with integrated-service digital networks.[8]

High-speed data-transmission phone lines will soon become popular in technologically advanced countries. This new standard is based on a single network that ends the needless duplication of seperate telephone and telex circuits and is capable of adapting itself to future changes. This type of network (ISDN) allows users to bring telex services to isolated spots. It can also accomodate improved data-transmission speed for fascimile, electronic mail, and other computer - related services. These are, in fact important services needed by developing countries in order to accelerate their socioeconomic development. It is, therefore, important to upgrade the local network to the level of digital networks.

There are specific advantages to introducing ISDN technology to local area networks :

1) Users can acquire new services directly though their lines, as shown in the proposed architecture.
2) Caller identification services allow business institutions to take orders over the phone and also

display automatically on screen customer files through a second channel, with the initial call still in progress, and update these files in consultation with their customers. The same technique can be brought to the factory floor with CAD/CAM systems for design alteration.
3) By 1990, ISDN will be a fact in the North, and therefore, it will become necessary for the South to update its local networks and its software and hardware systems to be able to connect to the North with the least cost.

One of the unique features of this network is that all expensive computers and data processing equipment can be located in another country. At the initial stage, it only requires the receiving terminals capable of collecting the data and information and displaying them in a manner useful to the user. An information generator and processor

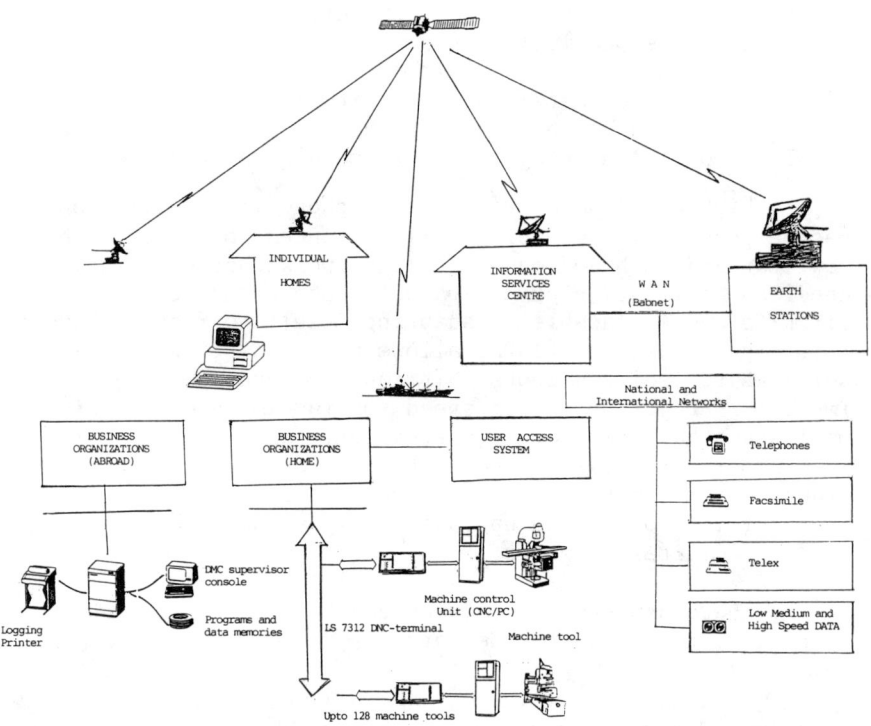

Fig. 3   Proposed Satellite Communication Network.

can also be kept at the other end as shown in the proposed architecture. Extra cost, expensive equipment, and manpower can be initially avoided since collaborators in developed countries can attend to these details. A modern factory can be completely automated in this manner (see Fig. 3) Scientists, engineers, and doctors can utilize this network to communicate with their counterparts or laboratories to upgrade their knowledge and exchange information. Thus, a network of this nature will serve as a hub for development and will motivate developing countries toward meeting their goals.

Conclusions

The present world situation makes it necessary for developing countries to build a technology - based economy. However, for such nations to achieve this goal, they need to be assisted and supported in the fundamental scientific research needed to upgrade their technological capabilities in order to accelerate their development strategies. It is clear that satellite communication technology can be used as a tool, easily and economically, to achieve this objective. The case for the use of this technology has still not been stated convincingly and forcefully enough to attract the development agencies that must willingly invest or lend money in this area. Whereas developing countries will have to alter their political and investment strategies, developed countries as well as venture capitalist, including banks, will have to take a closer look at investing in developing countries and seeing them not as a poor risk but as a future market.

The need also exist to demonstrate to both the user and investor the benefits that can be gained by commercializing space technologies, specifically satellite communication technology. It, therefore, can be said that with renewed discussions taking place either to remove or reduce third word debts, a new and viable approach can be taken in commercializing space technologies in developing countries with the emphasis on regional cooperation. However, international cooperation is needed in order to propagate advanced communication technologies thus developed to developing countries.

Cooperation among nations, public - and private-sector institutions will become particularly necessary if access to the use of satellite technology is to be improved significantly. The ongoing pilot projects sponsored by INTELSAT, INMARSAT, ESA and USTTI, and other programs organised by the UN such as ITU's training programs for manpower development and telecommunications policy changes, UNESCO's program for communications development, and the UN's space application program directed towards the peaceful use of space will become extremely useful for establishing the initial foundation for the network proposed in Fig.3.

For any future expansion in the field of satellite communication, the World Bank, and major Asian, African and Arabian banks should be encouraged by development agencies, as well as those who own such technology, to become involved in these development programs, encouraging and assisting developing countries in structuring their communication networks in such manner so as to bring revenue to the system.

Finally, since we now know that advanced technologies can accelerate the speed of socioeconomic development with the use of satellites, the highest priority must be given to roganizing technology-transfer programs via satellites using the unused satellite - time where possible. We close our presentation with the words of Arthur C. Clarke:[10]

> During the last decade, something new has come into the world. Two-dimensional communications networks are replacing vertical chains of command, in which orders moved downwards and only acknowledgements went upwards. We are witnessing the rise of the Global Family or Tribe, if you like. Its electronically linked members will be scattered across the face of the planet, and its loyalties and interests will transcend all the ancient frontiers.

References

1. Norbury, J.R., "Mobile Satcoms for the Future," Journal of Electronic Engineers, July/Aug. 1987

2. Carver, J.H., "Promoting Co-Operation in the Peaceful Users of Outer Space, "Proceedings of the COSPAR Workshop on Promotion of Space Research in Developing Countries, July 1984, 23-29

3. Sunaryo, R., "The Indonesian Experience in Space Research and How COSPAR Can Help in the Future," Proceedings of the COSPAR Workshop on Promotion of Space Research in Developing Countries, July 1984, 109-111

4.   "Project Share", INTELSAT pilot project, 1985/1986.

5.   Singh, I.B., Telecommunication and Development Prospects for 21st Century, <u>Proceedings of an International Forum on Telecommunication's for Development,</u> Oct. 1986, 52-63

6.   Goldschmidt, D., "Financing Telecommunications for Rural Development, "<u>Telecommunications Policy</u>, Butterworth, New York, 1984, 181-203

7.   Lundberg, O., "INMARSAT Spreading It's Wings," <u>Interdisciplinary Science Review</u>, Vol. 11, 1986, 50-63

8.   Kobayashi, K., "Shaping Communications Industry to Meet the Needs of the ISDN Age", Telecommunication, Vol. 53, 1986, 183-210

9.   "The Challenge of Change", International Symposium of Ministers of Communications, Vancouver, British Columbia, Canada, June 9-11, 1986

10.   "The Impact of Space Exploration on Mankind", Pontificiae Academiae Scientiarvm Documenta, the Vatican., 1-5, 1984

11.   Clarke, A.C., "Star Peace", Address on receipt of 10th Charles A. Lindbergh award, Paris, May 1987.

Low Traffic Density, Small Terminal Network, and Satellite Antenna Design for Communications in the Rural Areas

L. Bardelli,* F. Martinino,† and F. Rispoli‡
Selenia Spazio, Rome, Italy

Abstract

This paper deals with the applications of satellite communications techniques for improving the very poor telecommunications scenario in developing countries and to access isolated and dispersed users.

After a brief introduction to the current situation in rural areas of the world, the main characteristics that determine the strategy for communications problem-solving are summarized, and the most important requirements for both the earth segment and the space segment are outlined.

Some system considerations for satellite network design and outline of the main criteria that must match the rural areas requirements follow. In particular, two main concepts for satellite multiple access techniques are proposed: Single Channel per Carrier/Demand Assignment Multiple Access (SCPC/DAMA) and Code Division Multiple Access/Spread Spectrum (CDMA/SS).

Two examples of applications conclude the paper; the first one concerns a CDMA/SS network architecture design for 9600 b/s voice communication and TV programs distribution in African countries; the second one is relative to an on-board reconfigurable multispot antenna design to optimize the coverage of African regions.

Introduction

The communications environment of developing countries is quite different from that of the industrialized countries. Voice communications in rural areas are usually characterized by a limited number of telephone users spread across a large region. The situation is especially unsatis-

---

Copyright © 1990 by the American Institute of Aeronautics and Astronautics, Inc. All rights reserved.
* Space Division, Commercial Systems.
† Ground Division, System Engineering.
‡ Space Division, Antenna Department.

factory in Africa, where the telephone density is 0.7 telephone per 100 inhabitants, whereas the Asia-Pacific region, which constitutes 55% of the world's population, has only 17% of the world's telephones.

Domestic TV distribution is still very uncommon and data communications or services such as fax-simile, telefax, etc., are nearly nonexistent. The rural area users can be represented by isolated communities such as villages or perhaps remote commercial establishments (i.e., exploration camps in Canada, ranchers in Australia, offshore oil platforms in the North Sea) situated in places where telecommunications are generally not available. However, the trouble with rural areas is that they suffer poor geographic conditions, low income levels, low population densities, and underdeveloped social infrastructures that cause technical and economic problems.

From these brief considerations it is clear that one of the most important tools of the development of rural areas (especially in developing countries) is telecommunications service, in particular voice communications and TV programs distribution, to give then a proper link with industrialized countries.

In recent years the use of satellite technology has become more and more attractive [1] for improving the telecommunications scenario in these areas. This paper deals with the use of a satellite of the Olympus class to implement the incoming primary telecommunications services, and to access dispersed users.

Two main concepts have been studied for the terrestrial network: SCPC/DAMA and CDMA/SS. An example of the CDMA/SS with link budget and design has been worked out. An on-board reconfigurable antenna and relevant spots to cover African regions are illustrated in the next pages.

Rural Area Communications Requirements

The main aspects of rural areas context, that determine the strategy for communications problem solving, are the following:
1) very low population density;

2) almost total absence of telecommunications infrastructures and services;

3) nearly complete absence of other important social service such as electrical power distribution, roads, etc.;

4) existence of only a few large cities, with small villages often very far from the capital or other cities; and

5) need of rapid development in the communications which will affect the growth of the other areas.

A satellite system allows a communications network to be set up in shorter time than conventional terrestrial transmission system such as cable and/or fiber optics networks and terrestrial radiowave links. Moreover, in a country with vast territory and low population density, the setup of terrestrial networks with high grade penetration could easily become too costly. Because use of a satellite system can allow the user premises communications with a cost that is almost independent of the network penetration grade.

Today satellite technology gives a first fast step toward future development that will be based on subsequent terrestrial facilities whose implementation requires some infrastructure services that are still to be developed.

From this point it is easy to derive the principal requirements of the satellite network design for the space segment and earth segment. These requirements are summarized as follows:

space segment

1) dedicated payloads with high gain spot beams;

2) use of frequency bands for the uplink and downlink to avoid high rain attenuations while obtaining good antenna gains (typically Ku-band);

3) possibility of frequency and polarization reuse; and

4) channels bandwidth designed in order to accommodate efficiently telecommunications services.

earth segment

1) networks with no sophisticated common resource control;

2) low cost terminals;

3) low power consumption for use of alternative electrical source such as solar panels, thermoelectrical generators, etc.;

4) small aperture terminals to guarantee easy installation and maintenance;

5) possibility of an outcome of more than one telecommunications service using the same network (telephone and/or data communications and TV services); and

6) open network facilities to interface by means of the master station, (generally located in the capital), terrestrial network (i.e., public switching network, cable TV, TV radiowave links).

System Considerations

In the design of a satellite network for rural areas, it is important to consider the main requirements summarized in the previous paragraphs, as some of these can be real constraints.

At first, an evaluation of the required network resource must be made in order to define the number of circuits (bidirectional channels) needed; furthermore, a grade of service must be established as close as possible to the specific environment under consideration.[2]

The network resource can be represented by A, as expressed in Erlang for the required traffic, while the grade of service is Pb, given by the probability of blockage; the link availability is D and the bit error probability is Pe. Typical values of these parameters can be

$$Pb = 1\%$$
$$D = 99.5\% \qquad (1)$$
$$Pe = 10^{-3}$$

with a traffic intensity (A) that can vary from 10 to 20 Erlang. Assuming the basic Erlang B equation

$$Pb = \frac{A^N / N!}{\sum_{K=0}^{N} (A^K / K!)} \qquad (2)$$

where Pb is the probability of blockage, A the traffic, and N the total number of circuits available for the network; for a given traffic intensity A and a given probability of blockage Pb, the circuits number N can be derived.

The probability of blockage is the number of calls receiving a busy signal to the total number of incoming calls, moreover, the above model assumes that the calls arrive randomly and that all calls receiving a busy signal are lost. Furthermore, the peak hour calls number N that must be served can be derived from the intensity traffic value A:

$$N = A/A_o \qquad (3)$$

A_o being the source average traffic value that can be estimated on a statistical base.

From a satellite point of view a key aspect is the multiple access technique selection for sharing the total available network resource. From the considerations made in the previous paragraph the two main concepts demonstrate a good solution for the problem, and these are SCPC/DAMA and CDMA/SS.

In the first case the network resource is represented by the frequencies number that must be assigned to the system; for a given circuits number N, the frequencies number is 2N. In the second case the resource is the number of the codes that can be active at the same time for the system and that will be equal to 2N.

It is not our intention to carry out a comparison analysis between these two different access techniques and their relative performances. It should only be noted that the SCPC/DAMA is a typical system based on centralized control, while the CDMA/SS can be implemented by using a decentralized resource access control.

Satellite Network Architectures: A Example of CDMA/SS

The following system analysis refers to a hypothetical African satellite whose transponder characteristics could be similar to that of the Olympus 1 Specialized Service Communications System.[3]

It is assumed that one 27 MHz transponder is for TV distribution and only one-third of another transponder is for telephone communications.

Telephone Communications

The CDMA/SS multiple access technique is assumed. Moreover, it is assumed that the coded voice signal is at 9600 bit/s, with the requested maximum bit error rate equal to 10^{-3} for at least 99.5% of the time. In the case of the

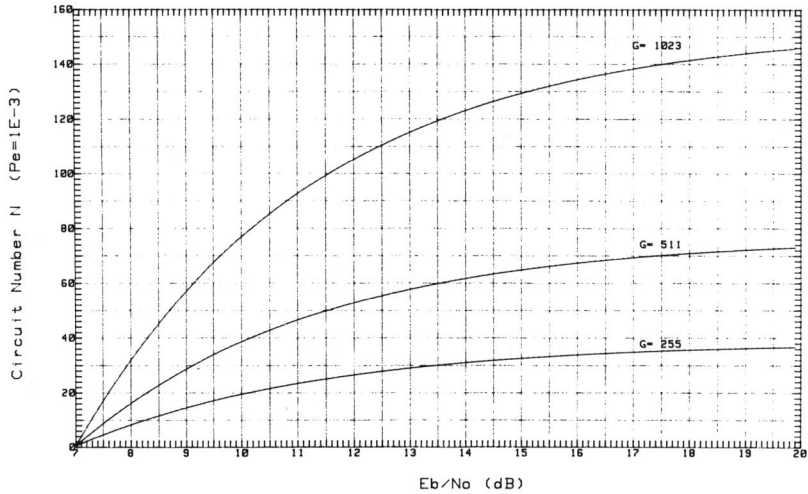

Fig. 1 Circuits number N vs Eb/No for different code gain values [with (Eb/No)e = 7 dB].

CDMA/SS technique an equivalent Eb/No ratio is defined as [4]

$$\left(\frac{E_b}{N_o}\right)_e = \frac{E_b/N_o}{1 + \frac{2}{3}\left(\frac{M-1}{G}\right)\frac{E_b}{N_o}} \quad (4)$$

where

M = number of simultaneously transmitted codes with the same power

G = code gain

$(E_b/N_o)_e$ = E_b/N_o value to have a given error probability P_e for one transmitted code (M = 1)

E_b/N_o = E_b/N_o value to have the same given error probability P_e for M simultaneously transmitted codes

Defining the desired transmission quality in terms of bit error probability P_e, then $(E_b/N_o)_e$ value is known and from

Eq. (4) it is possible to derive an expression that gives the number M of simultaneously transmitted codes as function of E_b/N_o and G values, that is

$$M = \frac{3}{2} G \left[\left(\frac{E_b}{N_o}\right)_e^{-1} - \left(\frac{E_b}{N_o}\right)^{-1} \right] + 1 \qquad (5)$$

because a two-way communications requires two codes, the number of telephone circuits N will be

$$N = \frac{3}{4} G \left[\left(\frac{E_b}{N_o}\right)_e^{-1} - \left(\frac{E_b}{N_o}\right)^{-1} \right] + \frac{1}{2} \qquad (6)$$

It has to be noted that, if

$$\frac{E_b}{N_o} = \left(\frac{E_b}{N_o}\right)_e$$

we have: $N = 1/2$

that is, a one-way transmission only is possible, while for $Eb/No \longrightarrow \infty$ we obtain the maximum of available circuit number N_{Max} for a given code gain G

$$N_{MAX} = \frac{3}{4} G \left(\frac{E_b}{N_o}\right)_e^{-1} + \frac{1}{2} \qquad (7)$$

Assuming the use of Binary Phase Shift Key (BPSK) modulation, an $(Eb/No)e$ of 7 dB is necessary for transmission without any codification.[5]

With reference to this situation the circuit number N for different values of Eb/No and code gain G can be evaluated; in particular in Fig. 1, Nc vs Eb/No for G = 255, G = 511, and G = 1023 is given. It is possible to note from Eq. (7) or from Fig. 1 that Nc is linear in G.

It is important to underline that

where

$$G = \frac{R_c}{R_b} = 2^\alpha - 1 \qquad (8)$$

R_c = chip rate

R_b = bit rate

$\alpha = \lg_2(L_c + 1)$

with L_c being code length; moreover, for the spread spectrum signal with BPSK modulation the required bandwidth B_s is:

$$B_s = G R_b (1 + \alpha) \qquad (9)$$

where α is the roll-off that is chosen about 0.8 to have a 9 MHz bandwidth with G = 511.

The approach proposed involves the use of a fully digital spread spectrum modem that gives a loss of about 2 dB for the use of hard decision on the chip in the baseband receiver.[6] Moreover, about 0.5 dB of implementation loss in the demodulation data-aided processes has to be considered. A total loss of 3 dB with respect to the ideal code gain can be assumed, a resulting performance is shown in Fig. 1 considering the curve with G = 256.

A capacity A of about 12 Erlang can be assumed as a typical value to give a good performance in terms of service quality. Assuming the Erlang B formula (last calls are cleared), we can evaluate the required circuits number N for a given probability of blockage Pb of the network.

Figure 2 shows the traffic capacity A vs the circuits number N for a probability of blockage Pb of 0.01 and 0.001, respectively. With 25 circuits it is possible to have a traffic capacity of about 13 Erlang for Pb = 0.001 and of about 16 Erlang for Pb = 0.01; in consequence it is assumed to consider N = 25. Figure 1 shows that for N = 25 it is

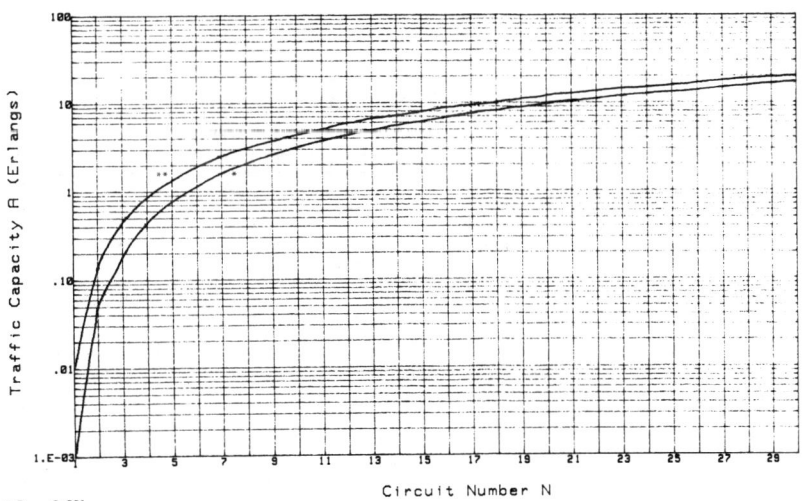

* P_b = 0.001
** P_b = 0.001

Fig. 2 Traffic capacity A as function of circuits number N for probability of blockage of 0.01 and 0.001 by Erlang B.

necessary to have an Eb/No value of 11.5 dB. Moreover, the number of peak hour calls as a function of Eb/No for Pb = 0.01 and Pb = 0.001 will be 600 and 500, respectively, for Eb/No = 11.5 dB.

From the above considerations link calculations can be carried out. Figure 3 shows the Eb/No value as a function of the Tx power for different values of the antenna diameter and for clear sky conditions.

It is important to note that we have disregarded the transponder intermodulation processes because we consider the Travelling Wave Tube Amplifier (TWTA) working in a linear zone, and we have also considered the available transponder power is shared up to three equal parts.

For a link availability of 99.5% the rain attenuation exceeded for 0.5% of the average year time, for the uplink and for the downlink, respectively. In particular with reference to the climate zones, as defined by Ref. [7-10], in which the world map is divided as shown in Fig. 4, it can be seen that the African continent includes six different values of rainfall intensity (R).

Table 1 depicts the climate zones with relative rainfall intensity R for 0.01% of time and the corresponding rain attenuation exceeded for 0.5% of time for downlink, while Table 2 shows the same for uplink. There are three climate zones (E, C, and K) in which the rain attenuation is under 2 dB, while for the climate zones P and N the rain attenuation is up to 8 dB.

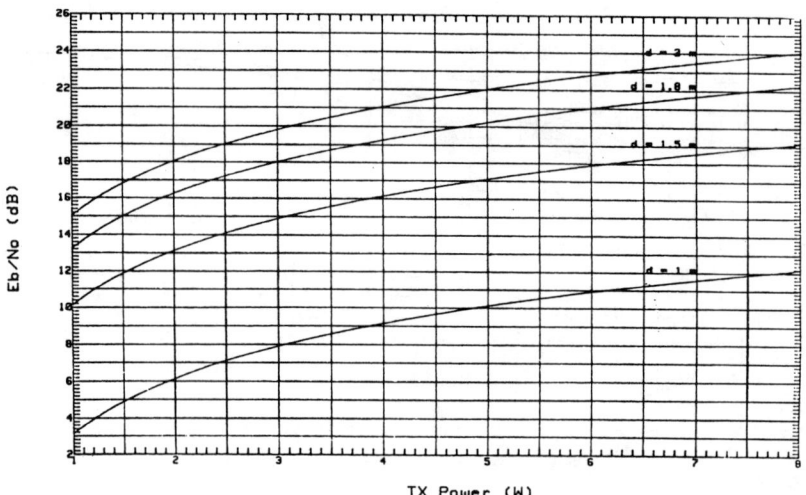

Fig. 3 Resulting Eb/No values vs Tx power for different values of the antenna diameter (clear sky conditions).

SATELLITE COMMUNICATION IN RURAL AREAS

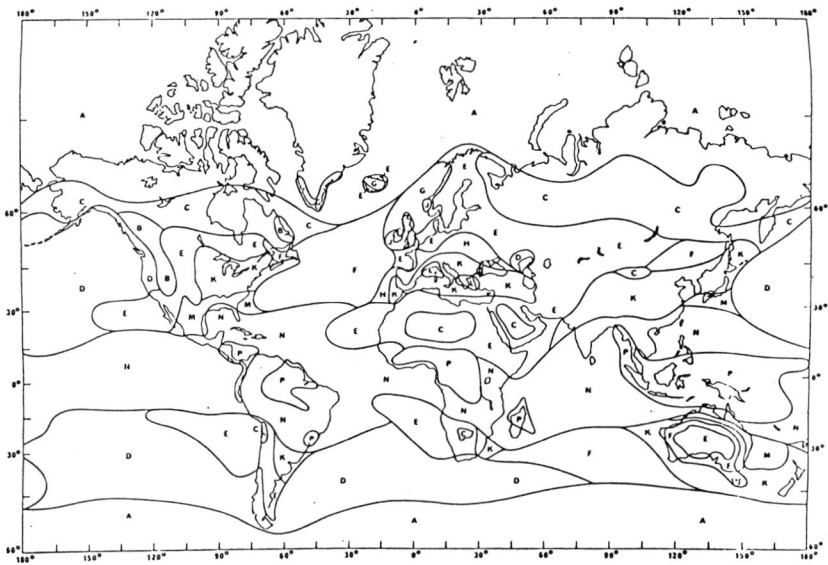

Fig. 4 Climatic zones.

Table 1 Rain Attenuation Values (0.5%)
For African Climatic Zones in the Down Link

Climatic Zones	Rainfall Intensity R (mm/h) 0.01%	Rain Attenuation (dB) 0.5%
E	22	0.7
C	15	0.4
K	42	1.4
P	145	6.8
N	95	3.8

In Fig. 5 the margins of rain attenuation as a function of the Tx power is shown for different values of antenna diameters. For climate zones in Africa with low attenuation (E, C, and K) an antenna with 1.8 m diameter and 2 W of Tx power is required, while for climate zones with high attenuation (P and N) an antenna with 2 m diameter and 4 W of Tx power is required.

Table 2 Rain Attenuation Values (0.5%)
For African Climatic Zones in the Uplink

Climatic Zones	Rainfall Intensity R (mm/h) 0.01%	Rain Attenuation (dB) 0.5%
E	22	0.9
C	15	0.6
K	42	1.9
P	145	8.1
N	95	4.8

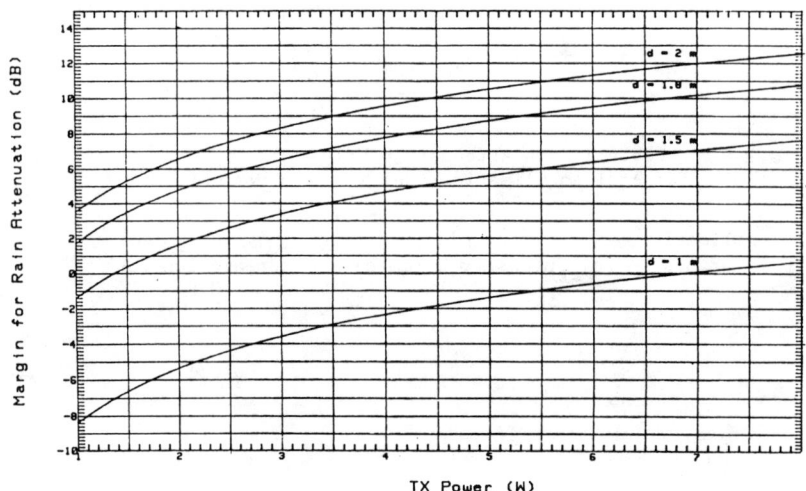

Fig. 5 Margin for rain attenuation vs TX power for different values of the antenna diameter.

It is preferred to have as low Tx power as possible to minimize the total terminal power consumption because the power supply is by solar panels and could be critical.

TV Program Distribution

For TV signal transmission it has to be assumed that all 27 MHz transponder bandwidth in FM modulation with a peak to

peak frequency deviation Δfpp of 13 MHz will be used.
The TV programs are transmitted from Capital Earth Terminals (CET) and are received in Small Earth Terminals (SET) and Main Earth Terminals (MET). In this case, the transponder saturation Effective Isotropic Radiative Power (EIRP) is transmitted by satellite.

On the receiving side the weighted signal to noise ratio is

$$\left(\frac{S}{N}\right)_{vp} = \frac{3}{2} \left(\frac{C}{N}\right)_t \left(\frac{Bn}{Bv}\right) \left(\frac{\Delta fpp}{Bv}\right)^2 E_v \, p \qquad (10)$$

where

$(C/N)_t$	= total carrier to noise ratio
B_N	= noise bandwidth (27 MHz)
B_v	= video bandwidth (5 MHz)
Δfpp	= peak to peak frequency deviation (13 MHz)
E_v	= emphasis coefficient (1.59)
p	= weighting coefficient (13.18)

Fig. 6 shows the weighting signal to noise ratio as a function of SET antenna diameter; for a 1.8 m diameter

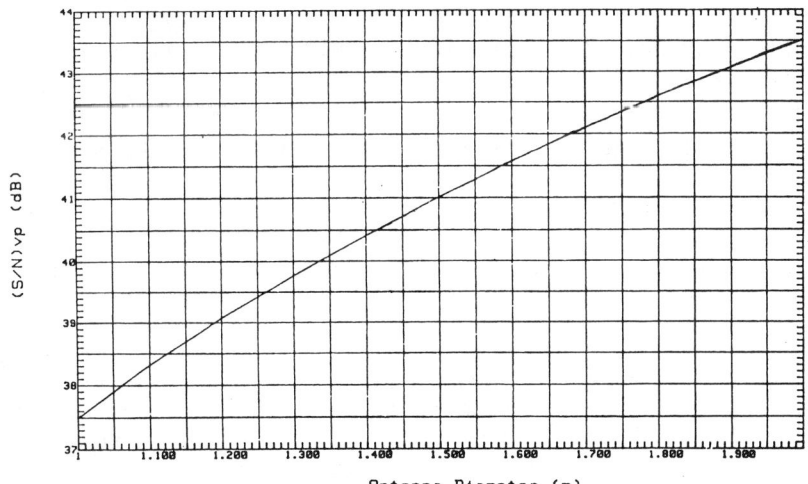

Fig. 6 Weighted signal to noise ratio as function of SET antenna diameter.

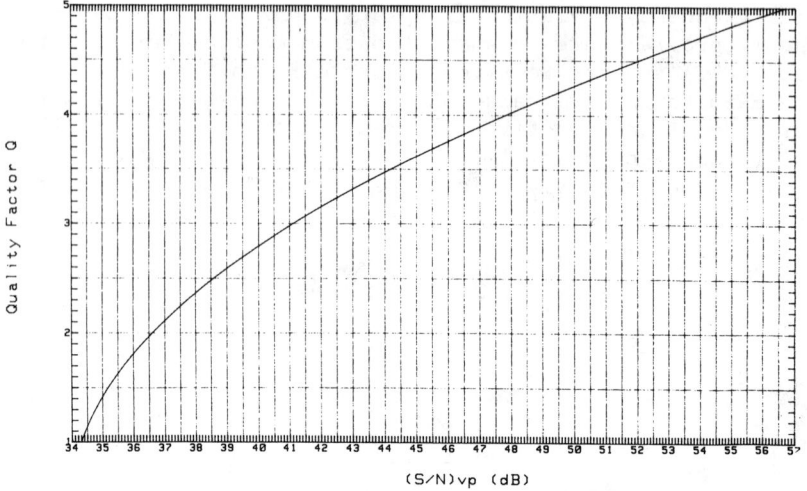

Fig. 7 Television quality factor Q vs (S/N)vp.

(S/N)vp of about 42.6 dB, and for a 2 m diameter a (S/N)vp of 43.5 dB are derived.

Fig. 7 shows the TV quality factor (Q) as a function of (S/N)vp; the quality factor indicates, for a given (S/N)vp, the subjective quality factor in reception for a mean user. For (S/N)vp value between 42.6 dB and 43.5 dB, the quality factor (Q) is between 3 and 3.5.

In Table 3 the means in terms of quality of the Q five levels are given; for $3 \leq Q \leq 3.5$ the quality is between quite good and good. The last results are valid only for clear sky conditions.

If a quite good quality (Q = 3) for at least 99.5% of the time,[11] is required a margin of about 1.5 dB for a 1.8 m antenna diameter can be achieved. These margin values could be suitable for climate zones E, C, and K, but not for the zones P and N where the rain attenuation is much higher. So with an antenna diameter of 2 m a quite good quality can be reached in the climate zones P and N for a percentage less than 99.5%.

It possible to conclude from the above preliminary network analysis and link-budget results, that considering a CDMA/SS access technique for a 1/3 of 27 MHz transponder bandwidth with code gain of 511, a network of about 300 SETs, two METs, and one CET can be realized using 25 simultaneous circuits and giving a probability of blockage of 0.001 or 0.01 with a traffic capacity of about 12 or 6 Erlang.

Table 3 Mean of Q Five Levels in Terms of
Quality and Degradation

Quality Factor Q	Quality	Degradation
5	very good	imperceptible
4	good	perceptible, but not disturbing
3	quite good	quite disturbing
2	poor	disturbing
1	bad	very disturbing

The antenna diameter can be equal to 2 m to have a safe margin and a better quality performance for TV program reception. The solid state amplifier will be 2 or 4 W depending on the particular climate zone under consideration. The number of voice modems for the CET could be 20, and 10 for each MET.

The most important characteristics of network under consideration are summarized in Table 4.

Figures 8-10 show the block diagram of the SET, the MET, and the CET. Every terminal shall be able to generate the network entry protocol for any one of the codes dedicated to other stations and in particular for the following:
1) the network access control is decentralized, however, the CET shall manage service channel codes with increased process gain and communication codes for network monitoring, billing, and master level system access control, to limit network saturation.

2) Every terminal shall evaluate, during idle time, the possibility to access the network via Bit Error Rate (BER) measurements, performed through the correlation scores obtained on the service channel code, transmitted by the CET.

The access procedure is performed through the following phases:
1) evaluation of link feasibility by the latest available channel BER measurement;
2) attempt of acquisition of the code assigned to the called terminal; and

Table 4 The Most Important Characteristics of the Network

Telephone Communications:

Access technique	: CDMA/SS
Coded voice rate	: 9600 bit/s
Code gain	: 511
SS signal bandwidth	: 1/3 27 MHz transponder bandwidth
Modulation	: BPSK
Traffic capacity	: 13 Erlang (Pb=0.001) or 16 Erlang (Pb=0.01)
Circuits number	: 25
Busy hour calls number	: 500 (Pb=0.001) or 600 (Pb=0.01) (tm=3 minutes)
Error probability	: 10^{-3}
Link availability	: 99.5%

TV Program Distribution:

Distribution type	: Broadcasting
Modulation	: FM
TV modulated/signal bandwidth	: 27 MHz
Peak to peak frequency deviation	: 13 MHz
Quality factor	: 3/3.5
Link availability	: 99.5% or less than 99.5%

Radio Frequency Subsystem Characteristics:

Antenna diameter	: 2 m
Tx power	: 2 W or 4 W
Receive System Temperature	: 350°K
Tx frequency	: 14 GHz
Rx frequency	: 12.5 GHz
IF telephone Tx/Rx frequency	: 70 MHz
IF TV Rx frequency	: (0.95 / 1.750) GHz

3) evaluation of BER index over the incoming code by the called terminal.

The billing information is sent to the CET which centralizes the billing accounting on a periodic polling-based procedure.

The CET performs continuous monitoring of the network terminals status polling for each SET/MET over the service channel. The CET is able to send broadcast freeze messages to prevent further terminals network accessing in order to

SATELLITE COMMUNICATION IN RURAL AREAS

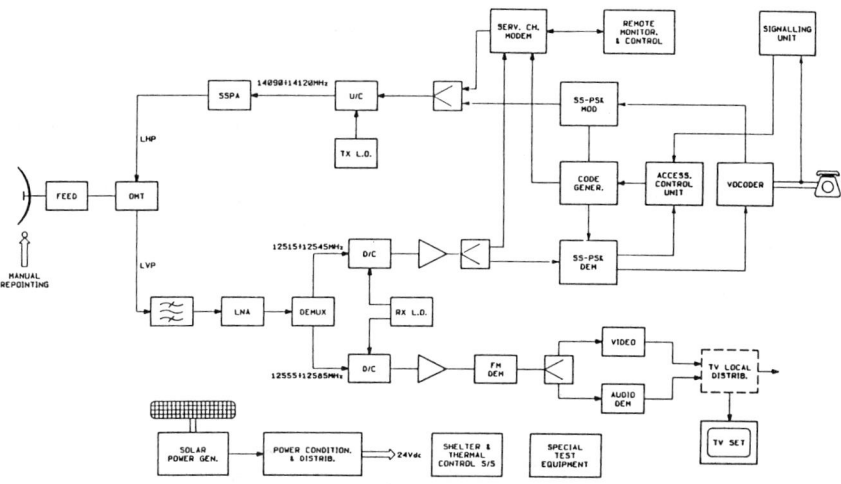

Fig. 8 Block diagram of the SET configuration.

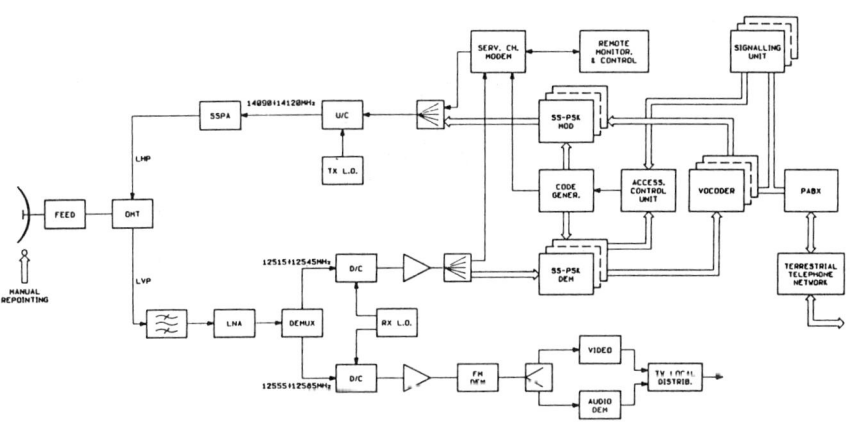

Fig. 9 Block diagram of the MET configuration.

avoid CDMA saturation. This kind of network monitoring is superimposed over the distributed network access control.

Beyond the monitoring and control functions supported by the CET, a remote terminal Radio Frequency (RF) power turn off capability is included to keep faulty stations from interfering with the normal network operation.

Satellite Antenna Design

A flexible antenna system should be used to generate narrow and wide shaped reconfigurable beams at channel

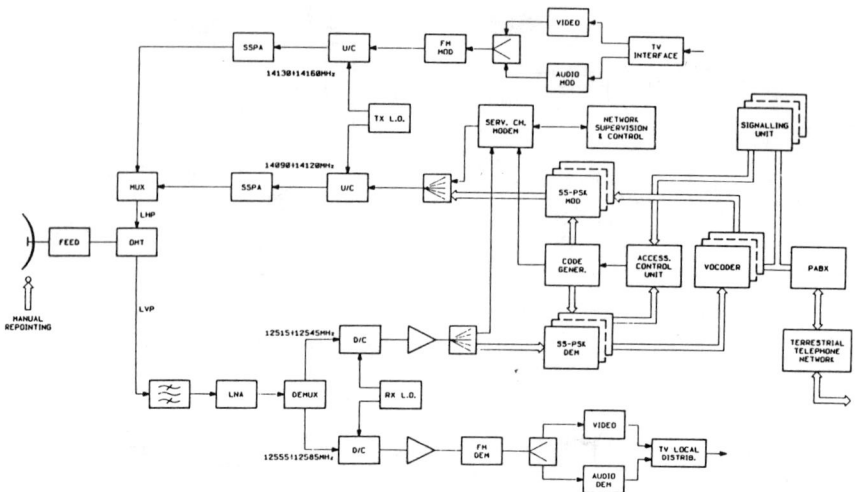

Fig. 10 Configuration block diagram of the CET.

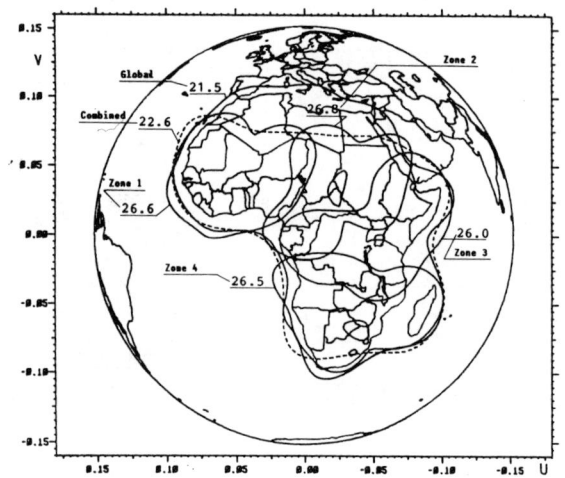

Fig. 11 Typical beam layout for African continent.

level, making it possible to allocate the payload resources according to the traffic patterns and users' needs.

Figure 11 shows a typical beam layout for the African continent. There are four zone beams (two for each polarization) partially overlapping, covering almost the whole continent, and exhibiting a gain of about 27 dBi.

Each couple of copolarized beams can be reconfigured into a nearly global beam for which the gain is 21 dBi.

The downlink is composed by an offset dual-gridded reflector for reusing the frequency spectrum through orthogonal linear polarizations and by two independent feed arrays.
The reconfiguration of the beams is achievable by properly connecting groups of horns with mechanically variable power dividers.

For the uplink it is possible to use either an elliptical beam antenna (G = 19 dBi), which allows a global coverage, or a scaled version of the downlink antenna with a multiple beam capability.

References

[1] Cacciamani E.R., Garner W.B., and Salamoff S.B., "The Emergence of Satellite Systems for Rural Communications", Proceedings of the IEEE, Vol. 72, No. 11, Nov. 1984, pp. 1520-1525.

[2] CCIR, Vol. 4, Annex 1 of Rep. 208-5, "Rural Telephone Communications".

[3] ESA, "Large Telecommunications Satellite Project, (L-SAT), System Performance Specification - SP-6-1-1" Appendix B to ESTEC Contract, March 1986, pp. 19-41.

[4] Pursley M.B., "Performance Evaluation for Phase - Coded Spread Spectrum Multiple Access Communication. Part I: System Analysis," IEEE Transactions on Communication, Vol. COM - 25, Aug. 1977, pp. 795-799.

[5] Feher, K., Digital Communications, Prentice Hall, 1983, pp. 156-159.

[6] Martinino F., Pugnaloni A., Saitto A., and Tripodi M., "Error Probability Evaluation for a Digital Spread Spectrum Modem used in Satellite Communications", - Alta Frequenza, Vol. VII, Jan. 1988, pp. 21-28.

[7] CCIR, Rep. 724-1, "Propagation Data Required for the Evaluation of Coordination Distance in the Frequency Range 1 to 40 Ghz".

[8] CCIR, Rept. 563-2, "Radiometeorological Data".

[9] CCIR, Rept. 721-1, "Attenuation by Hydrometeors, in particular Precipitation, and other Atmospheric Particles".

[10] CCIR, Rept. 564-2, "Propagation Data Required for Space Telecommunications Systems".

[11] CCIR, Rept. 642-2, "Recommendation 567-1, 568".

Payload, Bus, and Launcher Compatibility for Multibeam Mobile Communication Satellite Systems

Nizar Sultan*
Canadian Astronautics Limited, Ottawa, Canada

Abstract

This paper illustrates a methodology for first establishing the feasibility of fitting different satellite payloads to specific bus-launcher combinations, and then ranking the solutions. For a given payload, consisting of antenna systems and transponders and requiring some value for power-mass (P-M), several satellite buses and several types and sizes of launchers are available.

This paper suggests that the communication payload can best be finally optimized only after a detailed analysis is performed for the compatibility of the communication payload to that of all potential buses, the deployment and stowage of the antenna and feeds to all potential buses, and the stowed spacecraft to all potential launchers.

Such a methodology is demonstrated in this paper on a high-capacity maritime and aeronautical multibeam mobile satellite system. The feasibility of four different offset reflector antenna and feed configurations for stowage and deployment on five families of buses is analyzed from mechanical considerations. Similarly, the compatibility for all stowed spacecraft to the envelope of each of the different types of launchers is considered for Ariane and the shuttle.

I. Introduction

This paper illustrates a methodology for first establishing the feasibility of fitting different satellite payloads to specific bus-launcher combinations, and then

---

Copyright © 1990 by Nizar Sultan. Published by the American Institute of Aeronautics and Astronautics, Inc. with permission.
*Senior Staff Scientist; currently Senior Spacecraft Systems Specialist with Telesat Canada, Space Systems Department, Gloucester, Ontario, Canada.

ranking the solutions. For a given payload consisting of antenna systems and transponders and requiring some value for P-M, there are several satellite buses and several types and sizes of launchers. The selection of a specific set of options is dictated essentially by its feasibility, the most efficient bus utilization, and overall cost considerations. The problem is compounded by the fact that at the initial system design phase, many options for different P-M payload values can be generated as a result of various tradeoffs.

Such tradeoffs are performed[1-4] to optimize the payload for different requirements, system capabilities, performances, technologies, risks, costs, etc. In previous work[5] a simple and effective method of optimizing the payload for all potential buses was developed using the P-M diagram technique. In this paper this method is applied only after the mechanical feasibility of the payload to various buses and of the considered buses to various launchers has been established.

This methodology is illustrated on a multiple beam, frequency reuse, maritime and aeronautical mobile satellite system. First, various solid offset reflector multibeam antennas and feed arrays are outlined. Then their deployment on specific buses is considered, together with the feasibilities of fitting the various stowed satellites within different launcher envelopes. Finally, the results are given for the feasibility of different payloads to all buses, P-M capabilities and for the bus efficient utilization concept developed here.

II. Outline of Multiple Beam Antennas Considered

Four solid offset reflector antennas were considered at L-band.[6] Their geometries are outlined here to illustrate their impact on deployment and fitting the various buses available and on the compatibility of potential launchers to accommodate the stowed buses.

Two 2-m and two 3.4-m diam. solid reflector antennas are identical to those shown in Figure 1. Their apertures were limited to 3.4-m to avoid the need for elaborate deployment schemes, in order to be accommodated within available and cost effective launchers. The summary Table 1 below indicates a number of beams ranging from 6 to 52. Figure 1 shows a combination of spot and shaped beams generated by a 7-19 feed array occupying 0.96-1.65-m diam. Such antennas and feeds are considered for at least three

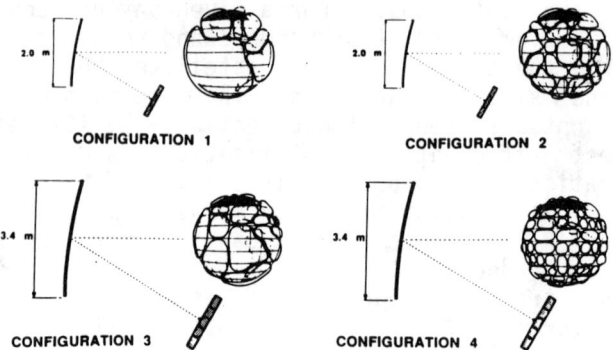

Fig. 1 Spot beam antenna configurations considered.

Table 1 Antennas characteristics

	ANTENNA CONFIGURATION			
	1	2	3	4
Focal length, f (m)	2.5	2.5	4.25	4.25
f/D	1.25	1.25	1.25	1.25
Reflector diam, D (m)	2.0	2.0	3.4	3.4
No. of beams	6	19	18	52
No. of transponders	4	7	11	19
No. of feeds	7	7	19	19

ocean geosynchronous satellites serving global maritime and aeronautical mobile users via global and spot beams.

III. Buses Payload Capabilities and Antenna Considerations

A bus is designed initially for a specific communication payload capability of dc power and mass. The payload is defined here as the antenna and the transponder systems. This implies that the antenna size and projection out of the bus are such as not to exceed the stability and structural design requirements. It also suggests a reasonable payload heat dissipation not exceeding the design limit.

This payload P-M design value can be parametrically varied by the bus manufacturers in two ways:

1) For a given stationkeeping condition, dc power, i.e., solar sails, batteries for eclipse capability, and power conditioning circuits, can be traded against the mass made available to the communication payload, and

2) For a given dc power condition the reduction of number of years of stationkeeping (SK), whether north-south (NSSK) or east-west (EWSK) from the design value, say from 10 to 7 years, can lead to savings in the fuel mass, and hence more mass is made available to the communications payload.

The manufacturers of five different buses were approached for the supply of their P-M parametric curves. These buses were:

1) Eurostar bus (BAe).
2) Aerospatiale bus.
3) RCA bus.
4) Olympus Bus (BAe).
5) Hughes 393 bus.

The buses' characteristics, as of 1986, are summarized in the tables below and shown in the following graphs. They represent a variety of conditions including

1) Lifetime (years y).
2) Stationkeeping (SK).
3) Percentage eclipse.
4) Power or mass margins.
5) Launcher: shuttle (STS) or Ariane (A).

These characteristics, some of which are shown in Figure 2-5, represent 52 potential buses for all conditions. Figure 6 shows them all combined. Twenty-one

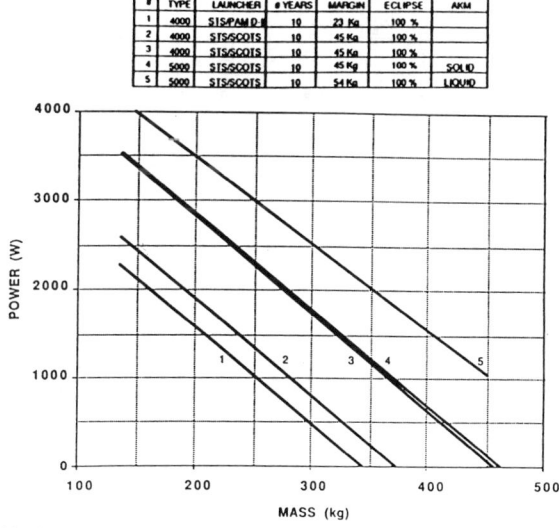

Fig. 2 RCA 4000 and 5000 buses typical power-mass parametrics curves for all lifetimes.

of these buses, shown in Figure 7, are for a 10-yr. SK and a maximum 3 degree drift at end-of-life (EOL).

Finally, if an STS launch cannot be considered, then effectively up to 17 buses are available for a 10-yr. SK and Ariane launch.

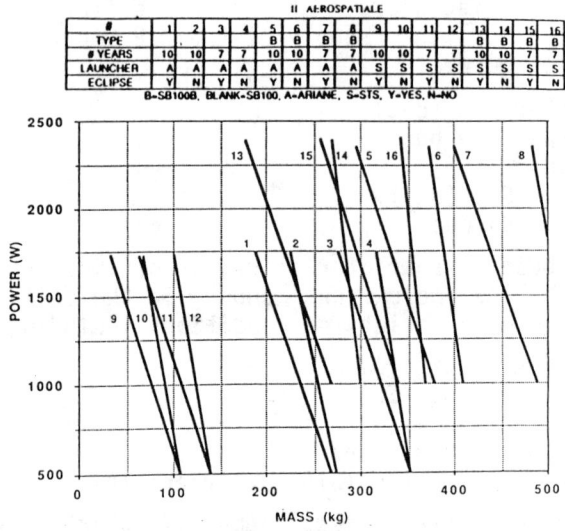

Fig. 3 Aerospatiale SB-100 and SB-100B buses typical power-mass parametrics curves for all lifetimes.

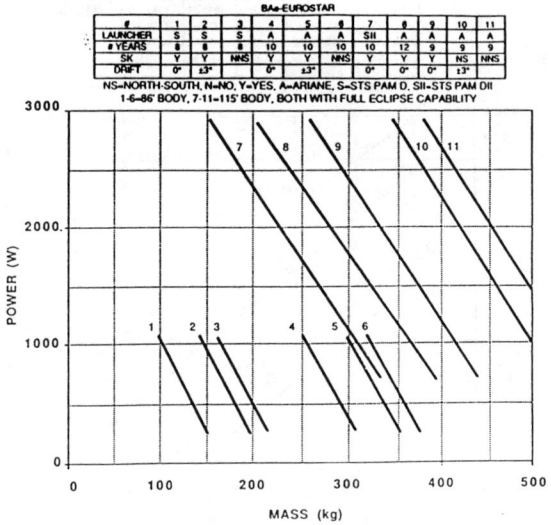

Fig. 4 B.Ae Eurostar buses typical power-mass parametrics curves for all lifetimes.

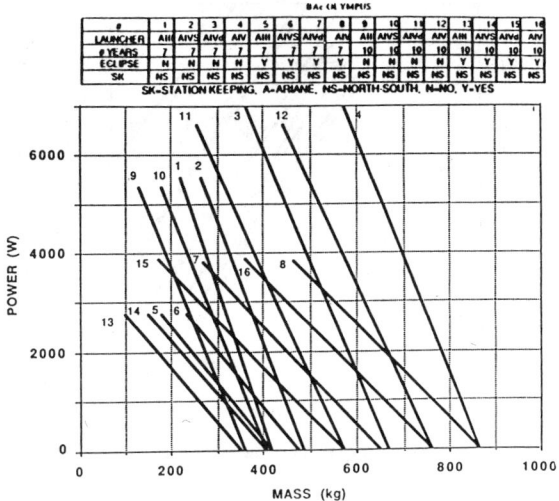

Fig. 5 B.Ae Olympus buses typical power-mass parametrics curves for all lifetimes.

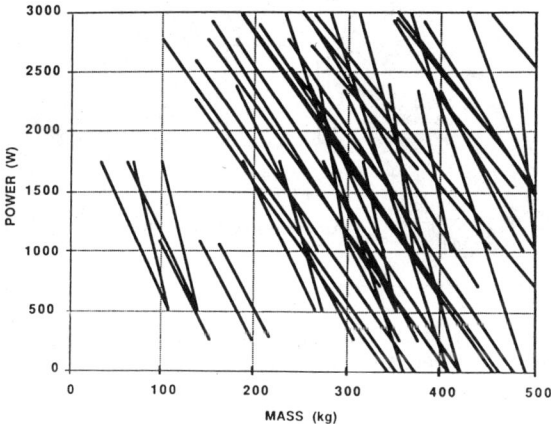

Fig. 6 All buses typical power-mass parametrics curves for all lifetimes.

The five buses have been examined with a view to the feasibility of using them to support the same mission. Only the feasibility of the geometry, with respect to fitting the resulting spacecraft into a launcher fairing, has been considered. Questions such as structural strength, stowed and deployed dynamics, and deployed SK are beyond the scope of this study.

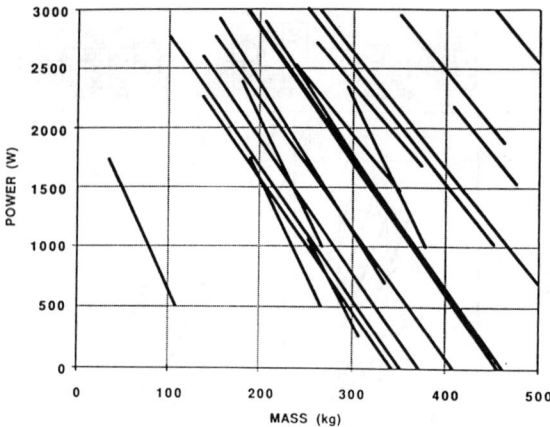

Fig. 7 All 10-yr. buses typical power-mass parametrics curves.

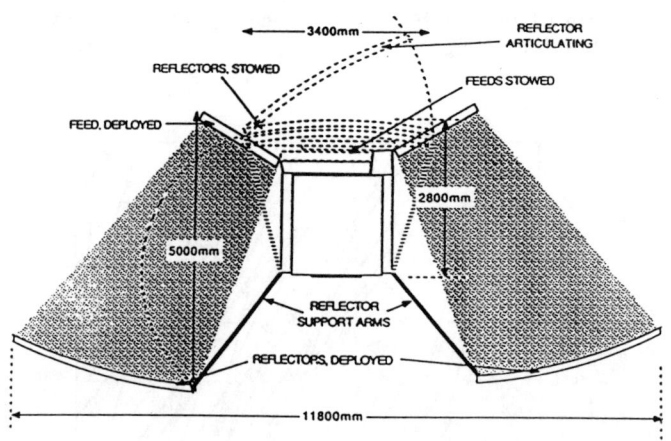

Fig. 8 Two 3.4-m diam. antennas and feed deployment concept on three axes-stabilized spacecraft.

In principle there is no limit to the possibility of "fitting" any antenna to any bus. However, two limitations were assumed. All articulation should be accomplished by means of "simple hinges" and the number of such hinges should be kept to the very minimum.

The factors driving these limitations are reliability in the first place and positional accuracy in the second.

Antenna deployment sequence is indicated in Figure 8 for two 3.4-m diam. antennas. Stowing sequence is in this order: feed no. 1 folds first, then feed no. 2, followed by reflector no. 1, and finally reflector no. 2.

Table 2 Compatibility of launchers with "twin antenna/bus" combinations

Launcher	Ariane								S.T.S.	
	Single launch				Dual launch					
	3	4 (1920) Type 01	4 (1920) Type 02	4 (2624) Type 02	Spelda type 01 Upper	Spelda type 01 Lower	Spelda type 02 Upper	Spelda type 02 Lower	Using PAM-D2	Axial launch
Bus										
2 meter antennas										
Eurostar	X	•	•	•	•	•	•	•	*	•
RCA 4000	X	•	•	•	•	X	•	X	•	•
Hughes 393	X	•	•	•	X	X	X	X	X	•
3.4 meter antennas										
Eurostar	X	•	•	•	•	•	•	•	X	•
RCA 4000	X	•	•	•	•	•	*	X	X	•
Hughes 393	X	X	•	•	X	X	X	X	X	•

Key
• indicates a fit
X indicates no fit

* indicates a case where there is a slight interference and detailed design would be required to resolve it

The antenna assembly mass, including each of the two L-band reflectors, the seven-element patch-type feed assemblies, and the C-band patch-type transmit and receive antennas, is estimated at 71 Kg.

IV. Payload/Bus/Launcher Considerations

One of the launchers, Ariane, provides the capability of placing up to two spacecraft directly into a geostationary orbit. There are two variations of the basic vehicle (Ariane 3 and 4). Ariane 4 has a number of variations in its payload fairing to allow for single and double launches of various sizes. Generally, of the various launcher options only some feasible cases are shown, as in Figures 9 and 10.

As an example, some effort in fitting the bus to launchers was devoted to the RCA buses. For other buses, such as Hughes 393 and Eurostar, the bus/launcher compatibility was arrived at in an approximate method by fitting the launch predeployment antennas on specific buses. Further work is needed in this area based on the availability of additional data.

With RCA buses two 2-m diam. L-band deployable reflectors, in addition to the feeds, may be launched on the shuttle transportation system (STS) or Ariane 4 rocket.

The STS option with PAM DII occupies about 18 ft. of the shuttle payload bay.

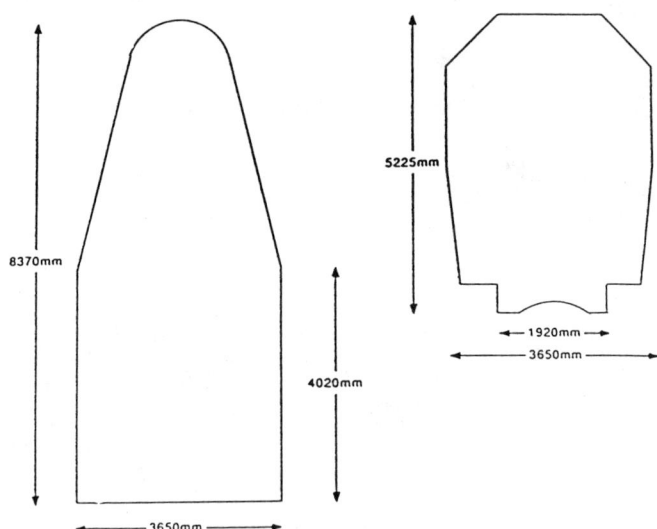

Fig. 9 Ariane 4 Spelda type-02 and dual launch upper fairing.

MULTIBEAM MOBILE COMMUNICATION SATELLITE SYSTEMS   159

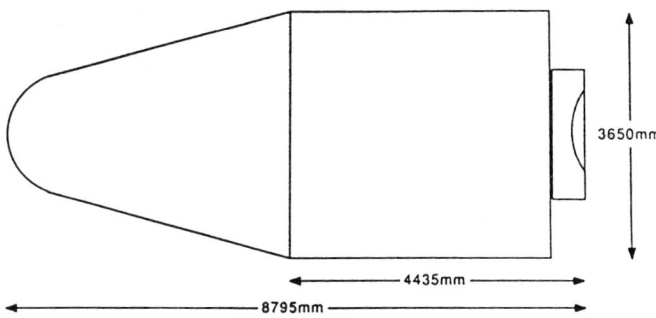

Fig. 10   Ariane 4 single launch 4 (1920) Type 01.

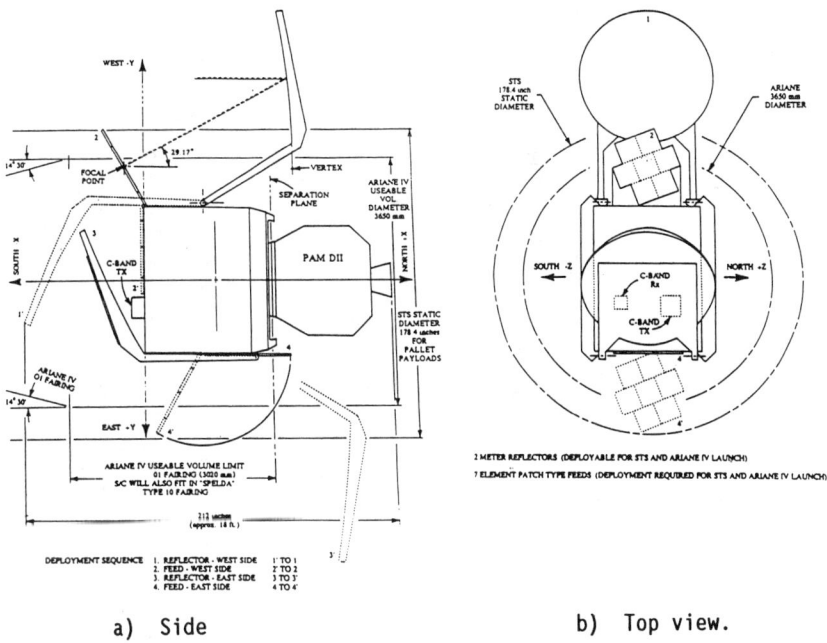

a) Side   b) Top view.

Fig. 11 RCA Ku-Band SCOTS configuration spacecraft (two 2-m diam. antennas).

The Ariane IV option (without PAM DII) may be launched within the "Spelda" or "01" fairings of this dual launch rocket.

The RCA buses are configured in the SCOT type geometry with one hinge in the reflector deployment and one in the feed deployment. In the stowed configuration this will fit the Ariane envelope, either as a single launch or in a

Table 3 Payload bus compatibility summary -
constant bandwidth, various lifetimes

Buses with various payload capabilities
Payload - bus compatibility summary

Antenna config-uration	Ocean	It#	BAe Olympus 1234567890123456	Aerospatiale 1234567890123456	BAe Eurostar 12345678901	RCA 12345	HAC 1234
1	A	1	***************1***	3******* ****	******	2*** ****
1	I	2	***************2***	3******* ****	******	1*** ****
1	P	3	***************1***	3******* ****	******	2*** ****
2	A	4	**************1**	3****** ***	2*****	*** ****
2	I	5	**************1**	3****** ***	2*****	*** ****
2	P	6	**************1**	3****** ***	2*****	****
3	A	7	2*** 1** **	***	3**	* ****
3	I	8	**** 2** **	1*** 3	***	* ***
3	P	9	3*** 2** **	*** 1	***	* ***
4	A	10	** 1* 3* *	*	2	
4	I	11	** 3* ** *	*	1*	*
4	P	12	** 2* 3* *	*	1	

1, 2 or 3 denotes best bus utilization in that order
* denotes other available buses
blank denotes bus non-compatibility

Table 4 Payload bus compatibility summary -
         constant bandwidth, 10 year lifetime

Buses with 10 years stationkeeping and eclipse capabilities
Payload - bus compatibility summary

Antenna configuration	Ocean	It#	BAe Olympus 1234567890123456	Aerospatiale 12345678901234567890123456	BAe Eurostar 12345678901	RCA 12345	HAC 1234
1	A	1	1***	* 3	**	2***	****
	I	2	2***	* *	**	13***	****
	P	3	1**	* 3	**	2***	****
2	A	4	1**	*	2*	3**	****
	I	5	1**	*	2*	3**	****
	P	6	3*	*	2*	3**	****
3	A	7	**	1		2	1**
	I	8	3*			3	2**
	P	9				2	1*3
4	A	10	1				
	I	11	2				1
	P	12	1				

1, 2 or 3 denotes best bus utilization in that order
  * denotes other available buses
blank denotes bus non-compatibility

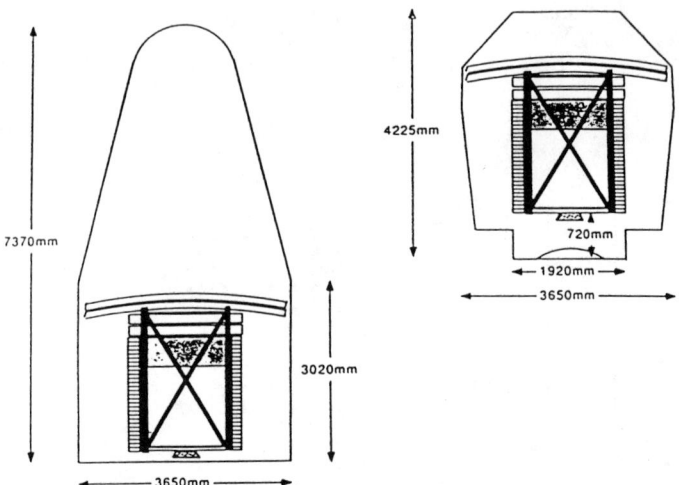

Fig. 12 RCA 4000 bus with 3.4-m diam. antennas in Ariane Spelda type-01 and dual launch upper fairing.

Spelda as part of a double launch, but not the STS-PAM envelope.

The configuration with two 3.4-m antennas is similar in concept to that of the 2-m case, but a second hinge is required for the reflector deployment so that the two reflectors can be stacked and used to provide mutual support through launch. The geometry allows the reflectors to be stowed in this way by the positioning of one feed slightly "ahead" of the other, while hinging both reflectors at the same level. Some details are shown in Figures 11 through 14.

Some preliminary work was performed on the compatibility of the two types of antennas to the five buses and to all Ariane and shuttle launches. Table 2 summarizes the results.

V. Available Buses and Their Utilization Efficiency

Before any compatibility analysis is performed, the total payload P-M requirements have to be defined for a set of parameters, such as different ocean satellites or antenna configurations, etc. as shown in Figure 15 for the case of constant global Erlangs traffic.

These results show that all the antenna configurations studied require payloads with nearly the same dc power, i.e., about 1.4 kW (for solid state amplifiers of 20% efficiency). However, the corresponding payload mass

MULTIBEAM MOBILE COMMUNICATION SATELLITE SYSTEMS

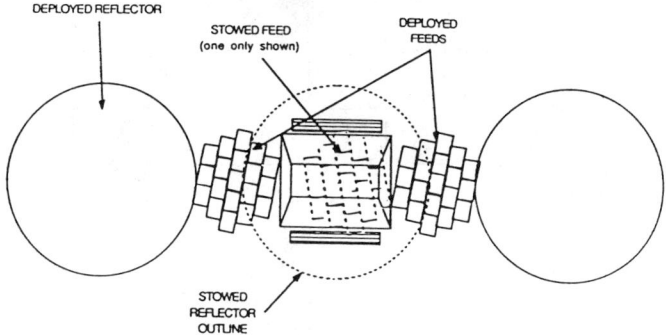

Fig. 13 Plan view for Eurostar with two 3.4-m antennas, showing stowed and deployed feed and reflector positions (reflector structure omitted for clarity).

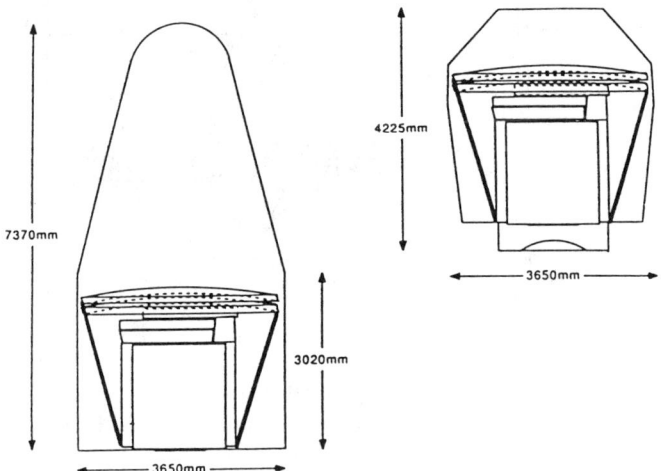

Fig. 14 Spelda Type 01 and dual launch upper fairing showing Eurostar bus with two 3.4-m diam. antennas.

varies significantly between 220 Kg and 460 Kg for antennas 1 and 4, respectively. The conclusion is that such a spacecraft is mass driven and has equal payload power requirements for all antennas.

The more significant results are shown in Figure 16, where the same payload power and mass curves are superimposed[5] on the bus payload capabilities for buses with different lifetimes. A visual inspection reveals which bus P-M characteristics can carry the payload. A more accurate and analytical approach is used below to study the potential compatibility of payloads to buses and the efficiency of bus utilization.

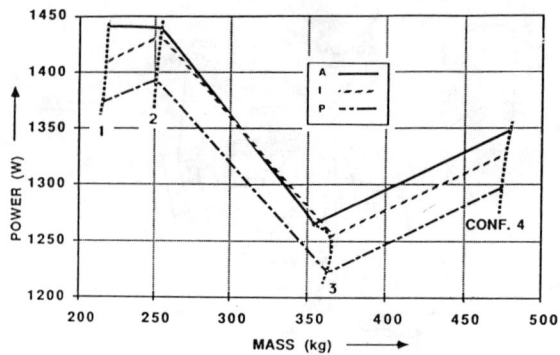

Fig. 15 Total payload mass and DC power (10% margin).

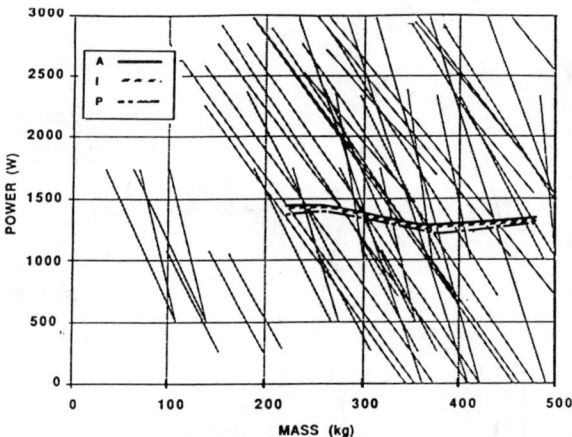

Fig. 16 All buses superimposed on mobile satellite payload power-mass diagrams.

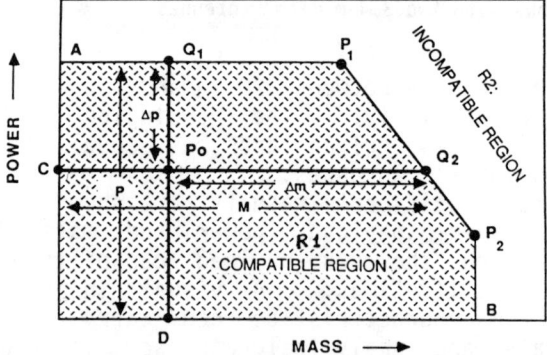

Fig. 17 Payload compatibility shaded area is region R1. Optimum bus utilization generally occurs when payload P-M is closest to line P_1P_2.

MULTIBEAM MOBILE COMMUNICATION SATELLITE SYSTEMS

Consider the P-M performance of a typical bus illustrated in Figure 17 by the P-M diagram.

The straight line P_1P_2 represents the tradeoff between reducing the dc power capability of a bus in return for increasing the mass available for a payload, and vice versa. The line segments AP_1 and P_2B are the outer limits for the bus power and mass capability, respectively. Together the three line segments define the boundary of the region within which any combination of power and mass will be available to the payload. For example, a payload whose characteristic power and mass requirements lie within region R1 can be supported by the bus, whereas if it lies in region R2 the bus is not available to the payload.

The efficiency of the bus utilization is an indication of how far away the payload P-M characteristics lie with respect to the bus capability boundary (within R1). P_1P_2 represents an optimum match between payload requirements and bus capability. In this case, the bus utilization is 1.0. In R1, the further away the payload PM characteristics are from P_1P_2, the lower the bus utilization efficiency becomes.

The bus utilization efficiency (BU) for a given transponder payload can be defined (Figure 17) as:

$$BU = [\ 1\ -\ 0.5\{\ p/p^2 + (\ m/m)^2\}\ ]\ X\ 100\%$$

where p/p = the fraction of nonutilized bus power, and
m/M = the fraction of nonutilized bus mass.

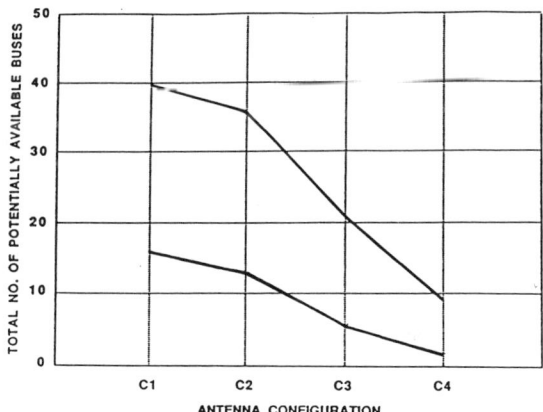

Fig. 18 Total number of potentially available buses of any lifetime and for 10 yr., all four antenna configurations.

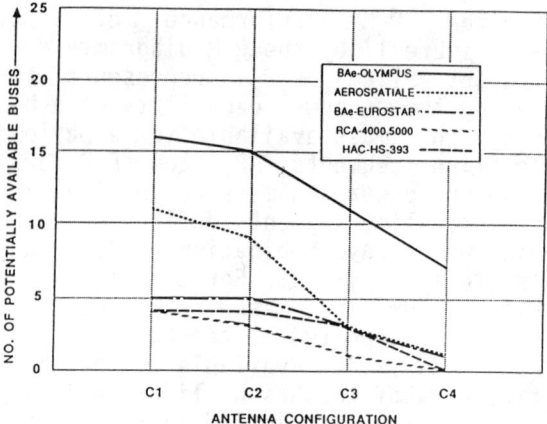

Fig. 19 Total number of potentially available buses of any lifetime for all four antenna configurations.

The compatibility analysis model has been applied to the 52 available buses and 192 different payload cases. Typical analytical results for only 12 cases (item nos. 1-12) for all buses and for buses of at least 10 yr. stationkeeping are presented in Table 3.

Table 4 illustrates the most efficient bus for four specific antenna payloads and the total number of available buses.

As can be seen, the general trend in the total number of potentially available buses for antenna configurations

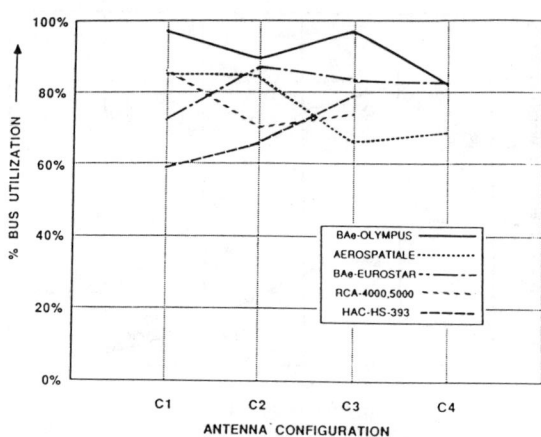

Fig. 20 Percentage bus utilization (any lifetime) for all four antenna configurations.

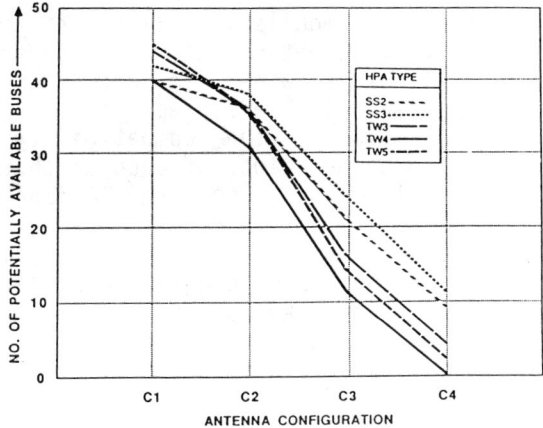

Fig. 21 Number of potentially available buses of any lifetime for different HPA types and all four antenna configurations.

going from 1 to 4 is decreasing, due to increasing payload mass.

Typical results are shown in Figures 18-20 for the number of available buses, in total and for each manufacturer, for any lifetime, and for 10-yr. lifetimes, together with the best bus efficiency. Figure 21 illustrates the versatility of the compatibility software: it can provide, for example, the effect of different high power amplifier (HPA) configurations on the number of available buses. Hence, the effects on the latter of different solid state HPA efficiencies and different travelling wave tube types are readily available.

VI. Conclusion

It may appear that the optimization of a satellite system consists of first reducing the communication payload, both in power and mass, in a tradeoff analysis of all potential parameters, then of looking for the most efficient bus and launcher.

The communication payload can best be finally optimized only after a detailed analysis is performed for the compatibility of:
1) The communication payload to that of all potential buses
2) The deployment and stowage of the antenna and feeds to all potential buses, and
3) The stowed spacecraft to all potential launchers.
This is demonstrated on a high-capacity maritime and

aeronautical multibeam mobile satellite system. The feasibility of four different offset reflector antenna and feed configurations for stowage and deployment on five families of buses is analyzed from mechanical considerations. Similarly, the compatibility for all stowed spacecraft to the envelope of each of the different types of launchers is considered, for Ariane and the shuttle.

At least 52 potential bus payload capabilities were found available. A software program was developed to analyze the communication payload compatibility with each of the 52 buses. The same software provides a bus utilization efficiency, i.e., a figure of merit for the bus utilization.

This results in a variety of options that are made easily and conveniently available for any change in the communication payload requirements or the technology considered.

Acknowledgment

This paper is based on work performed under the sponsorship of the International Maritime Satellite Organization (INMARSAT). Any views expressed here are not necessarily those of INMARSAT.

The author is grateful to Mrs. Edith Eburne for careful typing and preparation of this manuscript.

References

[1] Sultan, N. and Ng, P., "Capacity Optimization for Spot Beam Advanced Mobile Satellite Systems," IAF 37th International Astronautical Conference, Innsbruck, Austria, Paper IAF-86-339, Oct. 1986.

[2] Dachert, F., de Montlivault, J.L., and Coirault, R., "Optimization of a High Capacity Communication Satellite for Europe," AIAA 10th Communication Satellite Systems Conference, Paper 0698, Orlando, FL., Mar. 1984, pp. 664-672.

[3] Sultan, N., "Design Considerations for Multibeam Frequency Re-use Maritime Mobile Satellite Systems," Symposium on Commercial Opportunities in Space, Roles of Developing Countries, Taipei, Republic of China, Apr. 1987, pp. 240-242.

[4] Sultan, N. and Wood, P.J., "Adaptive Sub-Bands Channelization: Solution to Reconfigurability of Multibeam Frequency Re-use Maritime Mobile Satellites," AIAA 12th International Communication Satellite Systems Conference, Paper 88-0769, Arlington, VA., Mar. 1988, pp. 157-166.

[5]Sultan, N., Payne, W.F. and Carter, D.R., "Novel Approach to Optimization of Communication Payload for High Capacity Mobile Satellites," <u>IEE 3rd International Conference on Satellite Systems for Mobile Communications and Navigation</u>, IEE Publication No. 222, London, United Kingdom, June 1983, pp. 60-64.

[6]Sultan, N., "Planning of Advanced Maritime and Aeronautical Mobile Satellite Systems with Multibeam Frequency Re-Use," <u>IAF 38th International Astronautical Conference</u>, Brighton, United Kingdom, Paper IAF-87-480, Oct. 1987.

Rupture of the Spit of Sangomar – Estuary of the Saalum, Senegal

Amadou Tahirou Diaw*
University of Dakar, Senegal, West Africa
and
Nouhoum Diop†
Dakar Port, Senegal, West Africa
and
Yves-Francois Thomas‡
Ecole Normale Superieure, Montrouge, France

Abstract

The ancient mouth of the Saalum, named Lagoba, remains scarcely known. The sole sign of its existence has been, for long, the narrowing of the spit of Sangomar located some 2.3 km south of Jeefer.

This area has been submitted to a satellite follow-up for over 15 years. On February 27, 1987 strong northwest swells caused the spit to re-open at the level of Lagoba. The Spot imagery enables one to draw a bathymetric outline of Lagoba before (9 months) and immediately after (3 days) the rupture of the spit.

The aim of the present work is to discuss the conditions that favored this new rupture, notably the disappearance of the prelittoral bar.

Introduction

Saalum's estuary is located around 100 km south of Dakar (see Fig. 1). Centered at 13°49' N and 16°45' W, it presents a great diversity of landscapes: prelittoral bars, beaches, littoral spits, eolian forms, mangroves and tannes (saline soils submersible by tide), and is characterized

---
Copyright © 1990 by A.T. Diaw. Published by the American Institute of Aeronautics and Astronautics, Inc. with permission.
*Maître-assistant, Department of Geography, Ifan (Fondamental Institute of Black Africa).
†Engineer, Head of Department, Lighthouses and Buoyage Directorate.
‡Charge de Recherche, URA 141, Laboratory of Physical Geography Cnrs (National Center for Scientific Research).

by a great dynamics. Numerous infra-, inter-, and
supratidal sandy shapes are developing on its
western side and help channel its waters: shallows
in constant movement, and spit of Sangomar
migrating rapidly southwards. For economic reasons,
this area attracted the early attention of port
authorities: the banks obstructing the mouth and
the sandy spit of Sangomar have been submitted,
since the beginning of this century, to various
hydrographic and topographic surveys, as well as
aerial surveys and kinematic mapping tests.

Thomas and Diaw[1] presented a study on the
dynamics of the intertidal zone of Sangomar spit.
Referring to oral tradition, to the various works
describing the eroding phenomena that affected the
littoral of Sangomar (1914-1915, 1925-1930, and
1936), as well as to available imagery, one can
notice that the movements observed nowadays reached
an unprecedented scale with, for instance, a
southward progression rhythm of nearly 175 m-year^{-1}

This paper discusses the morphology of
prelittoral bottoms as observed before and after the
rupture of the spit noticed on February 27, 1987 at
Lagoba, or Joxaan some 2.3 km south of Jeefer. The
evolution of this part of Sangomar into a mouth of
the Saalum remains relatively unknown. While several
cuts have been mentioned for the spit, numerous
dates are given for these cuts, and there is
contradiction concerning the functioning period of
each of them. However, on the basis of the most
reliable works such as Minot,[2] Tromeur,[3] and Lefur[4]
one can:

1) Estimate that the Lagoba probably functioned

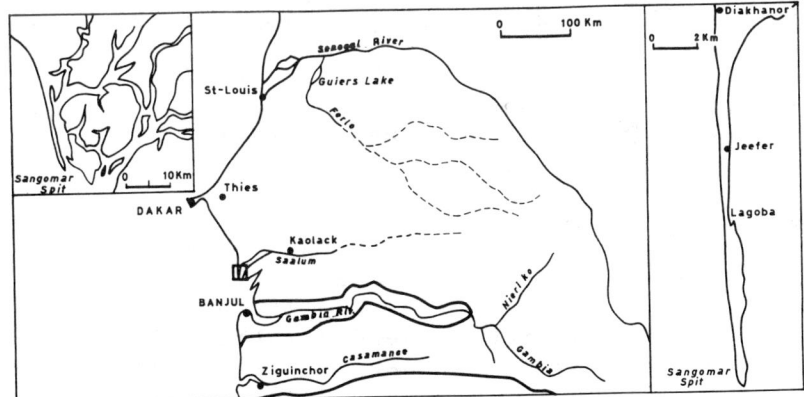

Fig. 1   Location map of the Saalum estuary and spit of Sangomar.

Table 1 Acquisition conditions of the Spot (HRV) imagery used in the study of Sangomar spit

Satellite	Spot-1	Spot-1
Reference Grid	022-322	022-322
Radiometer	HRV-2	HRV-2
Spectral mode	Multiband	Multiband
Ground resolution	20 X 20	20 X 20
View intake date	May 9, 1986	March 2, 1987
Time (Universal Time)	11h46'3"	11h33'40"
Latitude	14° 1' N	14° 1' N
Longitude	16° 33' W	16° 32' W
Orientation	9° 6'	8° 36'
Incidence	2° 18' (left)	20° 56' (right)
Solar rise	71° 18'	56° 8'
Solar azimuth	77° 54'	128° 7'

in 1860 by observing the outlines of the de Corbigny map and comparing them to those of the document established a few years later by Mage[5].

2) Notice that the Lagoba mouth functioned on August 28, 1928.

Material and Methods Used

Imagery

The selected images were acquired by the satellite Spot-1, with the n° 2 High Resolution Visible (HRV) instrument operating multispectral mode (XS).

The scenes are dated May 9, 1986 and March 2, 1987, i.e., 9 months before and 3 days after the rupture of the spit of Sangomar that occurred on February 27, 1987 (see Table 1).

Analysing method

A first image analysis shows that at each of the image taking dates hydroclimatic conditions were clement (see Table 2) and littoral water was sparsely loaded with suspended matter. Furthermore, the submarine beach has sandy bottoms that are not covered by vegetation and therefore have a reflectance that rarely varies. And last, the solar

Table 2 Climatological conditions noticeable during acquisition of the Spot (HRV) imagery used in the study of the rupture of Sangomar spit

Date	May 9, 1986	March 2, 1987
Temperature	24.4° C	24.6° C
Pressure	1012 hectopascal	1011 hectopascal
Relative humidity	72 %	90 %
Visibility	(1)	300 dam
Wind velocity	4 m. s^{-1}	3 m. s^{-1}
Wind direction	N - NW	N - NW

(1) measure not available

elevations during image acquisitions are important (71° in 1986 and 56° in 1987). All these conditions favor the carrying out of a bathymetric analysis.

In the study of bottoms, two spectral bands are used: the XS-1 band (λ= 0.50 - 0.59 µm) that is appropriate for the analysis of the morphology of infratidal environments in non-turbid prelittoral zone; and the XS-3 band (λ= 0.79 - 0.89 µm) suitable to distinguish the borders of the continental and the oceanic environments.

The analysing method used is the following:
1) The radiometric values were calibrated in terms of luminance. The equivalent Spot luminance L(λ) expressed by W.m - 2. Sr - 1. µm - 1 for a given λ band of an HRV radiometer is derived from radiometric values, i.e., from the digital numbers ND (λ) read on magnetic tape through the following ratio[6]:

$$L(\lambda) = \frac{1}{A(\lambda)} \cdot ND(\lambda) \qquad (1)$$

where $A(\lambda) = \alpha(\lambda) \cdot 1.3^{m(\lambda)} - 3$

The values of $\alpha(\lambda)$ and $m(\lambda)$ are respectively a coefficient of absolute calibration, i.e., a constant characteristic of a given HRV instrument at a given period for a gain of reference, and the gain actually used during acquisition (see Table 3).

2) The luminance due to reflection on the bottom was extracted. By substracting the mean sign

observable during high tide from the sign caught by the $L(\lambda)$ satellite, one obtains a value $L'z(\lambda)$ that only depends on the water depth z, the quality of waters, and the nature of bottoms. This operation therefore deducts the radiation backscattered by the ocean $Lw(\lambda)$ from possible surface effects such as specular reflection $Lg(\lambda)$ or reflection due to high tide breaks, i.e., the "froth" $Lc(\lambda)$, as well as molecular scattering (Rayleigh scattering) $Lr(\lambda)$ and scattering due to aerosols (Mie scattering) $La(\lambda)$:

$$L'z(\lambda) = L(\lambda) - \{[Lw(\lambda) + Lg(\lambda) + Lc(\lambda)].T(\lambda) + Lr(\lambda) + La(\lambda)\} \quad (2)$$

3) The weather conditions were corrected. The absorption by atmospheric gazes of the $L'z(\lambda)$ signal reflected by the bottom is corrected by means of the transmission term $T(\lambda)$ obtained by using the "5S" logicle[7] (see Table 3), which gives the value of $Lz(\lambda)$:

$$Lz(\lambda) = \frac{L'z(\lambda)}{T(\lambda)} \quad (3)$$

For the two images processed the values of $T(\lambda)$ are respectively 0.948 (May 9, 1986) and 0.945 (March 2, 1987).

4) The luminance measured under the ocean - atmosphere interface was calculated. $Lz(\lambda)$ is converted into its equivalent measured under the surface of the water body $Lz^-(\lambda)$,

$$Lz^-(\lambda) = \frac{Lz(\lambda)}{K} \quad (4)$$

Table 3 Values of λ in μm, $\alpha(\lambda)$, $m(\lambda)$, $T(\lambda)$, and f

Date	May 9, 1986	March 2, 1987
λ	XS-1	XS-1
$\alpha(\lambda)$	0.551	0.514
$m(\lambda)$	5	5
$T(\lambda)$	0.948	0.945
f	2.030	2.137

K being the transmission factor to the ocean - atmosphere interface (K ≃ 0.545).

5) To reduce the residual noise of the image and improve its visual aspect the elevated spectral constituents were suppressed by application of a low-pass digital filtering. The luminance file $Lz^{-}(\lambda)$ is convoluted by a 3 X 3 square mesh following type:

0.0625	0.1250	0.0625
0.1250	0.2500	0.1250
0.0625	0.1250	0.0625

6) The continent ocean limit was extracted by masking the convoluted image with the XS-3 band (near infrared), the absorption coefficient of the water in this part of the spectrum nearing 1.

7) In the marine field, the image was calibrated in terms of water depth, of coefficient of diffuse attenuation of waters, and of luminance of bottoms. Knowing that

$$Lz^{-}(\lambda) = Lf^{-}(\lambda, z \Rightarrow 0) \cdot e^{[-Kd(\lambda) \cdot f \cdot z]} \qquad (5)$$

f being a coefficient of optical path (see Table 3); $Kd(\lambda)$ the coefficient of diffuse attenuation of the water and $Lf(\lambda, z \Rightarrow 0)$ the luminance of the bottom estimated <u>under</u> the ocean - atmosphere interface when the water depth tends to 0.

Thus, one can give the following ratio:

$$z = p(\lambda) \cdot Ln\left[Lz^{-}(\lambda)\right] + q(\lambda), \qquad (6)$$

and extend z by using the least squares method thanks to calibration points of which the coordinates x, y, and z are known.

The preceding ratios illustrate the physical signifigance of both the slope $p(\lambda)$ and the intercept $q(\lambda)$ [8]

$$p(\lambda) = - \frac{1}{f \cdot Kd(\lambda)} \qquad (7)$$

$$q(\lambda) = \frac{1}{f \cdot Kd(\lambda)} \cdot Ln\left[Lf^{-}(\lambda, z \Rightarrow 0)\right] \qquad (8)$$

Table 4 Assessment parameters of the bathymetric model used, and characteristic values of the environment

Date	May 9, 1986	March 2, 1987
λ	XS-1	XS-1
N	9	8
$p(\lambda)$	-6.329	-2.384
$q(\lambda)$	27.259	9.557
r	0.945	0.962
Err	0.836	0.711
$Kd(\lambda)$	0.078	0.196
$Lf^-(\lambda, z \Rightarrow 0)$	74.230	55.102

Therefore, it is possible to obtain information on the quality of waters by extracting the coefficient of diffuse attenuation $[Kd(\lambda)]$:

$$Kd(\lambda) = - \frac{1}{f \cdot p(\lambda)} \qquad (9)$$

then, when z is known, and supposing that $Kd(\lambda)$ is invariable, it is possible to get information on the color of bottoms $[Lf^-(\lambda, z \Rightarrow 0)]$ for the points which the depth is given:

$$Lf^-(\lambda, z \Rightarrow 0) = e^{[f \cdot Kd(\lambda) \cdot q(\lambda)]} \qquad (10)$$

Results and Debate

Within only a few days, a major change occurred in the submarine and intertidal landscape: on February 27, 1987, there developed a breccia of nearly 4-Km width which created a second mouth in the estuarian complex of Saalum.

Retracing the unexpected and rapid evolution of prelittoral bottoms was made possible by the satellite approach that enables one to make a chronosequential approach, acquire real-time data, and obtain information of surfacic nature.

For each of the two images processed the values of the assessment parameters of the

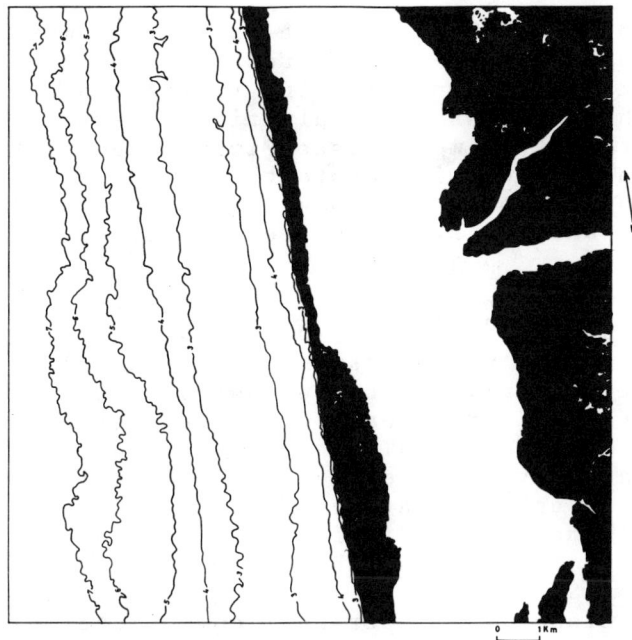

Fig. 2   Telebathymetric outlines of the prelittoral area of Lagoba obtained by interpreting the Spot HRV image of May 9, 1986; the isolines correspond to the signal bathymetry on the XS-1 band.

bathymetric model are given (N: number of calibration points, $p(\lambda)$: slope, $q(\lambda)$: intercept, r: correlation coefficient, and Err: standard estimation error), as well as characteristic values of the environment [$Kd(\lambda)$: coefficient of diffuse attenuation of the water, and $Lf^-(\lambda, z \Rightarrow 0)$: luminance of the bottom measured under the interface ocean-atmosphere (see Table 4)].

If one refers to Jerlov's standards for the classification of waters[9], the waters hemming the spit of Sangomar are liable to belong to the categories "Oceanic II" on May 9, 1986 and "Oceanic III" on March 2, 1987. These values seem to be acceptable since the waters "Oceanic II and III" are characteristic of the West African littoral.

And last, the luminance values obtained for a sediment full of water correspond to a value of the bottom reflectance in the XS-1 band, comprises between 13.4 % (March 2, 1987) and 15.3 % (May 9, 1986).

The coast-line and the isobaths comprised between -2 m and -7 m before and after the rupture were drawn according to the above described process and evaluation parameters of the bathymetric model (see Figs. 2 and 3).

The rupture is due to the conjunction of two kinds of factors: the immediate or meteorological factors and the static or geomorphological ones.

Hydrological and Meteorological Situation

A strong depression dug at 970 hectopascals, formed in the north-west Atlantic and related to a relative minimum centered on 30° N and 45° W, engendered a strong wave characterized by amplitudes of 2 to 3.5 m during its displacement. This harsh hydrologic situation was reinforced by spring tides that favored a successful attack against the supratidal field and the opening of a first breccia between the ocean field and the mouth of Saalum.

Fontana (1987, personal communication) thinks that resonance phenomena within the Sine-Saalum complex might have equally contributed to the amplification of the swell's energy.

Furthermore, the improvement of the rainy situation of the last few years probably raised the flushing effect of the ebb and slowed down the accretional trend.

Rupture of the Spit and Evolution of Bottoms

The study of the image acquired before the rupture (see Fig. 2) reveals deep changes in the distribution of isobaths at the level of Lagoba, in comparison with the situation observed in 1987 (see Fig. 3). This evolution is characterized by:

1) The disappearance of the prelittoral bar that is clearly observable in the image of May 9, 1986; and 2) the forming, in 1987, of an amphitheater-shaped prelittoral zone.

If little is known about the reasons why the prelittoral bar disappeared, the consequences of this disappearance are obvious: the role of the bar as an abutment was annihilated, thus enabling the swells to devote a more important part of their energy to the erosion of the Lagoba basin.

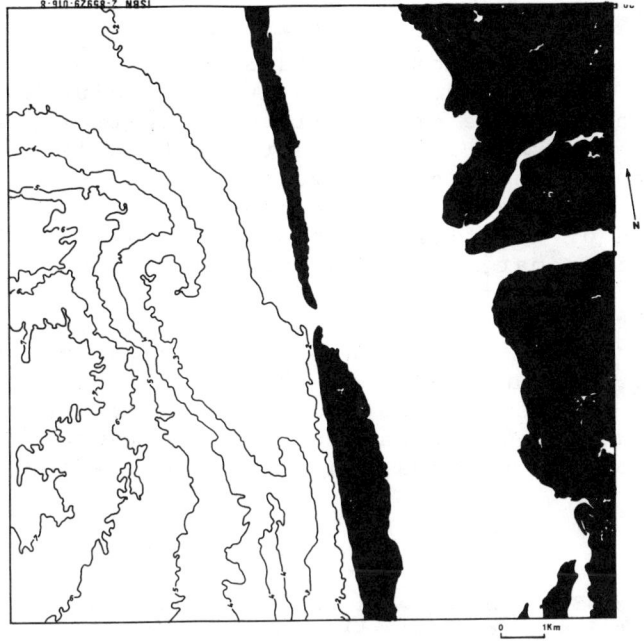

Fig. 3 Telebathymetric outlines of the prelittoral area of Lagoba obtained by interpreting the Spot HRV image of March 2, 1987; the isolines correspond to the signal bathymetry on the XS-1 Band.

Conclusion

The new situation created by the reopening of the mouth of Lagoba has already caused major transformations in the landscape.

In particular, the disappearance of large areas of the spit affected certain vegetal species as Rhizophora mangle L., Avicennia africana P. Beaw., Adansonia digitata Linn., and Phoenix reclinata Jacq.

The evolution of mangroves and tannes of various isles of the Sine-Saalum complex, notably in the isle of Dionewar, directly submitted to the effects of the opening of Lagoba.They are presently characterized by phenomena of over salinity which makes these environments even more unsteady and reduces the agricultural surface of the Niominka (A Senegalese ethnic group living in Saalum whose main activities are both fishering and agriculture.)

Preliminary observations by the Lighthouses and Buoyage Directorate, (Dakar Port) show remarkable changes of the tidal regime in the Saalum's estuary. Since the rupture of the littoral spit, the marine waters penetrate into the estuary through the mouths of Sangomar and Lagoba simultaneously. The lessening of the tidal wave in the estuary shortens the duration of the flow: at Kamatane station, 78 km upward Sangomar, the high tide occurs now 15 mn earlier than before.

If the Lagoba mouth is not obliterated soon, and if anthropogenic erosion by the thoughtless taking away of sandy material is not slowed down, the present-time situation would emphasize the changes observed now. The tidal regime, sedimentary transit, morphology of bottoms, and consequently the evolution

References

[1] Thomas, Y-F. and Diaw, A.T., "Kinematics of the littoral spit of Sangomar (Senegal)," French Society for Photogrammetry and Remote Sensing Bulletin, Vol. 114, Paris, 1989, pp. 14-16 and 27.

[2] Minot, A., "Minot report," High Works Ed., Dakar, 1934, 60 p.

[3] Tromeur, J., "Hydrographic Survey of the Saalum," Hydrographic Annals, Vol. 16, Paris, 1939, pp. 5-33.

[4] Lefur, A., "Hydrographic Mission in the West African Coast - Surveys of the entrance of Saalum," Hydrographic Annals, Vol. 4, Paris, 1950, pp. 125-139.

[5] Mage, E., "The Sine and Saalum Rivers," Maritime and Colonial Journal, Vol. 7, Paris, 1863, pp. 673-676.

[6] Cnes, and Spot-Image, "Spot data's users guide," Cnes and Spot-Image Ed., Toulouse, 1987.

[7] Tanre, D.. Deroo, C., Buhaut, P., Hermann, M., Morcrette, J. J., Perbos, J., Deschamps, P.Y., "Simulation of the Satellite Signal in the Solar Spectrum (5S)," Scientific and Technical University of Lille Ed., Lille, 1986, 150 + 115 p.

[8] Hallada, W.A., "Bathymetry Analysis Using Landsat-4 Thematic Mapper Data," Science Applications Research Ed., Riverdale, undated, 61 p.

[9] Jerlov, N.G., "Marine Optics," Elsevier Ed., Amsterdam, 1976, 231 p.

Saudi Arabia's Experience in Solar Energy Applications

Fahad S. Huraib*
King Abdulaziz City for Science and Technology, Riyadh, Saudia Arabia

Abstract

This paper presents the progress in solar energy research in Saudi Arabia with emphasis on the efforts of King Abdulaziz City for Science and Technology (KACST), which is a government research entity.

With the objectives to advance the development of solar energy technology as a viable, cost-competitive energy alternative, to introduce the solar applications industry to the Kingdom of Saudi Arabia, and to develop the Saudi human resources in the solar field, several major research projects have been undertaken by KACST Solar Programs, including: 1) converting solar energy into electricity for everyday use by the inhabitants of several rural villages, such as the Saudi Solar Village; 2) testing solar power as an energy source for space cooling and desalination, as is the case in the four Saudi universities' projects and the Yanbu water desalination project; 3) developing agricultural systems powered by solar energy to control the growing environment; 4) undertaking fundamental photovoltaic and solar thermal research; and 5) establishing high-technology laboratories for advanced solar energy research at the universities.

KACST Solar Programs is also engaged in the production of hydrogen gas using solar energy. This involves fundamental research, design, and construction of 10-kW and 350-kW photovoltaic/electrolytic hydrogen generation plants.

Several other projects are conducted in the Solar Programs, such as the Solar-Powered Highway Devices Projects, which consist of design, installation, and testing of solar systems to avoid potential road hazards on remote highways and tunnels.

Copyright © 1990 by Fahad S. Huraib. Published by the American Institute of Aeronautics and Astronautics, Inc. with permission.
* Solar Programs Director.

Introduction

The Kingdom of Saudi Arabia is one of the largest oil-producing countries in the world. It has continued to track the energy situation and market and to look far ahead when setting its energy policy.

One of the most important objective of the Kingdom's development plans is to spread out the economy base. Therefore, the Kingdom plans to extend the lifetime of oil availability and its reserves as long as possible to serve this purpose. The plan is to make more use of oil and its products to be the raw materials for petrochemical processes, which, in itself, is a major objective undertaken by the Kingdom.

As the Kingdom of Saudi Arabia is among the highest sun-intensive countries in the world with almost 3000 sun h per year, it was very natural to think of solar energy utilization and look to the development of its technology. The solar energy development plan was fully supported by people in the Kingdom and elsewhere. Moreover, during the 1981 International Conference on Renewable Energy, the Kingdom called for a vigorous promotion and development of new and renewable sources of energy.

In addition to tremendous oil savings that could be envisioned, there are other reasons that make the Kingdom so interested in solar applications. These are:

1) As the industrial base widens the electrical power demands expand. This usually occurs during daytime when the solar insolation is also high, which makes the demands and supply of electricity coincide.

2) Many of the Kingdom's villages are remotely located. For now, it is prohibitively too expensive to expand the national grid and electrify those villages.

3) The solar energy technology is still in the early developmental state. But it is not a complicated technology and, therefore, a country like the Kingdom with low technical resources can contribute and benefit a lot from the development of solar technology. In this area the Kingdom can rapidly increase the technological base and actively develop appropriate solar energy systems.

4) Solar power is clean and does not pollute the atmosphere, and its use does not affect the climate. In fact, the development of solar energy systems would be beneficial for developing countries of similar climate.

5) The abundance of low-cost utility power in the Kingdom discourages the industrial sector to get into the solar technology business at this time. This necessitates the Saudi government to take up the role of developing this new technology at least through its primary stages.

6) Saudi Arabia does not face an energy crisis now, but it believes that it must make a significant contribution to that search for energy alternatives.

For the reasons cited above, the Kingdom is now committed to several programs dedicated to this search for alternative energy technologies. The progress in solar energy research in Saudi Arabia is presented in this paper, with emphasis on the efforts of King Abdulaziz City for Science and Technology (KACST).

Three programs are currently underway at KACST, namely, the continuation of activities initiated under the Solar Energy Research-American/Saudi (SOLERAS) program (a joint Saudi-U.S. program), the Saudi-German program, and projects that are completely developed and conducted by KACST. These programs are briefly summarized in the following sections.

Joint Program with U.S.

SOLERAS was a unique program between the United States of America and the Kingdom of Saudi Arabia to address technical and economic issues for solar technology in each country. The vastly different cultural and political systems of these countries necessitated the development of a process to integrate these programmatic issues. In this sense the SOLERAS program accomplishments were not merely the result of scientific analysis but required a diverse set of management skills to fulfill all aspects of the program.

The solar energy program was one of the first major research areas developed by the Saudi Arabian National Center for Science and Technology (SANCST), which now has become known as King Abdulaziz City for Science and Technology, or KACST. SOLERAS was the first activity of the KACST Solar Program. It was funded by both countries: each country contributed $50 million for a five-year period, which was later extended to nine years. To put this in perspective, $50 million exceeded any other Saudi Arabian expenditures on solar research and any other international commitment to solar research by the United States.

Objectives

Projects included a broad scope of activities in the solar technologies, including centralized solar electric and thermal systems and decentralized applications for cooling and agriculture. These projects were designed to support the SOLERAS program objectives: 1) cooperate in the field of solar energy for the mutual benefit of the two countries; 2) advance the development of solar energy technology; and

3) facilitate the transfer of technology developed under the SOLERAS agreement.

Management Structure

The management plan developed to attain these objectives included the establishment of the SOLERAS Executive Board to ensure that both governments jointly exercised policy, financial, and management control over the complexities of a jointly funded research and development program. This eight-person board included senior officials from each of the following sponsoring agencies: 1) United States Department of Energy (DOE), 2) United States Department of Treasury, 3) Saudi Arabian KACST, and 4) Saudi Arabian Ministry for Finance and National Economy.

A four-person Project Selection Committee supported the Executive Board by providing technical and related advice.

Program Organization

A technical program plan was developed to specify the types of projects, determine the technical approach, and identify the solar energy technologies to be researched. The program addressed both the U.S. emphasis in research and development and Saudi interest in applications and testing.

Urban Applications

The objective of Urban Applications was to improve the quality of life for inhabitants in hot, arid environments by investigating solar applications for cooling. Since Saudi Arabia and the hot, arid regions of the United States have abundant sunshine and high cooling demands, there was strong incentive for investigation of solar cooling systems as alternatives to conventional cooling systems. The projects included both passive and active solar cooling designs as well as combinations of the methods.

Rural/Agricultural Applications

This area was given highest priority early in the U.S./Saudi initiative. Major funding was allocated to the development of an experimental solar village in Saudi Arabia. A major goal was to test solar technologies, particularly photovoltaics, to determine their capability for improving the quality of life in hot, arid climates. The potential for the integration of solar energy systems with controlled-environment agriculture was also a high priority.

Industrial Applications

The goal of this area was to use solar energy technologies in industrial applications requiring either thermal or electrical energy. The major project involved water desalination.

Resource Development Activities

Several activities supported resource development, including solar data collection and analysis providing quantitative data on solar energy resources to assist solar systems design. Universities in both the United States and Saudi Arabia received multiyear funding to conduct basic solar energy research in this area.

A major SOLERAS priority was the transfer of developed solar technology. This activity required the organization and sponsorship of major international technical solar workshops, annual short courses for Saudi and U.S. students in solar-related fields of study, and dissemination of project research knowledge and documents.

Accomplishments

The SOLERAS program goals were pursued through major engineering applications research projects, basic and applied science research projects, and technology transfer and resource development projects. In applications research, photovoltaics were used to generate electricity for remote villages, and solar thermal systems were applied as an energy source for space cooling, water treatment, and controlling growing environments in commercial greenhouses. Some of the most talented private corporate and university research groups in the United States and Saudi Arabia participated in the fundamental research projects. The technology transfer and resource development activities initiated broad international involvement in solar energy research and development.

The emphasis on open exchange and discussion of information also contributed greatly to the development of renewable energy. While some thought that such an approach was not favorable to a fledgling industry, SOLERAS identified and reviewed problems in an open, candid manner and systematically resolved the issues and developed more reliable, durable, and efficient systems. The intent of this stragtegy was to transfer technology to the general solar research community outside the contractors actually involved in SOLERAS projects. This approach allowed firms to improve their designs, systems, and components through review of the SOLERAS information base.

The desalination project conducted under Industrial Applications also illustrates the effect SOLERAS had on component and system refinement. The project's solar collectors provided by Power Kinetics Inc. were used as a test bed for design modifications and improvements developed during SOLERAS and have been incorporated in DOE's small community power experiments program.

The operation, maintenance, and performance results of the cooling field tests and solar desalination project enabled collector component manufacturers of reflective films and silvered mirrors to use the projects as test beds for new concepts. The 3M Corporation participated in the cooling field tests to develop reflective films with higher reflectance and better bonding characteristics. Falconer Glass Corporation was involved in the desalination project mirror testing to develop improved mirror edge protection solutions. Technical experiences reported in the SOLERAS documentation assisted not only participating firms but also the solar industry and related technologies.

Joint Agreement with Germany

The Kingdom of Saudi Arabia and the Federal Republic of Germany are together addressing several solar energy-related issues through a joint international research and development (R&D) program. The joint program, which was commenced in 1982 with the Solar Electric Stirling Engine Concentrator project, has expanded to include sizable projects devoted to the advancement of solar hydrogen technologies. The solar dish project, which succeeded in generating 50 kW of electricity from a single concentrator dish, is still considered the largest dish of its type in the world. The project budget was approximately 8 million Deutsche marks. The solar hydrogen projects will be presented in the coming sections.

Solar Hydrogen (HYSOLAR)

Introduction

The future potential of hydrogen as an energy carrier is unarguable, since hydrogen can be transported, stored, and burned using technologies already applied for natural gas. It can be produced by several processes using new and renewable energy sources such as solar energy. One method of generation is photolysis, in which solar radiation is used directly. This method has many problems related to separation and recovery,[2] so its potential for large-scale production of hydrogen is minimal. The simplest and cleanest method is electrolysis, in which solar energy is converted into electricity for electric decomposition of water.[3,4] Solar hydrogen generation will allow countries with high annual solar energy availability, such as Saudi Arabia, to export that energy to other, less sunny countries such as the middle and northern European and the northern American countries.[5]

It is argued by several, such as Nitsch,[6] that specific areas of the industrial and public energy consumption will be fulfilled during the next century by using hydrogen. This demands the development of hydrogen transportation, storage, and use technologies so that they can be handled by the public.

There are many problems that require extensive R&D effort before the solar hydrogen technology can be feasible both technically and economically. Such problems are requirements for immense solar collection areas, system matching,[4,7,8] storage systems,[9,10] distribution, use,[11] and cost. Those problems are being under major consideration in many research institutes and international cooperation programs, like HYSOLAR, around the world.

Objectives

The Saudi interest in solar hydrogen arose seven years ago when it was envisaged that hydrogen might become the main source of energy in the next century. The program objectives can be summarized as follows: 1) attain sufficient scientific knowledge for future commercial production and use of hydrogen in Saudi Arabia; and 2) facilitate the transfer of developed/acquired solar hydrogen-related technologies to the scientific community and the public of Saudi Arabia.

Program Management

The program is supervised by the Saudi Arabian-German Joint Committee for Cooperation in Science and Technology and managed by the KACST and the German Ministry for Science and Technology (BMFT).

The program is funded jointly with a total budget of 39.2 million Deutsche marks. The program duration is five years ending December 1989.[12]

Technical Plan

A technical plan was initially developed in 1983,[13] and revised after that during the course of the program. The plan consists basically of six tasks:

1) Demonstration of the reliability and safeness of solar hydrogen production through a continuous operation of the world's first 350-kW solar electrolytic hydrogen production demonstration plant that is to be constructed in the Solar Village near Riyadh, Saudi Arabia.

2) Design and construction of a 10-kW solar hydrogen production facility. This consists of a 10-kW photovoltaic power generation subsystem and three new and advanced electrolyzers. The project construction is completed in Stuttgart, Federal Republic of Germany.

3) Installation of a 2-kW laboratory test facility at the King Abdulaziz University, Jeddah, Saudi Arabia.

4) Investigation and research on photoelectrochemistry, advanced alkaline electrolysis, and fuel cells.

5) Study of the available techniques, equipment, and procedures for the use of hydrogen as fuel for future introduction to the public. The equipment under study is catalytic combustors, gas flame burners, steam generators, fuel cells, and internal combustion engines.

6) Building the professional community in the field of solar hydrogen technologies and training personnel for the monitoring, operation, maintenance, and repair of solar-powered hydrogen generation, storage, transportation, and use equipment.

Internal Program

The research and demonstration activities that are sponsored locally by the governmental agencies are mainly to demonstrate the technical and economic advantages of solar technology. The major projects' size, energy savings, and unit energy cost will be presented henceforth:

1) Photovoltaic (PV) Powered Highway Lighting and Warning Devices test undertakings generate approximately 800 MWh of solar electric energy annually. The two most significant demonstration projects are the lighting systems for two remote tunnels located in the southern mountains of Saudi Arabia. The total length of the tunnels is 712 m and the lighting system uses technology specially developed to use DC electricity from PV. The total projects budget is $4.5 million. The calculated cost of 1 kWh of electric energy generated is around $.10.

2) Domestic hot water systems are installed in several

housing compounds around the Kingdom. The average solar heating energy produced per square meter of collection area is about 11 MWh per year. One of the projects has 2244 m^2 of solar collection area and generates 25,000 MWh of useful heating energy per year. The calculated cost of 1 kWh of useful generated heating solar energy is around $.035.

3) PV-powered devices are installed for commercial production of electricity. Several of those projects are: 25 telecommunication stations in remote areas, two reverse osmosis desalination units, and dozens of cathodic protection systems for oil pipelines.

Conclusion

The Saudi Arabian government has actively supported the development work of some of the more promising solar technologies and will continue to do so in the future. Several solar demonstration systems that have been installed are operational, including the PV village power, PV water desalination, PV-powered high lighting, thermal water desalination, and passive solar buildings. We expect these systems to provide valuable field performance and reliability data for the practical users and to the solar communities in general.

With the continuous support and encouragement by the Saudi Arabian government for a high level of solar research involving international participants and Saudi Arabian universities, we believe that we can make a significant contribution in the future to the understanding of various solar technologies, their limitation and capabilities, and their use.

References

[1] Nitsch, J., and Voigt, C., "Strategies for the Penetration of Solar Hydrogen in Energy Systems," DGS, 4, ISF, Berlin, Federal Republic of Germany, 1982, pp. 961-980.

[2] Melvin, A., "The Impracticability of Large-Scale Generation of Hydrogen from Water Photolysis by Utilization of Solar Radiation," International Journal of Hydrogen Energy, Vol.4, 1979, pp. 223-224.

[3] Carpetis, C., Schnurnberger, W., Seeger, W., and Steeb, H., "Electrolytic Hydrogen by Means of Photovoltaic Energy Conversion," Hydrogen Energy Progress IV, pp. 1495-1512.

[4] Cox, K., "Hydrogen from Solar Energy Via Water Electrolysis," 11th IECEC, pp. 926-932.

[5] Bernnecke, P., Justi, E., Kleinwachter, J., and Rotzoll, R., "Solar-Hydrogen-Economy: Thermodynamic Optimization of its H_2-Transportation and Storing Tube Capacity," DGS, 4, ISF, Berlin, Federal Republic of Germany, 1982.

[6] Nitsch, J., "Large Scale Solar Hydrogen Production and Transportation," International Seminar on Solar Thermal Heat Production & Solar Fuels and Chemicals, DFVLR, Stuttgart, Federal Republic of Germany, October 1983.

[7] Freudenberg, K., "Solar Generator Performance with Load Matching to Water Electrolysis," Appl. Phys., Vol. A28, No.4, 1982, pp. 205-209.

[8] Freudenberg, K., "Solar Generator Performance with Load Matching to Water Electrolysis Longterm Averages and Range of Instantaneous Efficiencies," 4th E.C. Photovoltaics Conference, Stresa, Italy, May 1982, pp. 205-209.

[9] Iannucci, J., and Robinson, S., "Fixed Site Hydrogen Storage: I. Applications Impact," Journal of Alternative Energy Sources, Vol.2, 1981, pp. 3539-3605.

[10] Robinson, S., and Iannucci, J., "Fixed Site Hydrogen Storage: II. Comparison of Technologies and Economics," Journal of Alternative Energy Sources, Vol.2, 1981, pp. 3607-3625.

[11] Breele, Y., Gelin, P., Meyer, C., and Petit, G., "Technico-Economic Study of Distributing Hydrogen for Automotive Vehicles," International Journal of Hydrogen Energy, Vol. 4, 1979, pp. 297-314.

[12] KACST and BMFT Cooperation in Research, Development, and Demonstration for Solar Hydrogen Production and Utilization (HYSOLAR), "Agreement," Riyadh, Saudi Arabia, Feb. 1986.

[13] KACST and BMFT Cooperation in Research, Development, and Demonstration for Solar Hydrogen Production and Utilization (HYSOLAR), "HYSOLAR Program Description," Riyadh, Saudi Arabia, Feb. 1986.

The Saudi Center for Remote Sensing

Muhammad A. Tarabzouni*
King Abdulaziz City for Science and Technology, Riyadh, Saudi Arabia

Abstract

Satellite remote sensing is used throughout the world in many scientific disciplines, including geology, hydrology, cartography, land use and regional planning, and mapping and related disciplines. After the launch of LANDSAT 4, with its thematic mapper (TM) system and ground resolution of 30 meters, Saudi Arabia decided to become one of the first countries to directly receive the satellite data. On behalf of the government of Saudi Arabia, the King Abdulaziz City for Science and Technology (KACST), signed an agreement with the United States National Oceanic and Atmospheric Administration (NOAA) to access the data from Earth observation and meteorological satellites LANDSAT and NOAA. KACST built a receiving station, an image processing center, and a photograph processing laboratory. KACST also plans to sign an agreement with SPOT IMAGE of France to access data from the French Earth observation satellite Systeme Probatoire d'Observation de la Terre (SPOT).

The Saudi Center for Remote Sensing (SCRS) consists of a data acquisition system (DAS), a data processing system (DPS), and a photograph processing laboratory (PPL). The SCRS is able to receive, process, and analyze remote sensing data.

Introduction

The merging of remote sensing and space exploration has generated a great deal of interest and applications in a broad range of disciplines. Currently, we depend on space-borne sensors to provide information for tasks ranging from weather forecasting, mineral exploration, land use, and crop production forecasting to applications such as commercial fishing, range land monitoring, and pollution detection.

Copyright © 1990 by the American Institute of Aeronautics and Astronautics, Inc. All rights reserved.
* Director, Space Research Institute.

Definition of Remote Sensing

Remote sensing is the science of acquiring physical information about an object through the analysis of data obtained by a device that is some distance away from that object. Instruments used to gather this information are remote sensors. Your eyes, for example, act as remote sensors responding to light reflected from objects. Your brain acts as a computer, analyzing or interpreting the data gathered by your eyes.

Remote sensing instruments can be flown on high altitude aircraft and on orbiting satellites. Orbiting satellites are a relatively new research activity, in use only since the early 1960s. Some early applications of remote sensing are illustrated in Refs 1, 2 and 3. As shown in Fig 1, remote sensing incorporates two major processes--data acquisition and data analysis.

The elements of data acquisition are (a), sources of energy, such as the sun, which generate (b), energy that is propagated through the atmosphere, to (c), the Earth's surface. The Earth's surface reflects and absorbs some of the energy, depending on surface features. The absorption and reflection of energy is characteristic of materials and can be used to identify materials such as vegetation, bare land, water, and rock types. Sensing systems, (d), such as aircraft or satellites sensors, acquire the reflected energy and record it for analysis. This data, (e), can be recorded in pictorial or numerical form.

Satellites can, in summary, use sensors to record variations in the way the Earth's surface features reflect and emit electromagnetic energy.

The elements of data processing are (f), examination and interpretation of sensor data, using various computer-controlled viewing and analytic devices. Reference data, such as maps, inventories, statistics, and ground truth studies, may be used to augment the analysis. Information extracted from the sensor data and reference data include the type, extent, location, and condition of the various resources over which the sensor data were collected. This information, (g), is made available to users generally in the form of images, maps, tables, and written reports. Users, (h), apply this data to their particular areas of interest, forming decisions based on the information.

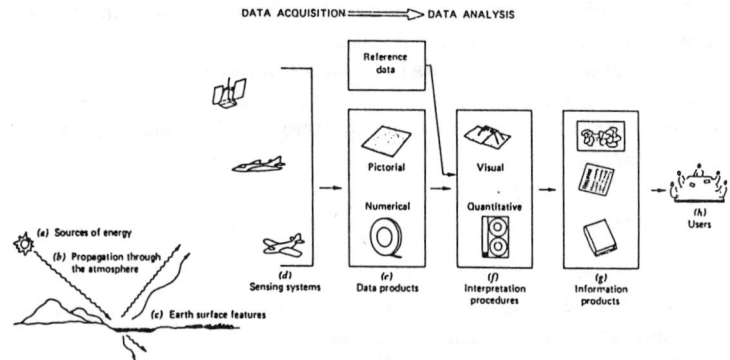

Fig.1   Electromagnetic remote sensing of Earth resources.

The SCRS uses these elements of data acquisition in various areas of interest. The data the SCRS uses is primarily acquired from satellite platforms.

LANDSAT

The United States National Aeronautics and Space Administration (NASA) launched LANDSATs 1, 2, and 3 on July 23, 1972, January 22, 1975, and March 5, 1978. They were placed into sun-synchronous orbits that crossed the equator in a southward direction at times ranging from 8:50 a.m. to 9:31 a.m. They acquired global coverage every 18 days (103-min orbit) at a nominal altitude of 918 km. All three have since been decommissioned.

LANDSATs 1 and 2 had two remote sensing systems augmented by a tape recorder system: 1) a three-channel return beam vidicon (RBV) system; 2) a four-channel multispectral scanner (MSS) system, as shown in Fig 2. The RBV images covered a scene 185 x 178 km with a ground resolution of 86 m. The MSS covered a 185-km-wide swath, with a ground resolution of 86 km in four wavelength bands: two in the visible spectrum at 0.5-0.6 micrometers (μm) (green) and 0.6-0.7 μm (red) and two bands in the reflected infrared at 0.7-0.8 μm and 0.8-1.1 μm. The tape recorder system was designed to record image data over areas of the world where there were no direct ground receiving stations.

LANDSAT 3 introduced two major changes in design: 1) a thermal channel was added to the MSS at Band 8, 10.4-12.6 μm with 234 m ground resolution; 2) the RBV system's ground resolution was improved by the use of a two-camera broad band system rather than the MSS, as shown in Table 1. Ground resolution was 30 m with a scene size of 185 x 173 km.

Unfortunately, the tape recorder system did not remain functional throughout the spacecraft's life, and the thermal band also developed operating problems and failed shortly after launch.

LANDSATs 4 and 5 were launched by NOAA on July 16, 1982 and March 1, 1984 respectively, with two sensor systems each: 1) an MSS system similar to

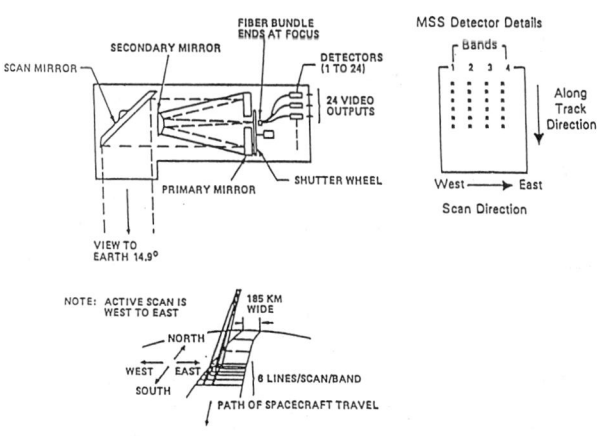

Fig. 2   Schematic of the LANDSAT multispectral scanner.

Table 1 Multispectral scanner and return beam vidicon characteristics

Spectral band	Spectral region, μm	Ground resolution, m
4	0.5 - 0.6	80
5	0.6 - 0.7	80
6	0.7 - 0.8	80
7	0.8 - 1.1	80
8	10.4 - 12.8	237
RBV-3 camera[1]	.48 - 12.8	80
	0.58 - 0.68	80
	0.69 - 0.83	80
RBV-2 camera[2]	.50 - 0.75	40

[1] LANDSAT 3 only
[2] LANDSAT 1 and 2 only

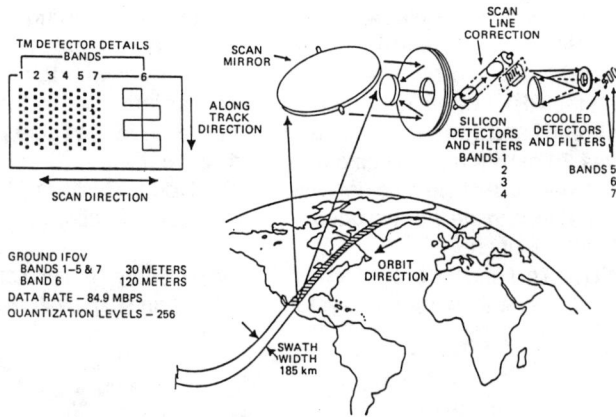

Fig. 3   Schematic of the thematic mapper.

that on the first three LANDSATs; 2) a TM system, as shown in Fig 3. They are in sun-synchronous orbits with a 98.2-degree inclination angle. They have repeating cycles of 16 days (98-min orbit) at a nominal altitude of 709 km. The ground resolution of the two systems is 57 m and 30 m respectively. They do not have tape recording systems.

TM is an advanced MSS with seven bands designed to maximize analytical possibilities for various applications. The seven bands are: 0.45-0.52 μm, 0.52-0.60 μm, 0.63-0.69 μm, 0.76-0.90 μm, 1.55-1.75 μm, 10.40-12.50 μm, and 2.08-2.35 μm, as shown in Table 2.

Bands 1-5 and band 7 of the TM have a ground resolution of 30 m and band 6 has a resolution of 120 m. The TM provides a substantial improvement in spatial, radiometric, and spectral resolution over the previous LANDSAT MSS systems. These improvements require a data transfer rate of 85 megabits per second, compared to the previous rate of 15 megabits per second. At these high data rates, the TM data cannot be recorded onboard and can only be received via direct transmission to a ground station or by a tracking and data relay satellite (TDRS).

LANDSAT 4 is still partially functional and LANDSAT 5 has exceeded its expected life. Two additional LANDSATs are planned for launch by the U.S. commercial organization EOSAT. These satellites will carry improved sensors and are expected to be operational in the early 1990s.

SPOT

The French Center National D'Etudes Spatiales (CNES) launched the remote sensing SPOT-1 on February 21, 1986. It was placed in a sun-synchronous orbit with a descending node of 10.30 a.m.. It acquires global coverage every 26 days. The payload of SPOT-1 consists of two identical high resolution visible (HRV) imaging instruments and two magnetic tape recorders.

The HRV instruments operate in two modes: a panchromatic (black and white) mode corresponding to observation over a broad spectral band of 0.51-0.73 µm; and a multispectral (color) mode corresponding to observation in three narrow spectral bands, 0.50-0.59 µm, 0.61-0.69 µm, and 0.79-0.90 µm.

A standard near-nadir SPOT scene covers 60 x 60 km. At its extreme oblique viewing angle of 27 deg, a scene covers 60 x 81 km with ground resolution of 10 m in the panchromatic mode and 20 m in the multispectral mode.

SPOT passes directly over the same area every 26 days. At the same time, by tilting the mirror during adjacent swaths it is possible to observe the same area from different angles in the acquisition circle.

In twin near-nadir mode, the two HRV mirrors are tilted 0.613 deg from the vertical line (away from each other). This is the "standard" mode.

In each HRV the mirror can be tilted so that the instrument can observe an area located up to 475 km from the track. This means that at the equator, the same area can be observed seven times in a full 26-day cycle. At 45 deg latitude, it is 11 times, and at 60 deg latitude, 18 times.

Table 2 Thematic mapper characteristics

Spectral Band	Spectral region, µm	Ground resolution, m
1	0.45 - 0.52	30
2	0.52 - 0.60	30
3	0.63 - 0.69	30
4	0.76 - 0.90	30
5	1.55 - 1.75	30
6	2.08 - 2.35	30
7	10.40 - 12.50	120

The mirror can be tilted in 91 steps of 0.6 deg each, up to a maximum of 27 deg to each side. Because of the curvature of the Earth, the maximum tilt respective to the vertical is 33 deg.

The probability of acquiring cloud-free images increases if an area is registered during successive passes, since weather changes rapidly. Each step of tilt equals roughly 9 km on the ground, and each full degree of tilt about 15 km. An E in front of the degree number means looking east, and a W in front means looking west.

The satellite can be programmed to constantly observe a given area for a certain period every time it passes within 475 km of the acquisition area. This gives a better guarantee of access to data than any other satellite system. This, in turn, means that it is possible to follow the course of development during an agricultural growth cycle.

SPOT's plane pointable mirror makes the registration of stereoscopic pairs possible, which provides the possibility of photogrammetrical processing for mapping and interpretation purposes. The mirror is controlled by an onboard computer, which is programmed daily by the mission and operations control center. The instrument can register images with angles of up to 27 deg in 45 stages of 0.5 deg. A stereoscopic pair that has been taken with an angle of 24 deg has a height/base ratio of one/one. Although the mirror can be tilted up to 27 deg, it is not recommended to use a tilt greater than 8 deg if good radiometry (i.e., comparable to near-nadir standard operations) is required. This restriction keeps the variation of the digital levels due to variation in projections to less than 2 per cent and also the reachable path on the ground to a width of approximately 200 km.

CNES plans to launch three additional SPOT satellites. SPOT 2 is to be launched in 1989 and SPOT 3 and 4 in the early 1990s and mid--1990s respectively.

NOAA Satellite Systems

The NOAA A through K satellite systems consist of operational, near-polar orbiting, environmental satellites. A prototype television infra-red operational satellite (TIROS), the TIROS-N, was launched in 1978. The NOAA 9, 10, and 11 satellites are currently in orbit. They provide images continually by constantly transmitting for reception by stations within range.

Table 3 Advanced very high resolution radiometer characteristics

Band	Spectral region, µm
1	0.580 - 0.60
2	0.725 - 1.10
3	3.550 - 3.90
4	10.300 - 11.30
5	11.500 - 12.50

Their sensor payloads reflect technological advances made during nearly 20 years of polar-orbiting weather satellite operations. The NOAA 9, 10, and 11 satellites operate in a near-polar, approximately sun-synchronous orbit at an altitude of 850 km and carry a wide range of sensors for environmental and meteorological data collection: 1) advanced very high resolution radiometers (AVHRR); 2) stratospheric sounder units (SSU); 3) data collection system ARGOS (DCS); 4) a space environment monitor (SEM); 5) microwave sounder units (MSU); 6) high resolution infrared radiation sounders (HIRS); and 7) search and rescue systems (SARSAT).

On these satellites the AVHRR sensor, being a low-resolution MSS, offers unprecedented possibilities for low-cost environmental monitoring over large areas.

The AVHRR sensors collect reflected and emitted radiation from the Earth's surface and atmosphere in four spectral regions covering both the visible infrared and the thermal infrared (see Table 3), with a nadir spectral resolution of 1.1 km. Frequency of reception from the NOAA system is 10 times daily, provided by a pair of satellites. In addition to monitoring atmospheric and meteorological parameters, the NOAA 9, 10, and 11 satellites, by their multispectral capability, have the potential of providing Earth resources-related data. Spectral bands 1 and 2 are comparable to bands 2 and 4 of the LANDSAT TM and offer a capability of monitoring vegetation biomass changes through linear combination of two bands. The thermal bands collect temperature data from the terrain and cloud tops and have the potential for detecting significant soil moisture changes in arid and semi-arid areas where vegetation cover does not have significant masking effect.

An important characteristic of the AVHRR sensor is its high radiometric resolution. The instrument samples the reflected/emitted radiation at 1,024 intensity levels (10 bit sampling), which is high compared to the 64 levels (6 bit sampling) of the LANDSAT MSS. The high radiometric resolution is advantageous because the signal/noise ratio of the sensor is relatively high and thus provides more precise data for discrimination of different terrain materials on the basis of their spectral reflectance and emittance.

Role of Saudi Arabia

Saudi Arabia has used LANDSAT information in the fields of geology, agriculture, mapping, environmental studies, land use, and city planning for more than ten years. This information has helped Saudi Arabia achieve its development goals.

Until January 1, 1987 however, Saudi Arabia did not have direct access to LANDSAT information via its own ground receiving station. Until that time the higher resolution data of LANDSAT 4 and 5 and of SPOT-1 were transmitted to a regional ground station outside Saudi Arabia, in France or Italy.

In 1982 Saudi Arabia decided to build a facility for remote sensing, and made plans to build a ground receiving station by signing agreements with NASA and NOAA and SPOT IMAGE to build a remote sensing facility at KACST in Riyadh. The facility was completed on September 26 1986. It provides the means to fully utilize LANDSAT, SPOT, and future satellite data.

Saudi Center for Remote Sensing

The SCRS is a complete system. It can acquire, process, and analyze remotely sensed image data from several satellites. It produces film, computer-compatible tape, and other output products for analysis.

Image data are also acquired from aircraft sensors in either digital or hard copy format. SCRS can also accept and process other image-input data of similar frequency, characteristics, and format, providing for future growth and expansion.

The SCRS is divided into three systems: DAS, DPS, and PPL.

Data Acquisition System

The DAS tracks, receives, and records data from the LANDSAT TM and MSS sensors, the SPOT HRV sensor in both the multilinear array (MLA) and panchromatic linear array (PLA) modes, and the AVHRR sensor of NOAA 9, 10, and 11 satellites. These data sets, which include both raw image data and telemetry, are recorded on magnetic tapes. These raw data tapes are transported to the DPS for processing.

The DAS capability to receive and record satellite data is provided by five major subsystems: 1) tracking and receiving subsystem; 2) antenna computer

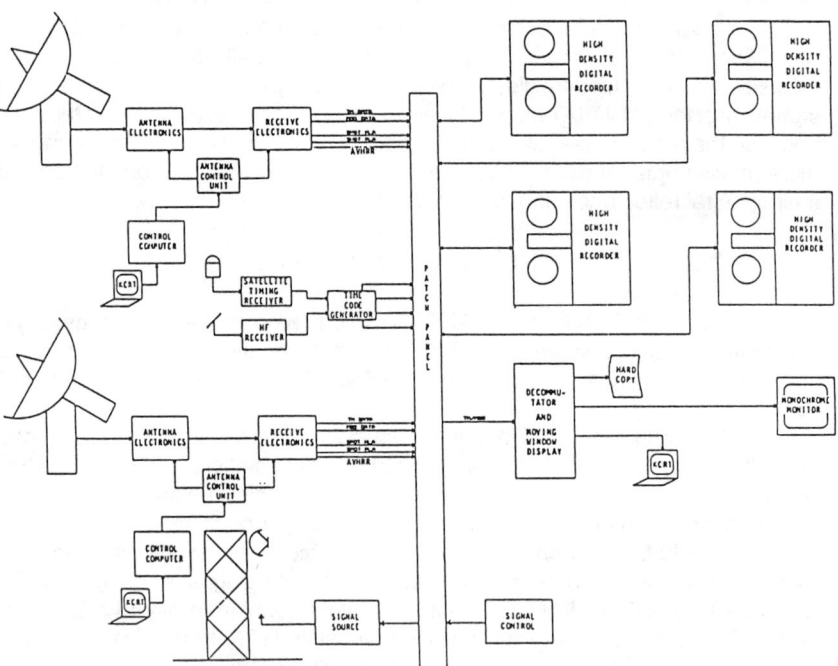

Fig. 4 Data processing system system requirements--daily throughput.

Fig.5   Coverage area of the Saudi ground receiving station.

subsystem; 3) timing subsystem; 4) recording subsystem; and 5) signal distribution and test subsystem.

The DAS is equipped with two 10-m S-/X-band antennas for acquiring LANDSAT and SPOT satellite data and one 3.8-m L-band antenna for acquiring NOAA satellite data. The S-/X-band antennas are mounted on pedestals approximately 7 m high. The L-band antenna is mounted on a pedestal approximately 2.4 m high (see Fig 4).

There are five S-/X-band receivers and four identical high density digital recorders (HDRR) to record the high data rates from LANDSAT and SPOT satellites.

NOAA television infra-red operational satellite (TIROS) AVHRR data, after being processed through the weather image processing system, are recorded on a 9-track, 1600 bit-per-inch Kennedy tape drive.

Magnetic high-density tape winding and cleaning equipment and tape degaussers are provided for tape preparation and maintenance.

The DAS is the most advanced station in the world, and with the two 10-m antennae the station can completely automatically track two satellites simultaneously 24 hours a day on S-band and X-band frequencies.

The ground receiving station can receive data over an area of 2406-million km^2, an area which encompasses the entire Arabian Gulf region and other neighboring countries, as shown in Fig 5.

Data Processing System

The DPS archives the MSS, TM, and HRV high-density tapes received from the DAS. The archived data is then screened to identify the scenes and create browse films. The DPS contains a data management module which tracks the acquired imagery, processes user requests for imagery, produces image catalogs, maintains a browse film library, and controls production.

The DPS processes the MSS, TM, and HRV raw data and generates high-resolution film and CCT products for image analysis and other uses. The products consist of bulk corrected, georeferenced, geocoded, and mosaicked data.

A bulk corrected image is defined as an image that has undergone both systematic radiometric and geometric correction. The systematic geometric correction removes known internal distortions introduced into the raw image data by the sensor mechanism, spacecraft motion, inaccurate altitude by sensor mechanism, Earth rotation, etc. The systematic geometric correction is made in the universal transverse mercator (UTM) map projection and does not require the use of ground control points.

A georeferenced image is defined as an image that has been bulk corrected and which geodetically registers to a map using ground control point information. The georeferenced image is also in the UTM map projection. Geocoded imagery is georeferenced imagery that has been further processed by rotation to provide alignment to true north. Mosaicked images consist of two or more neighboring geocoded images from the same sensor that have been merged into a single image.

AVHRR data recorded on CCT at the DAS can also be processed in the DPS to generate bulk corrected high-resolution film and CCT products. Georeferencing, geocoding and mosaicking are not performed on AVHRR data.

The DPS includes capabilities for performing a wide variety of interactive image analysis functions utilizing the image data types described above as well as image data produced by aircraft sensors in both digital and hard copy format.

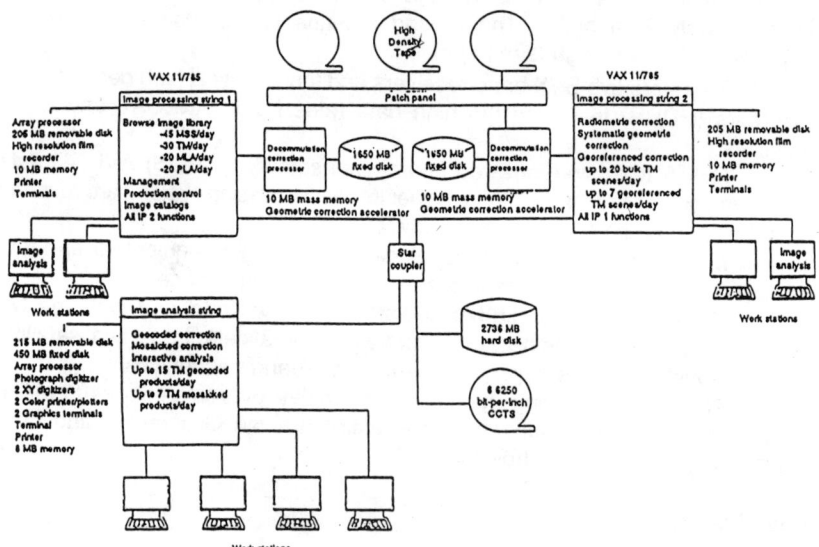

Fig.6   Data processing system.

The DPS centers around three VAX 11/785 computer strings. Two of these strings are identical and are configured mainly for performing the data screening and image processing functions. The third string is configured with peripherals required for image analysis. The data management functions may be performed on any string. The three computer strings share a set of disks and tape drives using a VAXcluster. The VAXcluster also provides high-speed communications between the strings (see Fig 6).

Photograph Processing Laboratory

The PPL provides photograph processing services to the SCRS from latent film products generated by the DPS.

PPL photograph processing functions and their related inputs and outputs are provided in response to written work orders prepared in the DPS data management subsystem or in response to special requests from the SCRS planning office.

The photograph products generated in the PPL include: 1) black and white microfiche browse film; 2) 241 mm color transparencies; 3) 241 mm black and white transparencies; 4) black and white prints and enlargements; and 5) color prints and enlargements.

The photographic processes used to generate these products include:1) the E6 positive color transparency process; 2) the Cibachrome positive color paper print process; 3) the RC negative black and white film process; 4) the RC positive black and white paper print process.

In addition to photograph processing, the PPL provides general services to other SCRS groups or activities, including: 1) archival storing of the standard film products generated in the course of daily activities; 2) monitoring the operation of the DPS film recording equipment and assisting in its maintenance; 3) assisting in the evaluation of system performance from the photograph processing end; 4) evaluating and monitoring film product quality and maintaining photograph process control; 5) supporting the SCRS user interface requiring the generation of requested photograph products and provision of special photograph support.

Station Status

The SCRS DAS is in full operation at the Riyadh site and regularly acquires LANDSAT 5 and NOAA 9, 10, and 11 satellite data. An archive of raw data tapes has been established and currently covers several acquisitions over the last year. On an average day, three or more NOAA and two or three LANDSAT 5 passes are acquired. This is done six days a week. SPOT data is acquired pursuant to an agreement with SPOT IMAGE that was signed in June 1989.

Until recently, the SCRS DPS was at a temporary site in Riyadh. General Electric (GE) is the prime contractor for the SCRS project and is responsible for integration of the systems and buildings. The SCRS and GE have jointly trained the Saudi personnel who operate the SCRS. Some production work has been done at the temporary site to verify the quality of the raw data archived in Riyadh at the DAS. The DPS was installed in August 1989 in a permanent facility, also

Table 4 Data processing system—system requirements, daily throughput[1]

SATELLITE	SENSOR	NO OF DATA ACQUISITION SCENES	NO OF DATA SCREENING AND MANAGEMENT SCENES	BULK CORRECTED SCENES	GEO-REFERENCED SCENES	GEOCODED SUBSCENES	MOSAICKED SUBSCENES	IMAGE ANALYSIS
LANDSAT 4,5	TM	60 AND	30 AND	20 OR	7 OR	15 OR	3 OR	AS REQUIRED
	MSS	60 AND	30	45	10	25	5	AS REQUIRED
SPOT	HRV	140	30 40	45 20	10 8	25 15	5 3	AS REQUIRED
NOAA	AVHRR	AS REQ.	AS REQ.	AS REQ.	AS REQ.	N/A	N/A	AS REQUIRED

[1] One day = 16 hours in two 8 hour shifts

in Riyadh, which was completed in mid-1989. The SCRS PPL was installed in a permanent site when the final work was done in late 1989.

Station Capabilities

The SCRS is capable of processing a large volume of data each day. Table 4 illustrates the system requirements for production during an operating day of two shifts of eight hours per shift.

The DAS is capable of acquiring 260 scenes each day: 60 MSS, 60 TM, and 140 HRV. This is equal to the number of scenes that occur within the SCRS acquisition circle on any given day.

The DPS is capable of screening a total of 115 scenes per day, distributed by sensor type as 40 MSS, 30 TM, and 40 HRV. Image processing requirements (bulk, georeferenced, geocoded, and mosaicked) are produced as requested by the users. On any given day, a request may be received for generation of any of the four types of imaging processes from each of the three types of sensor. The DPS is not normally requested to produce all four types of products from each of the three sensors each day, hence the "or" on the requirements chart between sensor types. Reading the requirements for TM image processing: 20 bulk correct scenes AND 7 georeferenced scenes AND 15 geocoded subscenes AND 3 mosaicked subscenes are required to be processed in one day for TM data.

Both NOAA data processing and image analysis product generation are performed as required by user requests, in addition to the above screening and image processing.

Conclusion

The SCRS is another example of the Kingdom's commitment both to mankind and to the peaceful use of space. The SCRS is the center of the remote sensing community in Saudi Arabia and the countries around it. It will serve as a local source of remote sensing data and a repository for both satellite data and related data such as maps and documents. Users will no longer have to acquire data from the United States or Europe.

The SCRS will be a laboratory for the examination and analysis of data and for working the data into products that can be applied to the problems and researches of government ministries and universities. The SCRS also offers the remote sensing community in the Middle East a center for training in the use and understanding of the data and methods used to process and analyze it. This transfer of knowledge and technology is vital to the continued technological development of the Kingdom of Saudi Arabia and its people.

Acknowledgments

The author would like to thank David Irwin for his technical and editorial assistance, Kazim Ali Zabin for the accuracy with which he typed the draft and final forms of the paper, and all of the people of the Space Research Institute and the SCRS at KACST for their unstinting cooperation while I wrote this paper.

References

1. Tarabzouni, M. A., "Computer-Enhanced Landsat Images for Ground Water Exploration in the Northern Arabian Shield: Hail Test Site," Master's Thesis, University of Tennessee, Knoxville, Tennessee, 1981.

2. Lillesand, T. and Kiefer, R., <u>Remote Sensing and Image Interpretation</u>, Wiley, New York, 1979, p. 3.

3. Landsat Data Users Handbook, NASA/GSPC, Greenbelt, Maryland, 1979, pp. 7-10.

Agricultural Applications of Remote Sensing in Hungary

G. Csornai*
FÖMI Remote Sensing Center, Budapest, Hungary

Abstract

   Agriculture is the most important economical factor in Hungary. Therefor a reliable, satellite data-based crop information system is very important. This paper provides an overview on the most important segments of a crop monitoring system, stressing the crop mapping and area estimation methods, results. Supported by the data processing system that was developed, results of crop mapping and area estimation using Landsat MSS bitemporal data and a traditional plus a novel method for a 0.6 million hectares area demonstrated that the novel method increased mapping accuracy by almost 10 percent. It also makes the survey more reliable at low cost. This and similar methods also suggest the more symmetric usage of digital image analysis and the features of geographic information systems.

Introduction

   Major environmental factors such as soils, climate, and terrain are very good for intensive agriculture in Hungary. The per capita cereal production is about 1.3 tons, which is more than that in the U.S. As about 30 percent of total land area is crop land, a reliable information system on the major crops is inevitable. The commodities production is based on the system of cooperative and state-run farms. The average holding of a farm is approximately 4000 hectares as opposed the 120 hectares in the rest of Europe.
   Because of the relatively large fields -- average size about 80 hectares -- the satellite data-based information system at the state level seems to be adequate.

Copyright © 1989 by the Institute of Geodesy, Cartography and Remote Sensing, Hungary. Published by the American Institute of Aeronautics and Astronautics, Inc. with permission.
*Head of Agricultural Applications.

The need for a remotely sensed databased information system has increased just recently because of the substantial cut to the state budget and staff running the traditional data collection system.

There have been efforts in using aerial photography for monitoring agricultural and horticultural farms. The use of color infrared (CIR) techniques gave good results, but the tecniques are not yet used operationally. However, there are some areas where the operational usage of CIR techniques are going to be introduced soon. These areas are soil survey for amelioration and crop cultivation error detection. This paper concentrates on agricultural applications of remote sensing by satellite, in Hungary.

Since 1981 the staff of the Hungarian Remote Sensing Center has focused development efforts on creating the basis of an efficient national crop information system (CIS) concentrating on the two major crops -- winter wheat and corn -- plus sunflowers and sugar beets.

Relying basically on Landsat data, first the assessment methods and the image analysis techniques had to be worked out together with the appropriate image processing systems and facilities. Though the development of crop assessment methods and an efficient data processing system is not finished yet, there are some preliminary results showing our original approach is worth pursuing towards the semi-operational stage in the near future. Currently the use of remotely sensed data is not operational, being only in the research and development phase.

Outline of a National Crop Monitoring System

Within any subtask of crop monitoring in Hungary the target area stretches from 0.5 to 5 million hectares, which is the area of one or more counties.

Whole crop monitoring comprises the following subtasks 1) detailed crop mapping, 2) area survey, 3) area estimation of the major crops sampling the area of the counties, 4) vegetation inhomogeneities detection and classification by the type and yield loss caused, 5) development stage assessment, 6) developmental evaluation, 7) condition assessment, and 8) objective yield forecast. A comprehensive CIS is certainly expected to provide usable information to either 1) the top agricultural administration and decision makers or 2) the farms or a set of farms. These are difficult requirements for the entire CIS, the data, the methods, and the spatial resolution to be used plus the timeliness, the data acquisition, and the processing capacity. It makes sense, however, to serve the needs of these two end users up to the point where their requirements are not controversial.

The previously discussed segments of crop monitoring generally comprise a series of consecutive problems to be solved. A subtask in that list normally makes use of the methods, techniques, results and processing systems provided by some of the previous subtasks in the list even if the underlying theories and models are quite different. This underlines the importance of the first three problems, which are subsequently discussed in more detail. Further detail is available on modelling the yield forecast and crop development.[1]

Previous Results and Approach

First, after having developed the basic digital image analysis procedures and tools, the adequacy of the methods and techniques of using Landsat multispectral scanner (MSS) monotemporal and multitemporal data was demonstrated.[2] A Landsat MSS time series was analyzed for a farm (approximately 6000 hectares), and an extrapolation was done for a 0.1 million hectare area of a unified land system on the Hungarian Great Plain. The classes used comprised four subclasses of corn based on their relative maturity. The accuracies were in the range of 85-95 % per pixel. The signature extension and classification were also quite good within the Hajdúság land system. The study also showed that multitemporal raw MSS data and multitemporal vegetation index data gave significantly better results than monotemporal data.

A later study[3] for a small area proved that the previous computer classification result itself had been almost faultless, showing "misclassification" on only the spots of the fields where inhomogeneities had really been present. Thus, in recomputing the classification accuracies, of 95-98% resulted. Besides, this provided a technique for the delineation of the in-field inhomogeneities, which is also important in crop development and production loss assessment. That showed the potential of our developed image analysis techniques, but the training and test fields were located within a very small area. Therefore, from the viewpoint of a country-wide survey it was a good indication only.

After having the preliminary results in 1983 that were actually the feasibility studies of processing methods and techniques rather than systematic, problem-oriented studies, we concluded that we should first improve image processing capabilities. This process resulted in an image analysis software system on a PDP-11/40 equivalent processor, where the color display was not perfectly integrated. Recently an effort has been made to integrate the image analysis, the reference data handling and geographic information, and modelling subsystems into a microVax compatible processor.[4]

Data processing system

The data processing system serves as a basis for the development of a remote sensing-based crop information system.[4] Although the subsystems are at different developmental stages and the whole system is also not yet sufficiently integrated, some features should be outlined.

The <u>image analysis subsystem (IAS)</u> has been the most important part as far as being an important tool in the agricultural land use mapping studies. It provides for major procedures, such as registration, geometrical correction, radiometric correction, supervised/unsupervised classification, and error assessment. The <u>reference data handling subsystem (RDS)</u> facilitates the input and data management of all the necessary ancillary data that are originally in map form, including ground truth, soils maps, meteorologic maps, and topographic maps. The data base system is of vector type with the attributes coupled. RDS provides a small but quite flexible set of features[5] that are generally expected from a vector-based geographic information system (GIS).

A separate subsystem for <u>geographic data modelling (GDMS)</u> has been devised. This division of the common GIS functions into two groups supports image analysis and modelling better[6], when different data sources are involved, creating a multidimensional problem of spatially distributed parameters. To emphasize this conceptual distinction between digital map handling and modelling using multidimensional spatial data, the term <u>geographic information and modelling system (GIMS = RDS + GDMS)</u> seems to be more appropriate. The IAS and GIMS are planned to be used in a fairly symmetric relation.

The <u>controller</u> is now analysis steps-oriented instead of representing an expert system terminal.

This new system is thought to give a framework for the development and semi-operational phase of the subtasks of crop monitoring.

Crop Inventory and Mapping in Hajdú-Bihar County in 1987

Starting from the earlier results a project was carried out for a county (Hajdú-Bihar, Approximately 0.6 million hectares area) to enable us to compare Landsat-based inventory results to that of the Hungarian Statistical Office for the first time in Hungary. The latter, traditional method is based on the obligatory reports of farms that are collected at the county level, where some modifications are done. These alterations are based on historical data and the special expertise of the staff. The statistics are then aggregated to the whole country at the ministry level. There is a tendency for county estimates to have a larger variance than the national estimate.

A county is the smallest unit for which the existing statistics collection method provides quite good, low-biased estimation. In our project[7] a wall-to-wall crop mapping approach was used instead of sampling frames.

The complexity of the area and the test of a novel classification method that was based on the previously digitized map of field boundaries were two genuinely new features of the project.

The Area, Data Used, and Processing System

Hajdú-Bihar County is located on the Great Plain in East Hungary. More than half of the county is arable land, and the county is important in the Hungarian crop production pattern. The soils are alluvial with a flat terrain, which explain why we selected the county as a test area. As this is also a test area for other investigations (e.g. monitoring the cultivation activities of farms) relatively much knowledge on the area has been accumulated.

Given the constraints of low cost, complete and cloud-free coverage by the Landsat MSS, we had to choose two unfeasible dates of April 19 and September 2, 1984. On the first date, only winter wheat and rye were seen. On the second date, only corn, sunflowers, and sugar beets were seen. It is believed that a single Landsat scene in the period of June 1–July 10 would result in the same or better mapping and estimation accuracies.

The <u>reference data</u> were used in an efficient form, first having the field boundary maps digitized (35 farms out of 100) by the end of 1987. These data were linked together with the crop codes of the selected fields from 1984. These latter data came from a nongeocoded data base.

These two processes resulted in a digital reference map. The stratification was accomplished on the basis of a soils map, geographical region classification, and a color composite of the Landsat data.

The processing system used was operated on the PDP-11/40 equivalent computer, but some programs were available at our microVax compatible processor. The ingegration level of the system was low, causing remarkable overhead costs for conversions and operator interactions.

Methods

Two methods were tested:

In our first approach similar series of steps and methods were used as in our earlier studies,[2] except the training and error estimation procedure was more efficiently supported by the GIS subsystem.[7] First, some selected areas from each stratum were clustered separately to find the spectral classes. This was followed by a complex cluster configuration assessment plus image visual investigation for the training samples to identify the spectral components of each class. The training was finished by building up the class statistics using the weighted sum of component statistics by the cluster occurrences in the given class. (This step assumes that class statistics can be derived as a mixture of probability density functions of the components.) If a test of the classifier (maximum likelihood) being built showed a high error rate, a retraining was done until an acceptable result was achieved.

In our second approach a novel method was used based on a reliable, good accuracy point-wise classification result map. Having the digitized fields' boundaries maps (DFBM) of the farms, first a thorough test on the change of boundaries was conducted. This uses a derivation operator to compare the actual Landsat data with the DFBM data base. (A common visual check was accomplished rather than a sophisticated automatic one being developed.) After this update in DFBM a reclassification was accomplished assigning the field to the class of the majority population within the field boundaries.

Discussion of Results

For the sake of simplicity four strata were carefully selected. The project revealed more technical errors and some theoretical methodological shortcomings. Some primitive

errors occurred during the creation of the digital reference map because of digitization errors and faulty coverage codes. This misled the classifier to a certain extent that was reflected in the misclassification tables (see Table 1). The resulting thematic map shows remarkable error in the sugar beet class because of the lack of reliable training samples for the particular category.
The ratio of reference area (training plus test) was kept in the 3-7% interval, which should have been lower in the strata.
  The accuracies in the strata were different, reflecting their homogeneities and complexities plus the availability of reference data.

Table 1 Misclassification tabla of Hajdúság (central) stratum

	Winter cereals	Corn	Sugar beets	Bare soil	Alfalfa, pasture	Settlement	Hit %
Winter wheat	3417		42	83	167	125	82
Winter barley	204						92
Corn		4005	241	193			83
Sugar beets	20	260	1560	20			78
Sunflowers							95
Silo corn	15	310		128		5	63
Alfalfa	10	79	17	3	207		60
Pasture	97				417	6	73
Green peas		38		78			64

Table 2 Ebes cooperative; comparison of the customary per poin-maximum likelihood (MXL) and the new classification

Categories	Fields	Area, hectares	Accuracy, %	
			MXL	Novel A-method
Winter wheat	18	857	92	99
Corn	12	785	95.6	100
Sugar beets	6	336	88.2	100
Sunflowers	1	10	95	100
Peas	1	55	96	100
Hybrid corn	3	74	88.8	100
Beans	1	92	90.5	100
Total	42	2209	93	99.6

As a whole the area estimation based on the crop map was in good correspondence with that of the Central Statistical Office for the broader categories of winter cereals, maize, alfalfa plus pasture. The distribution of cereals is: 1) wheat 87%: 2) rye 5%: 3) barley 8%.

The novel method was tested on more smaller areas. The results were good, as expected (see Table 2). The increase in correct classification is about 10% generally. This method misses the class only when the majority within a field is missed. This is very unlikely after having all the steps and data in the first approach corrected.

Conclusion

Though there is a need for substantial further research and development in setting up a national crop information service in Hungary, there have been some preliminary results as a result in crop survey techniques. The project demonstrated that the approach that was followed in the area survey of the major crops seems to be promising from the viewpoint of a national CIS. After eliminating some technical shortcomings, there are some problems that still need further research. A few of them are: 1) computer-aided methods of stratification, 2) efficient application of the statistical sampling frame when the aim is only an area estimation: 3) stronger computer support in training and test area selection: 4) the further exploitation of steady features in the thematic classification: and 5) more efficient support in error estimation. Complementary studies referring to Landsat Thematic Mapper, SPOT data (Satellite Pour l'Observation de la Terre), as opposed to Landsat MSS, could also improve our present methods to be able to switch into a semioperational phase on large regions in Hungary.

Acknowledgments

This work was supported by the State Board for Technical Development, the Ministry of Agriculture of Food, and the Hungarian Academy of Sciences.

References

[1] Ferencz, Cs., Tarcsai, Gy., Lichtenberger, J., and Hamar, D., "Some Results of Corn and Wheat Yield Forecasting by Satellite Remote Sensing," Proceedings of the 5th Symposium of Working Group Remote Sensing of the International Society of Soil Sciences, Hungarian Society of Agricultural Sciences, Budapest, Hungary, Apr. 1988, pp. 129-134.

[2] Csornai, G., Dalia, O., Gothár, Á., and Vámosi, J., "Classification Method and Automated Result Testing Techniques for Differentiating Crop Types," <u>Proceedings Machine Processing of Remotely Sensed Data,</u> Purdue University, West Lafayette, IN, July 1983, pp. 217-224.

[3] Csornai, G., Vámosi, J., Dalia, O., and Gothár, Á., "Vegetation Status Assessment and Monitoring in Agricultural Areas by Remote Sensing," 34th Congress os International Astronautical Federation, Budapest, Hungary, Oct. 1983.

[4] Csornai, G., "Agricultural Land Use Mapping Using Geographic Information System," <u>Proceedings of the 5th Symposium of Working Group Remote Sensing of the International Society of Soil Sciences,</u> Hungarian Society of Agricultural Sciences, Budapest, Hungary, Apr. 1988, pp. 272-275.

[5] Kiss, Z., "A Basic GIS: REBEKA," Users Manual, Budapest, Hungary, 1988.

[6] Bryant, N.A., and Zobrist, A., "IBIS: A Geographic Information System Based on Digital Image Processing and Raster Datatype," <u>IEEE Transactions on Geoscience Electronics,</u> GE-15, 13, 1977, pp. 152-159.

[7] Csornai, G., Dalia, O., Farkasfalvy, J., Nádor, G., Vámosi, J., and Zabó, P., "Research and Development Results on the Regional Crop Survey Methods," Institute of Geodesy, Cartography and Remote Sensing, Budapest, Hungary, Rept. 1987.

Yield Prognosis by the Productivity Criteria Using Spectral Signatures in the VIS, NIR and TIR Ranges

H. Barsch*
Pedagogical University Potsdam, Potsdam, GDR
K.-H. Marek†
Academy of Sciences of the GDR
H. Weichelt
Central Institute for Physics of the Earth, Potsdam, GDR
and
A. Gebhardt§
Academy of Agricultural Sciences of the GDR, Muencheberg, GDR

Abstract

Experimental investigations on the behavior of agro-geosystems have been accomplished under INTERCOSMOS cooperation for several years.

The object of the experiment "GEOEX-86" carried out in GDR 1986, was to verify former results under typical heterogenous soil conditions of the landscape of glacial lowlands and to expand the methodology of the determination of the agricultural productivity and yield formation. The tasks consisted of deriving the crop parameters from spectral signatures and of developing a computer-assisted agricultural crop monitoring system from remotely sensed data.

The technical equipment applied in the experiments consisted of 1) a spectroradiometer "MMR" (bands in accordance with the LANDSAT TM channels); 2) a field spectrometer "BSP 83" (40 channels between 0.4 - 1.1 um); and 3) a microwave radiometer "MWR" (λ = 2.8 cm). Parallel to this complex a four-channel aerial survey camera "MSK-4" was used to take multispectral photographs of a test site from an altitude of about 3,600 m. Also a number of biometrical and pedological parameters were determined, for instance, biomass, content of chlorophyll, content of water, leaf area index (LAI), and projective ground cover as well as soil moisture and content of organic matter.

Copyright © 1990 by the American Institute of Aeronautics and Astronautics, Inc. All rights reserved.

Introduction

Monitoring and estimating the state and development of agricultural crops are gaining increasing importance because of the intensified character of agriculture in the GDR. The tasks of remote sensing in this context have been addressed in a number of papers (Barsch et al.[1], Söllner et al.[2], Weichelt[5]). As a physical measuring technique probing an area, remote sensing is suited in particular to the acquisition of data from large areas used for agricultural purposes.

The Central Institute for Physics of the Earth and the Pedagogical University in Potsdam, GDR, have investigated the recognition of development and yield differences in agricultural crops by means of spectral signatures for several years.

These investigations were carried out as complex experiments under INTERCOSMOS cooperation between member states on test sites in the USSR, GDR, Poland, and Cuba in the years 1985-1988. In this paper results are presented that have been derived from measurements of spectral signatures in the test site "Falkenrehde" near Potsdam during the joint experiment "GEOEX-86" between the GDR and the USSR in 1986.

Area of Investigation

The size of the test site Falkenrehde is about 40 km^2 (see Fig. 1). It is situated 15 km northeast from Potsdam (the district capital) and has the landscape features that are typical for the internal lowland of the GDR (ground moraines of the last glaciation). Within the test site a 5-km^2 area was selected that at the southern end is bordered by a former glaciothermal channel. In the center of the lowlands where the groundwater table is elevated to surface level, there are marshy soils; the border, however, where the fluvial sands are conserved by the glaciofluvial valleys, features humus gleys. In the northern direction there is a ground moraine plateau with local diapirs of sand at domes. Here loamy underlaid sands are encountered that are partly slack or

Fig. 1 Test site Falkenrehde.

groundwater influenced and have a different top sand layer thickness. The catena of regional location categories is obtained by the agricultural location mapping. The considerable heterogenity of the soil cover in the test area is expected to give rise to a remarkable time-spatial differentiation in yield formation of agricultural crops.

Methods of Investigation

The measurement of spectral signatures was carried out for all the relevant phenological phases by a radiometric complex that can be operated either as an airborne version for the acquisition of flight track data (one dimensional) from low altitudes (50-150 m) or as a field version for data acquisition at the locations concerned. The radiometer complex consists of a modular multichannel radiometer (MMR) covering the VIS, NIR, SWIR, and TIR ranges and a microwave radiometer (MWR, developed and manufactured by the Technical

High School of Ilmenau) operating at a wavelength of $\lambda=2.8$ cm for measuring the emitted radiation. The channels of the MMR are in accordance with the bands of the LANDSAT TM. In addition to the spectral measurements, aerial survey flights were carried out by the multi-spectral camera MSK-4.

During the whole vegetation period, 1985/86, biometrical and pedological parameters were determined at 13 measuring points as terrestrial references. Investigations were carried out on winter wheat, winter barley, and sugar beets. In the next section the results on winter wheat are presented. The field measurements were performed at the points F, G, H, and I of the test field 511.

Results of Terrestrial Investigations

Because of the cool and rather humid weather at the beginning of the vegetation period, 1986, the phenological development and biomass production of wheat made best progress at measuring point I (location category D2a) until June, the location of which was dry and sandy and most distant from the lowlands. At this time the chlorophyll density was about 4.5 g/m^2 at measuring point I; the measuring points H and G, however, because of the strong moistening and therefore lower soil temperatures, were delayed in their development and had chlorophyll density values of 3.6-3.9 g/m^2.

From mid-June, during a dry and warm weather period and shortly before the phase of ripeness began, the mineral and moisture resources of the now heated humus rich soil became more and more decisive at measuring point G for the ear filling process.

The productivity of this area was then higher compared with those areas at the measuring points H and I and resulted in the highest crop yields during the final ripeness on the site. This location, which had inherently differentiated development of winter wheat, can also be clearly recognized by the spectral signatures.

For estimating the vitality of vegetation, a feature VM2 based on the remission values in

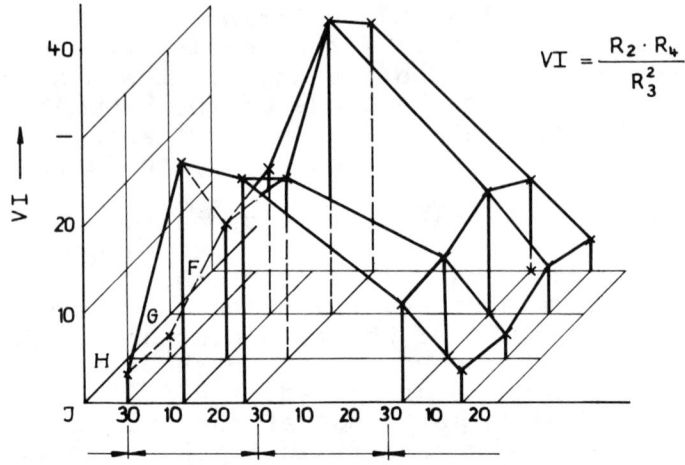

Fig. 2 Dynamics of the vegetation features VM2 at the locations of test field 511.

the green, red, and infrared spectral channels was computed by which the absorption of chlorophyll and the level of the infrared plateau was evaluated and was shown in preceding investigations to be most invariant against atmospheric effects (Weichelt et al.[4]).

Therefore, VM2 evaluates the photobiological activity and is computed by the following relationship:

$$VM2 = (K2 * K4)/K3 \qquad (1)$$

K with index stands for the remission values in the corresponding channel of the MMR. The water content of the plants is estimated using the ratio of channels K6 and K7:

$$WM2 = K6/K7 \qquad (2)$$

Figure 2 shows the dynamics of the vegetation feature VM2 at the four measuring points during the time interval April-July. The lowest values of VM2 were computed at the location H, which was situated on a slope. H can also be used for deriving the most unfavorable conditions of the location.

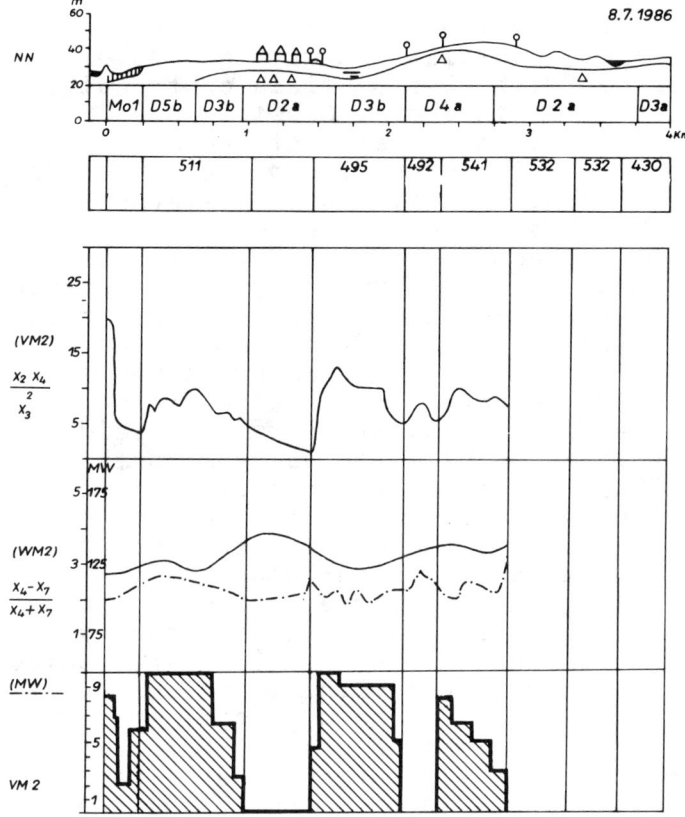

Fig. 3 Vegetation and water feature as well as clusters derived from profile 4 (July, 1986).

Analysis of Profile Measurements

As shown in an earlier study (Söllner et al.[2]), for the evaluation of the vegetation state it is convenient to consider other spectral bands which are influenced by the water content (MMR channels 6 and 7) or the object temperature (MMR channel 8) and matching both suitably to a criterion of productivity.

For this reason all spectral channels of the MMR and the microwave signal were used for the analysis of profile data. In doing so the profile data were separated in a number of different clusters by an unsupervised classification procedure (KMEANS algorithm cited in SPÄTH, 1975). Data processing was performed with single profiles and with groups of them in connection with several field plots cultivated with crops of interest. Considerable differentiations by means of the spectral signatures were found within fields of one crop as well as between fields of different crops.

The clusters extracted by the algorithm were evaluated by their centers. The estimation of vitality was made by using the vegetation feature VM2. Figure 3 shows the cluster result of the profiles dated from the July 8, 1986, and was arranged by increasing values of the vegetation feature VM2. It also includes the assignment of the clusters to the fields. Based on the location differences between the fields, significant variations in vitality were detected comparing equal agricultural crops on different fields. Within a field, however, the differentiation is rather poor.

Because of the limitation of the classified data set for the wheat fields, a better differentiation within the plots was achieved. In particular, the decline of productivity on the slope in the northern part of the field and on a dry sandy lens in the southern part of the field was clearly observed. The highest productivity was found to be in the lowlands.

The categories of all profiles extracted as a result of clustering were submerged to four groups of productivity. The basis for the estimation was the hierarchical classifier derived by Söllner et al., 1986, which in this case, however, was applied to the spectral features VM2 and WM2 derived from the cluster centers and to the radiation temperature measured in the TIR region.

Productivity category 1 features the maximum values of VM2 and WM2 and the minimum temperature value and therefore, can be assessed as most vital. Figure 4 shows the areal assignment of the productivity groups that were extrapolated by linking profile sections of equal productivity. By considering the classification results of the second aircraft flight on July 15, 1986, the rate of change of the vitality features can additionally be derived from the spectral data, which enabled

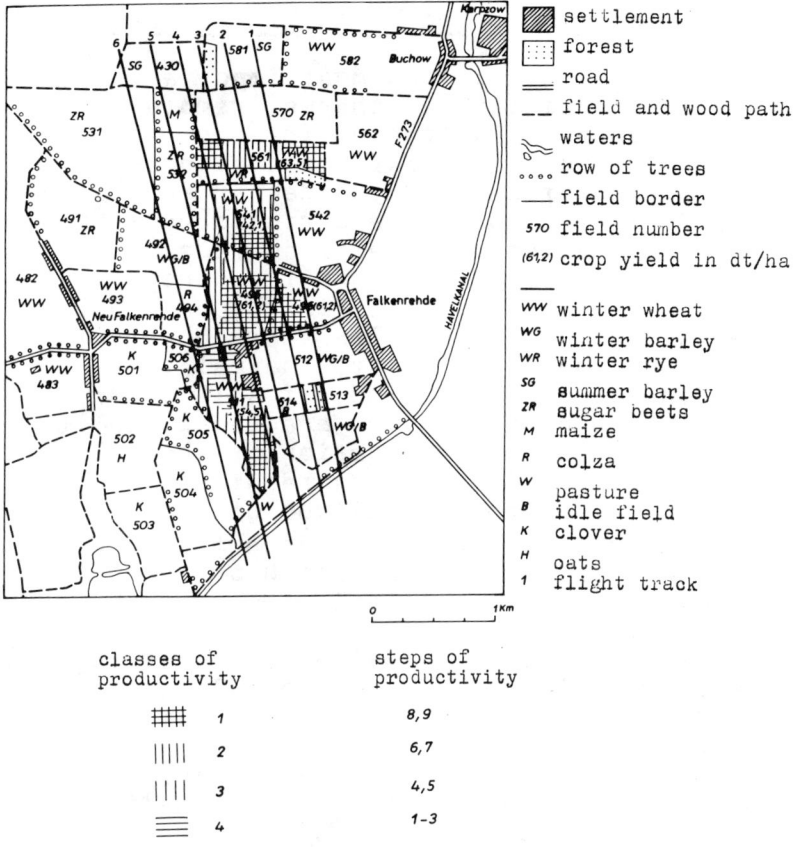

Fig. 4 Distribution of productivity classes measured spectrometrically on wheat fields (July 1986).

one to infer the speed of ripeness. In Figure 5 the combination of both results is demonstrated.

Therefore, a complex state assessment of the wheat cultures based on derived facts from the spectral data is feasible and turns out to be in good agreement with the terrestrial biometrical investigation and results. An estimation of the productivity of the wheat fields shows the results given in the lower part of Figure 4, which are in accordance to the values of productivity determined terrestrially. The computed yield prognosis values are about 10 % too high (see Table 1). The reason for this can be ascribed to spots of yield deficit at the field borders with soil compression caused by agricultural machinery.

Crop Monitoring Using TIR Imagery

Soil heat flux is the most important carrier of information in thermal remote sensing of bare soil. Special interest is paid to its dynamic part. Unlike the stationary geothermal heat flux from the deeper soil layers, it is caused by the periodic variation of absorbed insolation at soil surface.

For a given amplitude and frequency of absorption the dynamic heat flux depends only on thermal inertia, a complex thermophysical parameter of the soil. First, thermal inertia depends on soil moisture and, if soil is dry, on density and porosity. For practical use it is important that quantification of thermal inertia (i.e., soil moisture) requires measurement of diurnal amplitude of surface temperature and albedo. That means the remote sensing mission must be done both predawn and after solar noon and insolation must not be disturbed during this period by changing cloudiness. From the economic and meteorological points of view there arise serious limitations for operational use under European climatological conditions.

Much more promising is the combined use of VIS/NIR and TIR remote sensing for localization of highly moistured areas as a hint of soil compaction. These areas show a particular thermal behavior: despite higher absorptivity they appear at a minimum temperature because of evaporated cooling.

If the soil is plant-covered, the latent heat exchange (evapotranspiration ET) is decisive for plant canopy temperature. With

Table 1 Estimated and real crop yield of the wheat fields of test area

Field No.	511	495/496	541	561	Total
Estimated (decitons/hectares)	60.9	61.2	51.1	54.0	56.4
Real	54.5	61.2	42.1	-	-

Fig. 5 Areal distribution of change rate of the vegetation features.

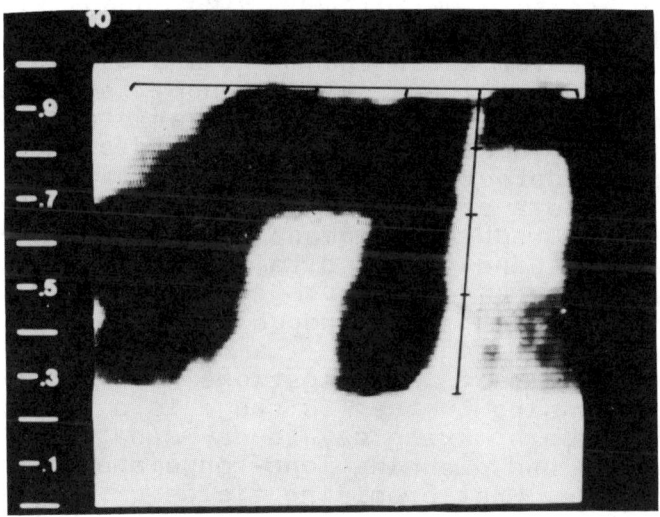

Fig. 6 Irrigation test plot of potatoes: a) panchromatic photography, b) TIR imagery with dynamic range of 10 K), c) scheme of irrigation treatments:
1   - without irrigation
2 - 6 - successive increase of irrigation of 5 mm
7, 8 - special irrigation program at high level

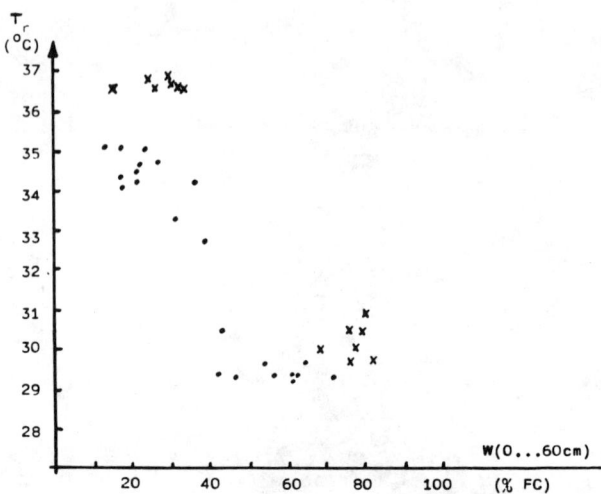

Fig. 7 Relationship between canopy temperature and spil moisture content W (in percent of field capacity) of the upper 60 cm of soil for winter wheat (.) and potatoes (x).

the exception of some environment a conditions (vapor pressure deficit and wind speed) ET depends on water supply of the plants and the conditions for water transport and diffusion. Thermal emissivity of plant canopy is therefore a sensible indicator for the water status and the phytopathological situation and can be used for early detection of stress conditions.

TIR data acquisition is made in the GDR by the thermographic equipment "AGA Thermovision 750" (wave band filtered to 3,6-5,5 um) mounted on helicopter Ka-26. Photographs in the VIS/NIR range are taken simultaneously with TIR imagery.

A number of investigations have shown that remote sensing in the TIR range is a valuable tool for both making day-to-day management decisions and planning long-range amelioration activities. Most promising fields of application are:

1.) Localization of highly moistured areas connected with soil compaction.
2.) Localization of underground drainage pipes.
3.) Identification of reduced evapotranspiration connected with water supply differences and

Fig. 8 Potato test plot with different infestations of nematodes (Globoders rostochiensis):
a) visual, b) TIR imagery, c) No. of eggs and larvae in 100 g soil material.

    phytopathological infestation (see Figure 6, 7, and 8).
4.) Monitoring of areal distribution water and localization of water run off.

5.) Micro- and mesoclimatological characterization of agricultural areas (cold air fluxes and influence of wind obstacles).

    Summing up, TIR remote sensing in combination with VIS/NIR imagery is a valuable addition to remote sensing methods for agricultural monitoring purposes.

References

[1] Barsch, H., Marek, K.-H., Söllner, R., and Weichelt, H.; "Investigation of Spectral Characteristics of Natural Objects in Test Sites of the GDR," <u>25th Congress of the International Society of Photogrammetry and Remote Sensing, Commission VII</u>, Rio de Janeiro, Brazil, 1984, pp. 75-84.

[2] Söllner, R., Marek, K.-H., Barsch, H. and Weichelt, H., "On the Condition of Agricultural Objects from Spectral Signatures in the VIS, NIR, MIR and TIR Wavebands," <u>Symposium on Remote Sensing for Resources Development and Environmental Management</u>, Enschede, Netherlands, Aug. 1986, pp. 321-324.

[3] Späth, H., *Cluster-Analyse-Algorithmen zur Objektklassifizierung und Datenreduktion.* München, R. Oldenburg Verlag, Wien 1975.

[4] Weichelt, H., and Herr, W., Zur Vorverarbeitung multispektraler Daten (Preprocessing of multispectral data)," *Vermessungstechnik*, Berlin, 35, 1987, 8, pp. 270-272.

[5] Weichelt, H., Beiträge zur in situ Messung von Spektralcharakteristiken natürlicher Objekte (Contributions to the insitu-measurement of spectral signatures of natural objects). *Publications of the Central Institute for Physics of the Earth*, No. 90, Potsdam, GDR, 1986.

Measures for Minimizing Radiation Hazardous to the Environment in the Advent of Large-Scale Space Commercialization

S. Nataraja Murthy*
Indian Space Research Organization, Bangalore, India

Abstract

The nature of hazardous effects from radio frequency (RF), light, infrared, and nuclear radiation in the advent of large-scale space commercialization on human and other biological species is presented. RF radiation hazards, in the order of severity, result from transmissions from ground to space, microwave energy beams from Space Solar Power Satellites (SSPS) to ground, and excess power flux densities on ground due to space emissions.

Measures suggested include using interlocks to cut off radiation towards ground, off-pointing of microwave energy beams in case of attitude failure, and limiting the "satellite of-axis gain data rate product" in case of space transmissions towards Earth. Hazards due space-borne lasers can be reduced by using reflective materials on buildings and reflective clothes by personnel. Underwater colonies can reduce high-power laser hazards. Nuclear power satellites when worn out are suggested to be deposited at some stable points in the solar system so that they may not enter the atmosphere as nuclear debris. Other measures suggested are creation of shelters in buildings and in cities.

I. Introduction

With the advent of large-scale space commercialization the risks associated with the commercialization process and its defense have increased. Radiation hazards are only one of the risks associated with space commercialization, but they may cause damage not only at one spot but also distributed over

---

Copyright © 1990 by the American Institute of Aeronautics and Astronautics, Inc. All rights reserved.
*Head, Frequency Management Cell, and Safety Officer.

large parts of the globe. They may also be spread over time, causing damage that may be sustained for quite a long time.

RF/microwave radiation hazards for personnel working in the vicinity of transmitting aerials are well documented.[1,2,3] In view of large-scale space commercialization the number of such RF links have been estimated, the areas affected are calculated and the risk to environment has been defined in this paper. Measurements recently made near one of the high-power Telemetry, Tracking and Command (TTC) antennas regarding power densities at different elevation angles have been used in calculating the risk. The measures to be taken to off-point the microwave beam from a solar power satellite have been highlighted. Though the risk from RF radiation from a number of spacecraft to the global population is at present negligible, the conditions under which the risk can be significant have been worked out, bringing in a new concept of satellite off-axis gain data rate product.

Standards are available for ground-based laser safety.[7,8] In view of the wide application of laser beams in the space commercialization and space defense networks, the hazards for ground environment are estimated for a malfunction in spacecraft orientation, and measures are recommended for global safety.

Hazards due to nuclear weapons[4] have been studied earlier. Risks due to radioactive material released from ground to nuclear reactors have also been reported.[5] But hazards from space-based nuclear plants proposed for the 21st century[6] have not been studied to the necessary extent. Measures for minimizing these, such as depositing space nuclear debris in some stable points in the solar system before their fallout, are suggested in this paper. In case of their inevitable fallout to ground, countermeasures are suggested on both a short-term and a long-term basis.

II. Hazards from RF Radiation

RF radiation hazards can be listed according to their severity in primarily three applications: 1) high-power radiation from Earth station/ground station antennas for commanding or communicating to spacecraft, including antennas of small domestic satellite transmitting terminals, 2) solar power satellite systems discussed for large space power generation and transmission to Earth by microwave beams, and 3) continuous transmission of RF radiation from the satellites causing large power flux densities on the Earth's surface. Figure 1 depicts the RF radiation hazards in the advent of large-scale space

SPACE COMMERCIALIZATION: RADIATION HAZARDS

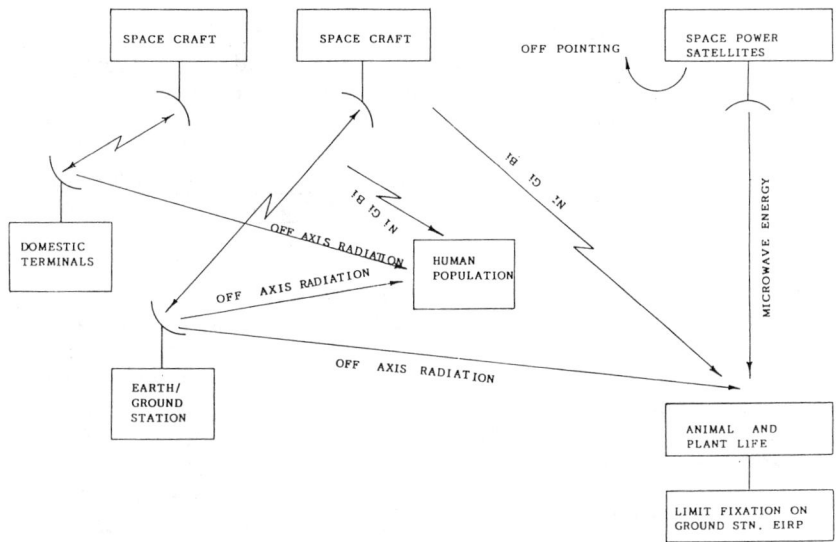

Fig. 1   Hazards from RF/microwave radiation and countermeasures.

commercialization and the measures suggestted to minimize the same.

A. Biological Effects of RF Radiation

It is common knowledge that high-power density RF radiation causes eye reddening and skin burns. Medium-power density continuous radiation is known to cause cataracts and disfunctioning of the testis. However, recent investigations have revealed that, at millimeter waves, microwave fields have resonant interactions with molecules that modulate a host of calcium dependent intracellular enzyme mechanisms. Bioelectromagnetic research is producing clear evidence of joint cell surface interactions of tumor promoters with RF modulated fields. As these interactions take place at very low levels of RF radiation, the primary stages of ecological species will be affected.

The biological effects are also related to the frequency of the radiation. While there is relatively less harm with radiation up to 3 MHz, the most harmful region is 3-100 MHz. From 1 GHz 300 GHz, where most satellite communications are present or are planned for future, the effects are almost the same. Current technologies use microwaves up to 30 GHz as in the Japanese National Space Development Agency (NASDA) network. The millimeter range starts from 30 GHz and upwards. The Italians plan to have a beacon at 50 GHz, and the technology is being upgraded to obtain satisfactory performance at 90 GHz (W-band). Millimeter waves may hold the future of satellite information transfer, whose risk for continuous exposure is no less than that of microwaves.

B. RF/Microwave Radiation from Earth Antennas and Domestic Picture Phone Communication Links

From the total number of spacecraft launched by different space agencies[11] untill 1986, the most conservative estimates of spacecraft to be available on different orbits are put at 16,000 by 2050. The transmitting Earth stations for manning and communicating to these spacecraft will be almost doubled, and the antennas used at any given time can be estimated at about 50,000. These are primarily directed at the spacecraft, but some energy is radiated towards ground through the side lobes and back lobes. (If a Tracking and Data Relay Satellite System (TDRSS) type of network is used, then the number of ground stations are reduced, but higher power densities from the transmitting antennas are required for information transmission.) There are also domestic communication terminals planned to be operated for transmission of picture phone/voice signals.

Assuming that an average whole body specific absorption rate of 0.4 W/kg is applicable for biological species, the geographical areas affected by the radiation are estimated as circles of about 1-3 km radius. Measurements made at an ISRO Telemetry, Tracking and Command (ISTRAC) ground station near a 2-kW radiating, 10-m-diam antenna at 2 GHz frequency at a distance of 60m have shown that the power densities are of the order of 0.3 mW/cm^2 at 6 deg elevation, while the continuous safe exposure limit according to East European standards is 10 $\mu W/cm^2$. Observations made on rate of beating of hearts in pithed frogs showed that there was much increase at an average intensity of 0.003 mW/cm^2 corresponding to 3 $\mu W/cm^2$.[10]

C. Exposure to Microwave Radiation from Space Solar Power Satellites

The SSPS system would collect about 8500 megawatts (MW) in space, meant to be beamed to ground through a micro-wave beam at 10 cm wavelength. The power received on ground is estimated at 5000 MW. The site on Earth surface receives energy at a power density of around 20 mW/cm^2 at the beam center, 0.1 mW/cm^2 at 18 km away. The Microwave beam directing system and phase control achieved by means of a pilot signal beamed from the center of the receiving antenna would control the pointing accuracy of the beam. During partial failure of the beam pointing system the coherence of the microwave beam would be lost, the energy dissipated, and the beam spread out so that the intensity reaches 0.18 mW/cm^2 around a larger area such that bird and animal species would be subjected to radiation hazard.

D. Continuous Transmission of Information from Spacecraft.

The power densities on the Earth surface due to transmission of information from satellites, which at the moment can be neglected, can take a significant value because of the high information transmission rates proposed for the future. A NASA study on the NASA Spacecraft Communications and Tracking Project prepared by the advanced TDRSS working group visualizes that because of the aggregate video and emergent data requirements associated with a space station base and its payloads the projected data rates exceed 500 mega bits per second (MBPS). Other possible drivers for higher data rates of 5-10 giga bits per second (GBPS) include solar flare programs for brief periods. In addition, developments in holography for 3D visual application may raise data rates of space stations for micro gravity and life sciences experiments to a few giga bits by the year 2000.

According to the recommendations of Consultative Committee for International Radio (CCIR) and International Telecommunication Union (ITU) radio regulations the power flux densities permitted on the surface of the Earth from any individual spacecraft is -154 dBW/4 KHz.m^2 or -190 dBW/m^2.Hz.

Suppose the number of satellites of a particular information handling class is Ni with an off-axis gain Gi and with an information rate of Bi bits per second. Then the threshold power density of $10\,\mu W/cm^2$ is exceeded when $10 \log \Sigma\, [(Ni.Gi.Bi)/2]$ is greater than 190 dB assuming half the satellites illuminate the surface of Earth and only one tenth of them are activated at a given time. By 2050 if this limit is reached alternate techniques of modulation shall have to be used to limit the power flux densities on the earth surface.

E. Measures for Minimizing RF Radiation Hazards

Measures for minimizing long-term exposures of the environment to RF radiation can be stipulated as follows:

1) To strictly adhere to a pointing angle limit from zenith to 5 deg elevation and provide interlock circuits to switch off transmission if the elevation falls below 5 deg in case of ground-based transmitting antennas pointing towards low Earth orbit (LEO) satellites.

2) To specify the effective isotropic radiated power (eirp) power density limits for domestic satellite picture phone/telecom terminals in the side lobes and back lobes.

3) To insist on various administrations to exercise strict control over the power spectral densities at the Earth surface due to spacecraft transmissions.

4) To adopt modulation schemes such as spread spectrum techniques and adopt improved coding schemes to reduce the transmitting eirp of the spacecraft.

5) To limit the Earth surface power flux density due to domestic satellite television transmissions in individual countries to -144 dBW/4KHz.m^2 as in the case of the other transmissions above 5 deg elevation angle.

6) To resort to sectorial blanking, i.e., shutting off radiation in the azimuth sectors of denser populations by providing wire mesh diffraction fences around the antennas or keeping the antennas in sites naturally shielded throughout the azimuth.

7) To affect automatic deorientation of the microwave power transmitting antennas of SSPS towards a non-Earth oriented direction in the event of beam failure.

8) To limit Σ ($N_i.G_i.B_i$)/2 product through international agreements for protection of the environment.

9) To off-orient dwelling areas in a house away from transmitting aerials.

III. Hazards From Light and Infrared Radiation

Space commercialization is expected to give rise to a strong defense network that houses most advanced types of lasers with large power carrying beams aboard the spacecraft. Among different types, promising ones for space applications are 1) a hydrogen fluoride (HF) laser operating at a 2.7 - µ wavelength, and 2) a free electron laser (FEL) operating in

SPACE COMMERCIALIZATION: RADIATION HAZARDS

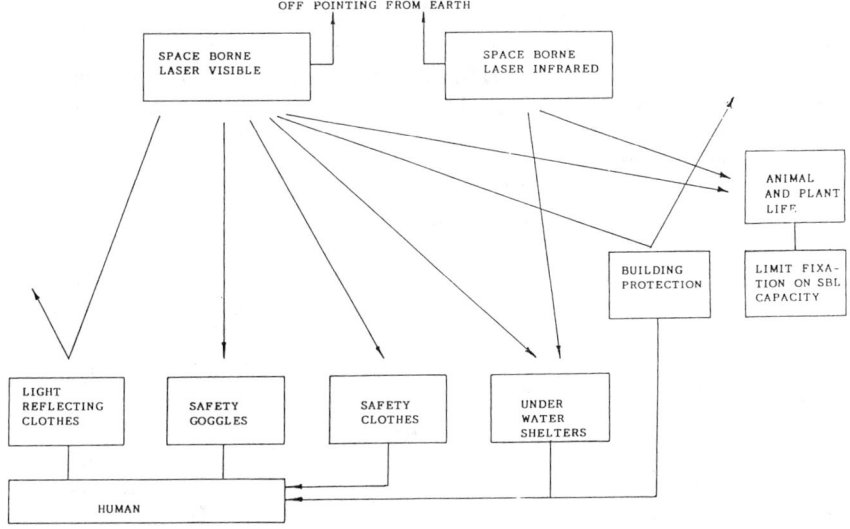

Fig. 2. Hazards from space-borne lasers and countermeasures.

a large wavelength range of 1 mm to 100 nm. Figure 2 represents the hazard from light and infrared radiation in the event of large-scale space commercialization and measures to be taken to minimize the effects.

A. Biological Effects of Light and Infrared Radiation

Laser beam exposure can result in tissue damage from any of the following types of mechanisms, i.e., the thermal effects, photochemical effects, acoustic transients, and chronic exposure.[7,8]

Thermal-related effects present the major type of damage from laser beam exposure. The extent of thermal damage is principally dependent on the tissue, i.e., the most effective absorber of the particular laser radiation wavelength, the strength of the irradiance of the incident laser beam, the size of the irradiated area, and exposure duration. Given the vast energy ranges of lasers, the effects on tissue can vary from no effect, simple tissue reddening, steam generation, charring, or even explosive tearing or ripping where radiant exposure levels are high. While direct exposure to laser radiation is an effective hazard for an intensity of 1 W/cm^2 for short durations, chronic exposure levels of 0.08 W/cm^2 can cause cataract formation in the eye. The skin sustains severe damage wherever energy densities approach several joules per square centimeter.

Protective standards for the exposure of the eye and the skin for the laser are given by the Occupational Safety and Health Administration (OSHA) based on standards of American National Standards Institute (ANSI). The maximum possible exposures for direct exposure by different species of environment range from 3×10^{-3} J/cm^2 to 10^{-2} J/cm^2.

B. Laser Radiation from Space

The lasers will be incorporated on large space platforms. Powers of the order of 20 MW are feasible on the platforms by 2050 or even earlier. An HF laser provided with a reflector diameter of 10 m and operating at a 2.7-μ wavelength can provide an intensity of 25 kW/cm^2 at a distance of 2000 km. The beam spread is of the order of 0.32 rad. For reaching a power density of less than 1 W/cm^2 the spread can be assumed as 30 m (about 15 beam spreads). In case of misalignment in tracking, there will be colossal tissue damage for animal and plant life and a human population of about than 10,000 will be blinded when the beam looks towards densely populated countries like India, assuming that a sweep is made by the laser for an arc of 2000 km.

C. Measures for Minimizing Light and Infrared Radiation Hazards

Measures for minimizing light and infrared radiation hazards can be stipulated as follows:

1) The laser beam should be off-pointed from the spacecraft-Earth direction within a few milliseconds of the occurrence of misalignment.

2) Earth inhabitants shall put on laser safety goggles and shall dress with highly reflective material under a constant threat of activation of the laser.

3) All buildings shall be provided with reflecting mirrors to direct the beam away from the horizon.

4) Underwater colonies will help in absorbing the mechanical shocks and thermal energy.

5) The capacities and wavelengths of space-based lasers shall be fixed to reduce the effects on the environment.

SPACE COMMERCIALIZATION: RADIATION HAZARDS

IV. Hazards from Nuclear Radiation

Nuclear power is being suggested not only for long-range space propulsion, but studies indicate[6] that nuclear power may be used for driving commercial payloads of space agencies. Among them is a 100 KWe-class nuclear power subsystem that can be expanded to power levels of 1 or 2 MW. It is called a multi-megawatt (MMW) solar power (SP) nuclear reactor satellite. The engineering subsystems are meant to cater to several space industries and laboratories. The feasibility studies have indicated the following options:

1) A fast liquid-cooled reactor coupled with an advanced thermoelectric converter.

2) An in-core thermionic system with a pumped sodium potassium coolant.

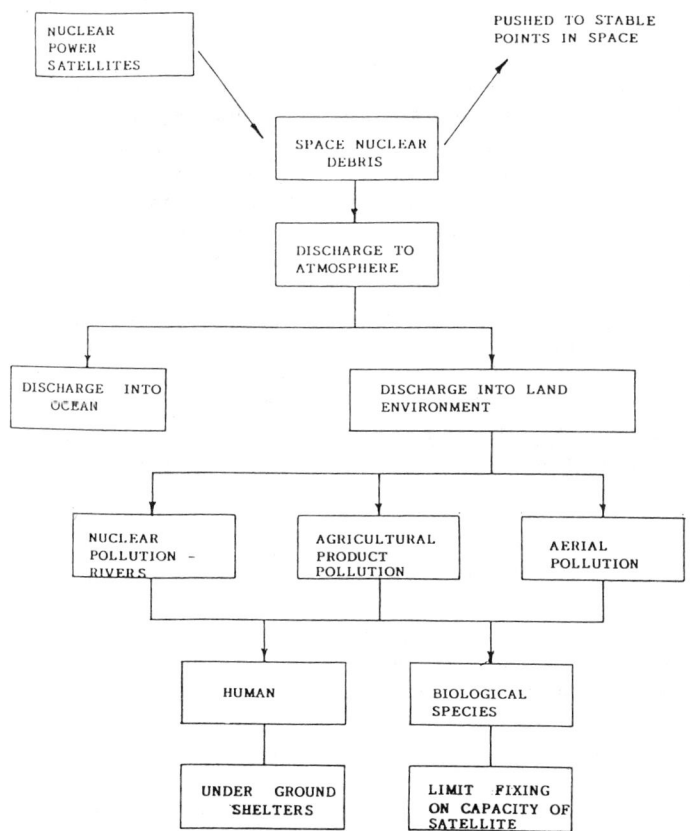

Fig. 3   Hazards from nuclear radiation and countermeasures.

3) A relatively low temperature fast reactor coupled with a free piston Stirling engine.

Hazards from such satellites start right from the launch site, as at the moment the success rate of launch combined with successful operation of the satellite is surely not 100%. Even while in orbit there may be some instances where the fuel may melt, as in the case of the ground-based sodium-cooled fast breeder Enrico Fermi reactor in 1966. Any part of the reactor that is susceptible for vibration damage causes the release of radioactive material. Figure 3 represents the hazards from nuclear radiation due to large-scale space commercialization and measures needed to minimize the effects.

A. Biological Effects of Nuclear Radiation

When an MMW nuclear power satellite has spent its useful life in orbit and starts reentering the atmosphere, in the event of the shield breaking off the nuclear debris gives rise to many types of residual radiation. Out of these gamma radiation and beta radiation are important to consider. Beta radiation produces incapacitation in ecological species. After discharge to atmosphere and transport down from winds transfer of beta and gamma radiation to ground results in irradiation of the population by three important routes:
1) external radiation from the radioactive material, 2) internal irradiation from inhalation of the same, which is subsequently suspended into the atmosphere, and 3) internal irradiation from the consumption of contaminated foodstuffs. Excessive absorption of radioactive iodine results in effects on thyroid glands. Parts of biological systems more susceptible to radiation damage are 1) bone marro, which produces red blood cells ; 2) the lining of the intestines, which loses its capacity to reline the intestines; and 3) brain and muscle cell damage. Beta radiation affects the skin if the exposure is more than 600 rads. During the fallout of the radioactive material more beta than gamma radiation is emitted and at short ranges, the beta dosage can be as much as 100 times that produced by the associated gamma radiation.[4] Thus a man with contamination only on his hands is likely to suffer more from the effects of beta radiation on the skin of his hands than from whole body radiation from the associated gamma ray exposure. The effect on other biological species is more pronounced and has been studied extensively. Exposure above 150 rads is believed to be risky for humans and above 50 rads for biological species on a scaled down proportion.

B.   Measures for Minimizing Nuclear Hazards

Measures for minimizing nuclear hazards include the following:

1) The space-borne nuclear reactor shall be effectively shielded so that when they reenter the atmosphere they can fall as large lumps that can be tracked from ground radars; thus, the position of the fall can be determined and the population can be alerted accordingly.

2) Each nuclear power satellite will have to be provided with a rocket motor that can place the worn-out satellite in some stable point in the solar system so that the location can be used as a nuclear dumping site.

3) In case of nuclear debris entering into atmosphere, short-term and long-term^5 countermeasures will have to be taken. Short-term actions are to be designed to limit early health effects, which include evacuation, sheltering, and issuance of stable iodine tablets. Sheltering has to be done on a large scale as more and more areas are prone to risk. Every home shall be provided with a shielded room against gamma radiation. The room shall be preferably underground. It is worthwhile to earmark some underground colonies, passages, and spaces for sheltering large masses. In large cities the underground spaces already available shall be expanded. Long-term actions include relocating personnel to noncontaminated areas for some time periods, banning contaminated foodstuffs, and decontaminating vast areas by vacuum cleaning, sand blasting, and soil removal in agricultural lands.

4) The capacities of the nuclear power satellites shall have to be limited to reduce the effect on environment.

V. Conclusion

Large-scale space commercialization leads to hazards due to RF/microwave, light and infrared, and nuclear radiation. While microwave radiation effects are more felt on the environment due to ground transmitting aerials and solar power satellites, their effect is likely to exceed the threshold power density of $10 \ \mu W/cm^2$ in about 50 yr due to direct radiation from the satellites.

Infrared and light radiation from lasers shall be off-oriented from the Earth when the tracking accuracies are not met. Also, buildings on the ground shall have to be provided with reflecting material, and people shall wear safety goggles and reflecting clothes for minimizing hazardous effects. Underwater colonies help in quicker dissipation of heat and mechanical energy.

Nuclear power satellites shall have to be shielded properly and shall have to be deposited in some stable points in the solar system when they are worn out. Elaborate tracking of nuclear debris and countermeasures are necessary if the debris falls to Earth.

More than any other measures, restraint by man in commercialization of space helps in preserving a healthy life on earth and in space.

VI. Acknowledgments

The author is grateful to Professor U.R.Rao, Chairman, Indian Space Research Organisation, for kindly suggesting corrections to the paper. Thanks are due to K.V.Venkatachary, Director, and P.N.Jayaraman, Deputy Director, ISTRAC, for suggesting RF radiation level measurements for safety reasons.

References

[1] Skolnik, M. I., Radar Hand Book, McGraw-Hill, Newyork, 1970, Section 29-26.

[2] Shankara K.N., "Microwave Radiation Hazards", Symposium of Bio-Medical Engineering, Osmania University, Hyderabad, India, June 1978.

[3] Marha, K., "Microwave Radiation Safety Standards in Eastern Europe," Institute of Electrical & Electronic Engineers, Transactions on Microwave Theory and Techniques, Feb. 1971, pp. 165-168.

[4] McNaught, Nuclear Weapons and their effects, 1st ed., Vol.IV, Brassey's, London, England, 1984.

[5] Clark, M.J., and Kelly, G.N., "Radiation Exposure of UK Population from Routine Discharges from Civil Nuclear Installations", Nuclear Energy, Vol. 21, Aug.1982, pp. 275-288.

[6] U.S. Department of Defense, "Antimissile and Antisatellite Technologies and Programs", 1st ed., Noyes, Park Ridge, New Jersey, 1987.

[7] Mallow, Chabot, Laser Safety Hand Book, Van Nostrand, Reinhold, Newyork, 1978.

[8] Sliney, David, and Wolbarsht, Myron, Safety with Lasers and Optical Sources, Plenum, New York, 1980.

[9] Ross, Adey, "Effects of Microwaves on Cells and Molecules," Nature, Vol. 332, June 2, 1988, p.401.

[10] Foster, K.R., and Pickard, W.E. "Microwaves: the Risks of Risk Research," Nature, Vol. 330, Dec. 10, 1987, p. 531.

[11] Kessler, D.J., "Report on Orbital Debris: Issues on Current Status in USA", Space Frequency Coordination Group, 3rd annual meeting, Nov. 1982, p.34.

[12] Sinclaire, W.K., "Risk, Research and Radiation Protection," Radiation Research, Vol. 112, 1987, pp. 191-216.

Remote Sensing Activities in Japan

Keiji Maruo*
Remote Sensing Technology Center of Japan, Tokyo, Japan

Abstract

The launch of the first Japanese Earth observation satellite, Marine Observation Satellite (MOS-1), February 19, 1987, is well worthy of special mention in the remote sensing activities in Japan. The second Japanese Earth observation satellite, the Earth Resources Satellite-1 (ERS-1), is scheduled for launch with the H-I rocket in 1991. The next generation of Japanese Earth observation satellites will be the Advanced Earth Observing Satellite (ADEOS), which is planned for launch in 1993 by the H-II rocket. Landsat data have been received and processed at National Space Development Agency (NASDA)'s Earth Observation Center (EOC) since 1979 and supplied to Japanese and overseas users via Remote Sensing Technology Center of Japan (RESTEC). Direct receiving of data from Satellite pour l'Observation de la Terre (SPOT) has begun by NASDA in October 1988.

I. Introduction

This paper first reviews the Japanese Earth observation satellites in operation and also in the developing stages. Since NASDA completed the ground receiving systems for Landsat in 1978, the utilization and research of remote sensing data are being carried out by NASDA, RESTEC, and government agencies with the participation of national research institutes, universities, public utilities, and corporations as well as private companies related to the development of image processing systems, aerial surveys, mineral resources, etc. Training courses in remote sensing technology for researchers of developing countries have been conducted by RESTEC under commission from the Japan International Cooperation Agency since 1977.

Copyright © 1990 by Keiji Maruo. Published by the American Institute of Aeronautics and Astronautics, Inc., with permission.
*Managing Director.

The purpose of this paper is to outline remote sensing activities in Japan focusing on topics such as Earth observation satellite development, ground receiving stations, data distribution, and international cooperation.

II. Earth Observation Satellite Development

Japan has several remote sensing satellite programs. They are the MOS-1 program, the ERS-1 program, and the ADEOS program.

A. Marine Observation Satellite-1 Program

1. Status and Planning of MOS-1

Marine Observation Satellite-1, the first Japanese Earth observation satellite, is an experimental satellite the purpose of which is to establish fundamental technologies for Earth observation systems in order to implement practical observation of the Earth (primarily the ocean). The preliminary design was carried out in 1979, the basic design was completed in the middle of 1981, and the critical design was completed in June 1983.

MOS-1 was launched by N-II launch vehicle from Tanegashima Space Center, NASDA, on February 19, 1987. Figure 1 shows the MOS-1 configuration and Table 1 shows MOS-1 orbit details.

2. Mission Instruments

MOS-1 has three sensors and a data collection system transponder (DCST). These sensors are the multispectral electronic self-scanning radiometer (MESSR), the visible and thermal infrared radiometer (VTIR), and the microwave scanning radiometer (MSR).

a) Multispectral Electronic Self-scanning Radiometer

The MESSR is an electronic scan-type radiometer with a 2048 elements charge coupled device (CCD). It has two bands in the visible region and two bands in the near infrared region. The instantaneous field of view (IFOV) is 50 m. It is equipped with two optical systems, and each optical system can scan approximately a 100-km width (for a total of 185 km) and has two gain modes (normal gain and high gain).

b) Visible and Thermal Infrared Radiometer

The VTIR is a mechanical scan-type (45 deg tilted scan mirror) radiometer having one visible and three thermal infrared bands for observation of sea surface temperature. The swath width is 1500 km, and the IFOV is 900 m for the visible band and 2.7 km for thermal infrared bands.

c) Microwave Scanning Radiometer

Table 1 MOS-1 orbit details

Item	Value
Altitude	908.7 km
Semimajor axis	7,286.9 km
Inclination	99.1 deg
Eccentricity	0.004
Period	6,190.5 sec
Coverage cycle duration	17 days (237 revolutions)
Revolutions per day	14-17/1 revolutions
Time of descending node equator crossing	10:00--11:00 a.m. (local time)
Ground trace accuracy (deviation at the equator)	20 km

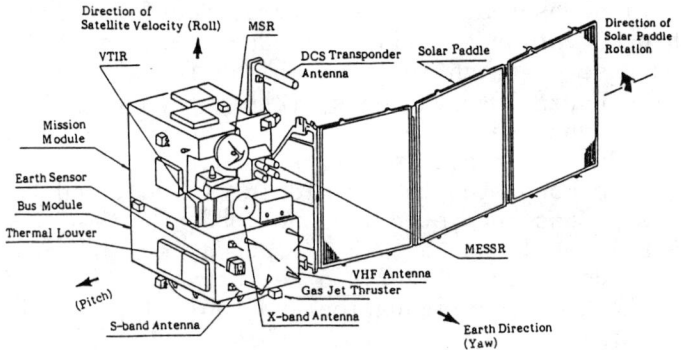

Fig. 1 External view of MOS-1.

Table 2 Characteristics of MOS-1 observational equipments

Item	Observation equipment				
	MESSR	VTIR		MSR	
Objective	Sea surface color and land use	Sea surface temperature		Watervapor content liquid water content ice snow distribution	
Observation wavelength, μm	0.51 ~ 0.59 0.61 ~ 0.69 0.72 ~ 0.80 0.80 ~ 1.1	0.5~ 0.7	6.0 ~7.0 10.5~11.5 11.5~12.5	-----	
Observation frequency, GHz	---	---		23.5~0.2	31.4~0.25
IFOV in km	0.05	0.9	2.7	32	23
Radiometric resolution	39 dB	55 dB[a] (Alb.=80%)	0.5 K	1 K	1 K
Swath width, km	100 (each optics)	1500		317	
Scanning method	Electronic scanning	Mechanical scanning		Mecanical scanning	

[a]Note: S/N ratio excluding quantization noise.

The MSR is a conical scanning microwave radiometer having two channels, which are 23.8 GHz and 31.4 GHz. MSR is effective for observation of water vapor content, liquid water content, and ice and snow distribution. MSR has a swath width of 317 km, and the IFOV is 23 km for 31.4 GHz and 32 km for 23.8 GHz. The characteristics of MOS-1 sensors are shown in Table 2.

3. Data Collection System Transponder

The DCS transponder is designed for basic experiments in order to determine platform location by the Doppler effect. It receives signals in the 400 MHz band from the data collection platform and retransmits them to the ground station by 1.7 GHz.

4. History of Operation

MOS-1 was launched on February 19, 1987, from Tanegashima Space Center, NASDA. Within the first two days, three-axis attitude was established.

In may 1987, transition to the experimental operation phase was declared, which was substantially the mission operation phase for the satellite. Since then MOS-1 has been in operation in a rather regular pattern.

5. MOS-1 Ground Station

Figure 2 shows the coverage of MOS-1 receiving round stations around the world, which are as follows:

1) Japan: EOC, NASDA (Hatoyama), Tokai University, Space Information Center (Kumamoto), and Showa-Base (Antarctic).

2) European Space Agency (ESA): Mas Palomas (Spain), Fuchino (Italy), Kiruna (Sweden), and Tromso (Norway).

3) Thailand: Bangkok, National Research Council of Thailand (NRCT).

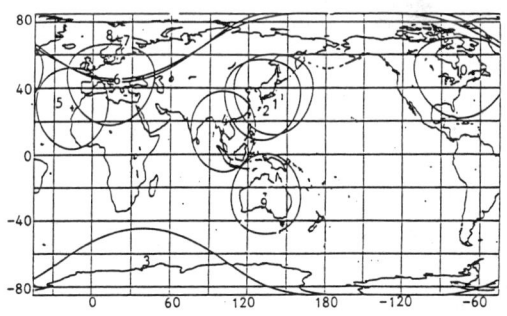

1.EOC 2.Kumamoto 3.Shouwa Station 4.Bangkok, Thailand
5.Maspalomas,Spain 6.Fucino, Italy 7.Kiruna, Sweden
8.Tromso, Norway 9.Alice Springs, Australia
10.Gatineau,Canada

Fig. 2   MOS-1 receiving stations.

Table 3 Development schedule of the ERS-1

Item	Japanese fiscal year							
	1984	1985	1986	1987	1988	1989	1990	1991
Design	Preliminary design		Preliminary design review ▲		Crotocal design review ▲			Pre shipment review ▲▲ Launch
			Basic design	Detail design				
Engineering model			Manufacture-assembly	Test				
Proto-flight model					Manufacture-assembly		Test	

Table 4 Main characteristics of ERS-1

Shape	Body approx. 1 m x 1.8 m x 3.1 m box type Synthetic aperture radar approx. 12 m x 2.5 m Solar cell paddle approx. 8 m x 3.4 m
Weight	Approx. 1.4 tons
Attitude control	Three-axis stabilized zeromomentum
Design life	2 yr
Launch vehicle	H-I (2-tage)
Launch site	Tanegashima Space Center
Orbit	Type Sun-synchronous subrecurrent orbit
	Altitude Approx. 570 km
	Inclination Approx. 98 deg
	Period Approx. 96 min
	Recurrent period 44 days
	Local time at 10:00 ~ 11:00 a.m. descending mode

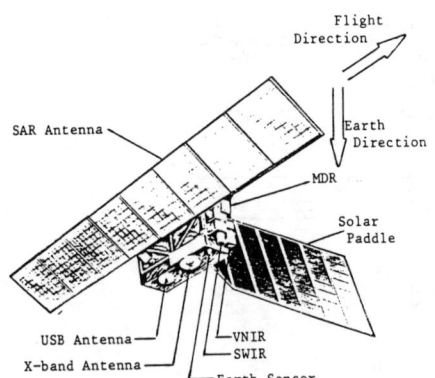

Fig. 3 Outlook of ERS-1.

4) Australia: Alice Spring, Commonwealth Scientific and Industrial Research Organization (CSIRO).
5) Canada: Gatineau, Canadian Centre for Remote Sensing (CCRS).

6. MOS-1b

The development of MOS-1b, which is a successor satellite of MOS-1 and has the same mission instruments, has been completed. MOS-1b was launched by the H-II rocked on February 7, 1990.

B. Earth Resources Satellite

1. Status and Planning of ERS-1

The primary purpose of ERS-1, which is the second Japanese Earth observation satellite, is to establish the fundamental technology of remote sensing from space by synthetic aperture radar (SAR) and optical sensor (OPS). It is primarily designed to survey and monitor nonrenewable resources, land use, agriculture, forestry, environmental protection, prevention of natural disasters, surveillance of coastal regions, etc.

The development of the satellite was shared between the Science and Technology Agency (STA) and the Ministry of International Trade and Industry (MITI). NASDA conducted the system studies of ERS-1 in 1980 and 1981 and started the research on sensors and other devices in 1981.

ERS-1 is currently in a proto-flight model (PFM) developing phase and is scheduled to be launched in FY 1991. Table 3 shows the development schedule of ERS-1. The main

Table 5 Main specifictions of mission subsystems

Synthetic aperture radar	Swath width	75 km	
	Resolution	18 m x 18 m (3 looks)	
	Off nadir angle	35 deg	
	Observation frequency	1275 MHz	
Optical sensors (OPS)	Swath width	75 km	
	Resolution	18 m x 24 m	
	Number of observation bands	Visible near infrared band	3
		Shortwave infrared band	4
		Stereoscopic band	1
Mission data transmitter (MDT)	Carrier frequency	8 GHz band	
	Modulation method	QPSK	
	Electric power amplifier	TWT	
	Antenna type	Mirror-surface corrected-shape beam antenna	
Mission data recorder (MDR)	Data recording rate	30 Mega bit per sec x 2 channel	
	Recording time	20 min	
	Operating time	2000 h (maximum)	

characteristics of ERS-1 are shown in Table 4, and the configuration of ERS-1 is shown in Figure 3.

2. Mission Instruments

ERS-1 will carry the following mission instruments:
1) Synthetic aperture radar (SAR).
2) Optical sensor (OPS) using the CCD system, with a visible and near infrared radiometer (VNIR) and a short wave infrared radiometer (SWIR).

The main specifications of mission instruments are shown in Table 5.

c. Advanced Earth Observing Satellite

1. Status and Planning of ADEOS

The main objectives of ADEOS, which is the next generation of Japanese Earth observation satellites, are as

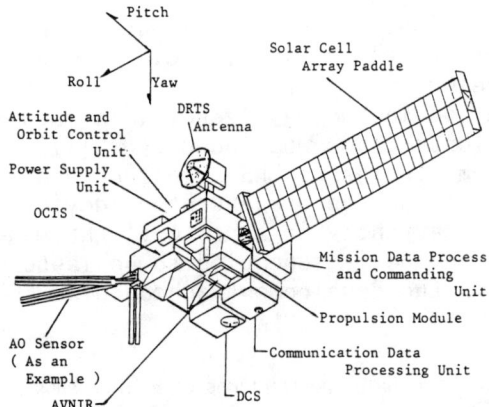

Fig. 4   Outlook of ADEOS satellite.

Table 6   Major characteristics of ADEOS

Total weight	Approx. 3.2 ton
Sensors	OCTS, AVNIR, POLDER, NSCAT, TOMS, IMG, ILAS, RIS
Design life	3 yr
Launch vehicle	H-II
Launch target	Early 1995
Weight of bus instrument	Approx. 1550 kg
Available power	3.5 kW (EOL)
Attitude and orbit control	3-axis stabilized (zero-momentum)
Data rate for direct transmission, data relay and data archiving	Max. 100 Mega bit per sec

follows:
1) To continue and advance the Earth observation technology that was developed in the MOS-1 and ERS-1 programs.
2) To develop advanced optical sensors for sophisticated Earth observation.
3) To contribute to the international community by supplying data and an opportunity for sensors selected through the announcement of opportunity (AO sensor) as additional sensors onboard.
4) To conduct experiments on Earth observation data relay using the Engineering Test Satellite-VI (ETS-VI) and the Experimental Data Relay and Tracking Satellite (EDRTS) to enhance global observation capabilities.
5) To develop a modular type of satellite to achieve the basic technology that is necessary to develop the future Japanese polar orbiting platform.

The conceptual design and preliminary design of ADEOS are currently being performed, and ADEOS is planned to be launched by the H-II rocket in FY 1993.

2. ADEOS System

The major characteristics and outlook of ADEOS are shown in Table 6 and Figure 5, and Figure 6 shows the Earth observation system featuring the ADEOS satellite. Observed data will be transmitted to the ground stations not only directly from ADEOS but also through the EDRTS.

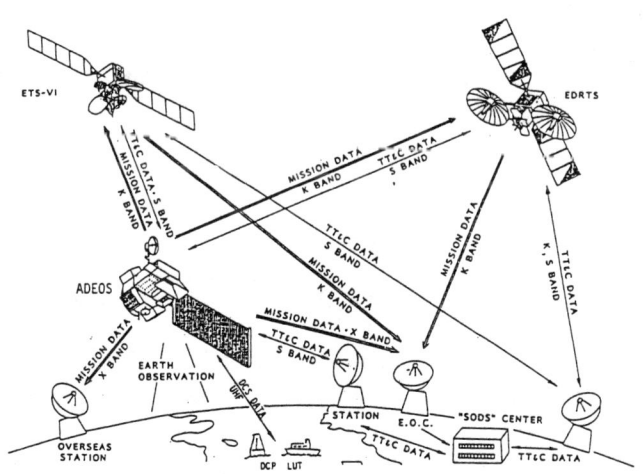

SODS : Space Operations And Data System
E.O.C : Earth Observation Center
TT&C : Telemetry Tracking And Command

Fig. 5   Earth observation system featuring ADEOS satellite.

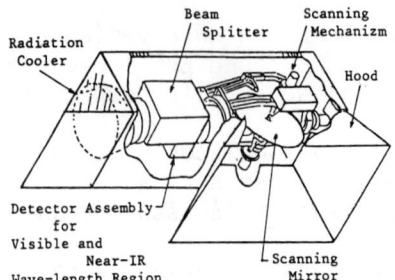

Fig. 6 Outlook of OCTS.

Table 7 Main parameters of ADEOS nominal orbit for operational observation mode

Orbit	Circular, sun-synchronous, recurrent orbit
Mean altitude	796.61 km
Orbit inclination	98.60 deg
Nodal period	100.92 min
Repeat cycle	41 days
Subcycle	3 days
Orbit/day	14 + 11/41 revolution
Local time at descending mode	10:30 a.m. ± 15 min

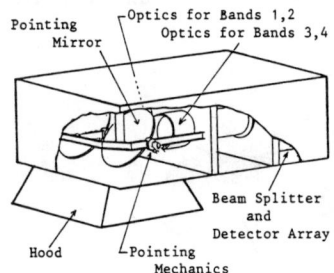

Fig. 7 Outlook of AVNIR.

Since ADEOS will have a mission data recorder, damped data also will be transmitted to specified stations to be determined. Downlink for the mission data transmission will be compatible with that of ERS-1. Main parameters of ADEOS nominal orbit for the operational observation mode are shown in Table 7.

3. Mission Instruments

ADEOS will carry two advanced optical sensors: 1) the ocean color and temperature scanner (OCTS), and 2) the advanced visible and near infrared radiometer (AVNIR). The main specifications and outlooks of OCTS and AVNIR are shown in Table 8 and Figures 7 and 8.

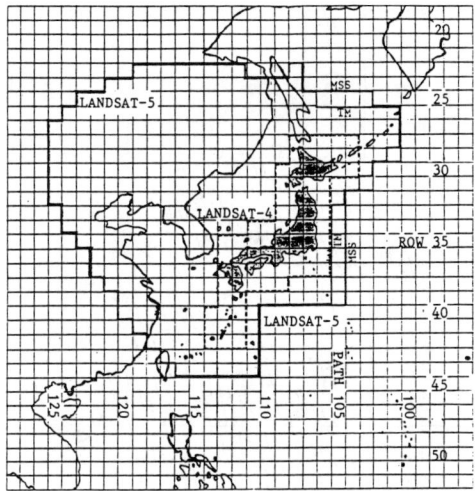

Fig. 8 Landsat coverage for FY 1987 (descending).

In addition to these two sensors, plans call for additional AO sensors to be installed to provide the international community to install their sensors onboard and to contribute to the progress of the Earth observation system.

The announcement of opportunity (AO) for participation in ADEOS program was issued in January 1988, and the six AO sensors have been decided in September 1989.

III. Reception, Processing, and Distribution System of Remote Sensing Data

A. Ground Station

In October 1978, NASDA completed the construction of the EOC for the reception and preprocessing of Landsat data. Operation begun in January 1979.

The EOC is located at Hatoyama, Saitama Prefecture, about 50 km NW of Tokyo, lat 36.00 N and long 138.12 E. EOC receives and preprocesses the data from Landsat-5, MOS-1, and French SPOT and produces computer compatible tape (CCT), photographs, and floppy disk data. Figures 9 and 10 show the antenna coverage of Landsat-5 and MOS-1.

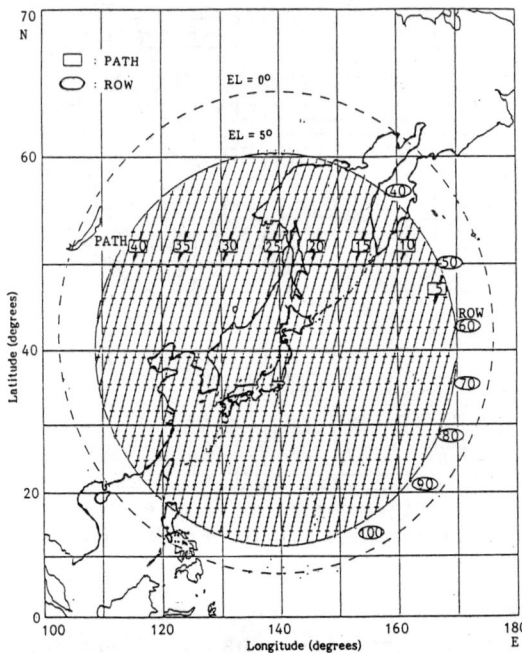

Fig. 9 Nominal Orbit trace of MOS-1, path and row center positions and coverage at EOC (descending).

Table 8 Specification of OCTS and AVNIR

Specification of OCTS

Item	Specification (goal)
Wavelength region	Visible: 6 bands; middle IR: 1 band; near IR: 2 bands; thermal IR: 3 bands
Scan angle	Approx. ± 40 deg
IFOV	0.85 mrad (1.7 mrad for mid. IR)
Tilting angle	± 20 deg (5 deg step)
Scanning system	Mechanical mirror vibration
Optics	Reflection type
Weight	Approx. 180 kg
Power consumption	Approx. 240 W

Specification of AVNIR

Item	Specification (goal)
Wavelength region	Visible: 3 bands; near IR: 1 band (resolution 16 m) Panchromatic band: 1 band (resolution 8 m)
IFOV	Approx. 20 μrad
Scanning angle	Approx. 5.7 deg (ground distance 80 km)
Pointing angle	Approx. ± 40 deg
Scanning system	Electronic scanning (push-bloom)
Optics	Reflection type
Weight	Approx. 200 kg
Power consumption	Approx. 250 W

B. Data Distribution System

The RESTEC was authorized by NASDA as the remote sensing data distribution office of Landsat-5, MOS-1, and SPOT to general users in Japan and foreign customers. It commenced Landsat data distribution in July 1979, started to distribute Landsat MSS floppy disk data for personal computer users in spring 1984, and commenced to distribute MOS-1 data on a test basis from August 1978 and on a routine basis from November 1987.

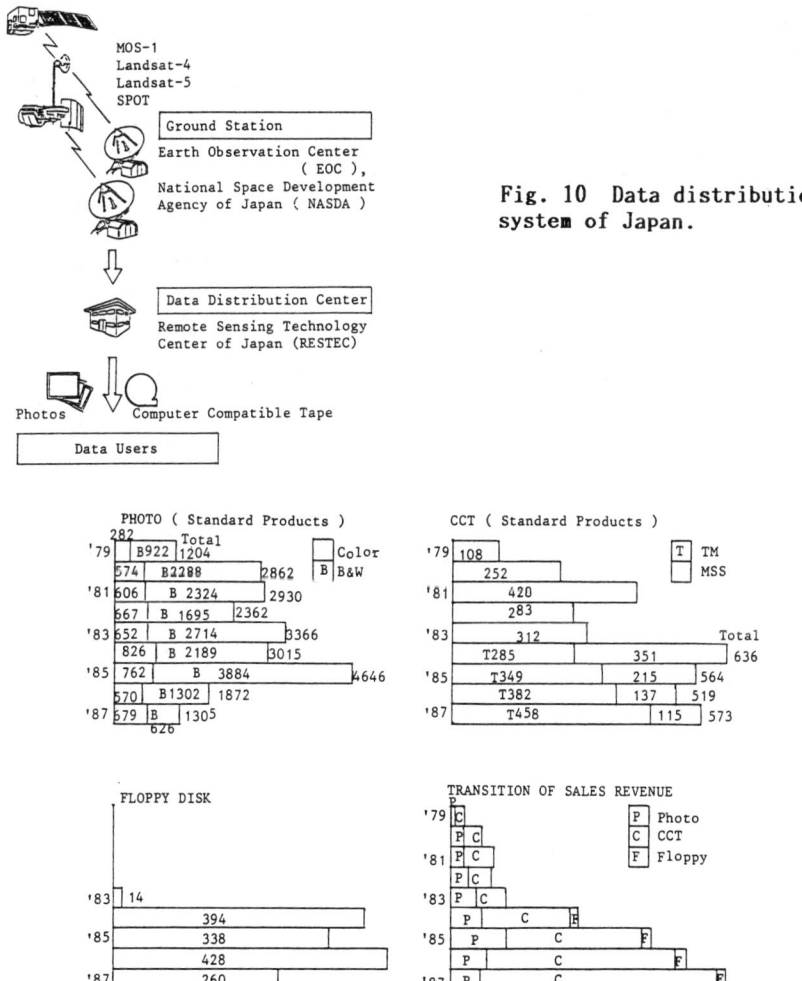

Fig. 10 Data distribution system of Japan.

Fig. 11 Transition of Landsat data distribution from 1979 to 1987.

A short history of remote sensing data distribution is as follows:
1) January 1979--EOC started reception from Landsat.
2) July 1979-----RESTEC started to distribute Landsat Multi-spectral Scanner (MSS) data.
3) June 1981 --- RESTEC started to distribute Landsat Return Beam Vidicon (RBV) data.
4) Aug./Sept. 1982-- EOC temporarily stopped reception due to upgrading of ground stations.
5) March 1984 -- RESTEC started to distribute floppy disk data on a test basis.
6) March 1984 -- Free distribution to NASDA's principal investigator system was terminated at the end of FY 1983.
7) April 1984 -- NASDA commenced to received Thematic Mapper (TM) data at Hatoyama, EOC.
8) April 1984 -- NASDA commenced to distribute MSS floppy disk data on a routine basis.
9) October 1984--RESTEC started to distribute TM data on a test basis.
10) May 1985 ---- RESTEC started to distribute TM data on an operational basis.
11) May 1986 ---- RESTEC commenced SPOT data distribution as the exclusive distributor in Japan.
12) August 1987---NASDA commenced MOS-1 data distribution on a test basis.
13) November 1987-NASDA commenced MOS-1 data distribution a routine basis and floppy disk data of MESSR.
14) April 1988----NASDA commenced the floppy disk data of Landsat TM and MOS-1 VTIR and MSR.

Figure 11 shows the flow of data distribution in Japan.

IV. Marketing of Remote Sensing Data in Japan

Marketing of Landsat remote sensing data in Japan is as follows: First, the transition of Landsat data distribution of CCT, photograph, and floppy disk data from 1979 to 1987 are shown in Figure 12. Photographs sharply decreased from 1986 and CCT maintained almost the same level in recent years, but CCT of TM has increased continuously Sales revenue have set good records and have increased steadily in recent years.

In Japan there are four major user categories, which are government organizations, industry, university and schools, and nonprofit foundations. Figure 13 shows user profile by type of organization on photographs and CCT from 1979 to 1987, and Figure 14 shows user profile by type of application field.

Figure 14 shows that the major application is land, which includes land information, land use, forestry, agriculture, geological structure, geology topography, mineral resources, energy, water resources, and land environment. Table 9 shows Landsat standard products distributed to categorized users in FY 1987.

RESTEC started to distribute MOS-1 products from August 1987 on a test basis and from April 1988 on a routine basis. Table 10 shows first year operation/distribution of MOS-1 standard products to categorized users. The number of products in this table includes the free distribution of the MOS-1 cooperative verification program, which has been carried out by NASDA.

RESTEC started the distribution of SPOT data for Japanese users in Japan on May 15, 1986, under the authorization of SPOT IMAGE, who is the sole distributor of products derived from SPOT satellite operated by Centre National d'Etudes Spatiales (CNES) in France. Under the

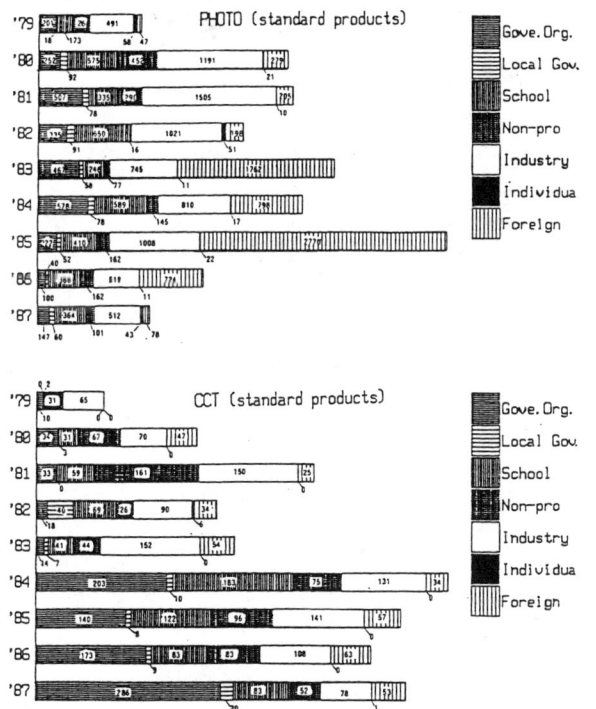

Fig. 12 Customer's profile of the type of organization.

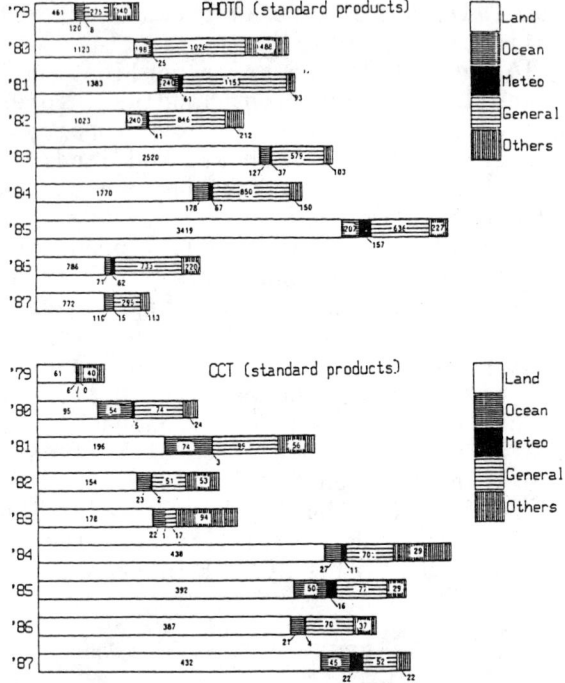

Fig. 13 Customer's profile by application field.

Table 9 Landsat standard products to categorized users in Fy 1987

Category of user	Number of users	Number of delivered data/Share, %							Quick look	Sales value share, %
		B/W	MSS Color	CCT	Floppy	B/W	TM Color	CCT		
1 Government users	28	57/15.0	20/6.4	36/31.3	0/0	48/19.6	22/6.0	250/54.6	232/75.6	51.3
2 Local government	39	10/2.6	19/6.0	10/8.7	21/8.1	3/2.1	28/7.7	10/2.2	2/0.7	3.0
3 Educational institute	63	174/45.7	51/16.2	31/27.0	117/45.0	91/37.1	48/13.2	52/11.4	11/3.6	10.2
4 Nonprofit foundation	24	21/5.5	5/1.6	15/13.0	11/4.2	10/4.1	65/17.8	37/8.1	19/6.2	7.6
5 Private company	168	95/30.0	183/58.3	20/17.4	103/39.6	46/18.8	188/51.5	58/12.7	11/3.6	17.8
6 Individual	12	11/2.9	30/9.6	0/0	8/3.1	0/0	2/0.5	1/0.2	26/8.5	0.7
7 Foreign country	20	13/3.4	6/1.9	3/2.6	0/0	47/0.7	12/3.3	50/11.0	6/2.0	9.4
Total	354	381/100	314/100	115/100	260/100	245/100	365/100	458/100	307/100	100.0

Table 10 First year operation/distribution; MOS-1 standard products to categorized users

Category of users	MESSR						VTIR				MSR				Quick look	
	B&W		Color		CCT		B&W		CCT		B&W		CCT			
	sale	MVP	sale	MVP	sale	MVP	sale	MVP	sale	MVP	sale	MVP	sale	MVP	sale	MVP
1 Government users	58	13	25	6	18	39	24	8	30	47	2	1	28	85	184	0
2 Local government	6	0	30	0	3	0	7	0	4	0	0	0	0	0	0	0
3 Educational institute	99	12	59	13	42	31	40	27	5	21	9	3	9	7	0	0
4 Nonprofit foundation	9	0	16	5	20	7	9	12	7	9	2	5	0	5	0	0
5 Private company	9	0	74	5	10	5	6	2	3	2	3	2	12	2	19	0
6 Individual	52	0	245	0	0	0	2	0	0	0	0	0	0	0	0	0
7 Foreign country	17	0	10	0	8	17	0	0	1	6	0	0	1	1	0	0
Total	250	25	459	29	101	99	88	49	50	85	16	11	50	100	203	0
	275		488		200		137		135		27		150		203	

Note: Left side number show the sales volume, and right side numbers show the free distribution for MVP.

Table 11 SPOT data distribution during April 1987 to March 1988

Photo	B/W	Panchro	18 frames	CCT	Panchro	68 scenes
		Multi	32 frames		Multi	75 scenes
	Color		79 frames			
Total			129 frames			142 scenes

Table 12 Data type number of products distributed in FY 1987

Satellite	Sensor	Photo	CCT	Floppy	Q.L.[a]	Total
MOS-1b[b]	MESSR	763	200	NA[c]	112	1075
	VTIR	137	135	NA	36	308
	MSR	27	150	NA	55	232
Landsat	MSS	695	115	260	105	1175
	TM	610	458	NA	202	1270
SPOT	HRV-PA	18	68	NA	-	86
	HRV-XS	111	75	NA	-	186
Total		2361	1201	260	510	4332

[a]Q.L. = quick look. [b]MOS-1 includes MVP. [c]NA = not available.

Table 13 Categorized users of products distributed in FY 1987

User	MOS-1[a] pht./dig.		Landsat pht./dig.		SPOT pht./dig.		Total pht./dig.	
Government	137	247	147	286	35	53	319	586
Local government	43	7	60	20	2	0	105	27
University, etc.	262	115	364	83	3	10	629	208
Foundation	58	48	101	52	58	38	217	138
Industry	101	34	512	78	1	42	614	154
Individual	299	0	43	1	30	0	371	1
Foreign user	27	34	78	53	0	0	105	87
Total	927	485	1305	573	129	143	2361	1201

[a]MOS-1 includes MVP.

contract between SPOT IMAGE and RESTEC, RESTEC distributed the unenhanced SPOT data in Japan. The sales record during the first year, which was from May 15, 1986, to March 31, 1987, was 44 frames of photographs and 65 scenes of CCT, which was better than expected.

Table 11 shows the second year of SPOT data distribution, which was almost three times greater than the previous year's record. The user's profile by application field on SPOT data is mostly land, which had a 90% share.

Table 12 and 13 show data type and number of products for satellites (MOS-1, Landsat, and SPOT) distributed in FY 1987 (April 1987 to March 1988).

Table 14 Students enrolled in the remote sensing training course from 1977 to 1988

Country	No. of student	Coundtry	No. of student	
Afganistan	2	Nepal	5	
Argentina	6	Pakistan	5	
Bangladesh	7	New Guinea	1	
Brazil	5	Panama	1	
Burma	1	Peru	2	
China	6	Philippines	9	(1)[a]
Egypt	1	Saudi Arabia	1	
India	9	Singapore	3	(1)[a]
Indonesia	13 (2)[a]	Sri Lanka	7	
Iraq	1	Thailand	15	(2)[a]
Jordan	1	Turkey	1	
Korea	9	Venezuela	3	
Malaysia	8			
		Total	122	

Note: The table excludes the MOS-1 training course for 20 students from Thailand.
[a] The numbers in parentheses reflect the special training course for Asian Pacific countries held in 1986.

We had two favorable events in 1988. One was MOS-1 data reception in the Bangkok station in August, and the other was SPOT data reception in the Hatoyama station (EOC) in October. These events should provide good circumstances for satellite data distribution in the future.

V. International Cooperation in Remote Sensing

A. Seminars and Training

Japan is taking part in the United Nations (U.N.) Space Application Program promoted by the UN Committee on the Peaceful Uses of Outer Space. In September 1980, Japan cosponsored the "U.N. Seminar on Remote Sensing Applications to Land Use Planning" with the U.N. for United Nations Economic and Social Commission for Asia and Pacific (ESCAP) members.

Japan also held the "Japan Seminar on Remote Sensing Applications" in Tokyo in March 1982 for the Asian and Oceanian researchers and managers who engage in the application of remote sensing in the fields of agriculture and forestry. The Association for Science Cooperation in Asia (ASCA) seminar on comprehensive utilization of remote sensing technology sponsored by the STA was held in February 1987 in Tokyo. Also, the International Symposia on the MOS-1 Verification Program, the Advanced Earth Observation Satellite, and Earth Science were held in Tokyo in 1988.

Since FY 1977, a "Group Training Course in Remote Sensing Technology" sponsored by the Japan International Cooperation Agency (JICA) has been held every year in Tokyo for members of developing countries. The purpose of this training course is to introduce recent research results on computerized digital analysis and to present current trends of advanced analyzing technology. In spring 1988 the 12th training course finished with 10 participants after two months, and the accumulated number of participants is now 122 from 25 counties.

Besides these regular training courses, special training courses for the MOS-1 receiving station in Bangkok, Thailand were held in February and October 1987 on commission from NASDA for 20 participants of ground station operators and MOS-1 data users. Table 14 shows the participants enrolled in the remote sensing technology training courses from 1977 to 1988.

B. Participation in International Meetings

In the field of remote sensing, Japan is participating in multilateral meetings, for example, the Committee on Earth Observation Satellites, the meeting of Remote Sensing from Space of the Summit of Industrialized Nations Working Group on Technology, Growth and Employment, etc. Through these meetings, Japan exchanges information of remote sensing activities with other nations and contributes to cooperation within international frameworks.

Furthermore, Japan has cooperation programs on remote sensing with the U.S., Canada, France, Australia, West Germany, and ESA under bilateral agreements.

C. Joint Research on Remote Sensing Technology with ASEAN Countries

Japan started a three year joint research program on remote sensing technology with the Asian countries (Thailand, Indonesia, and Malaysia) in 1986 in order to utilize MOS-1 data in tropical areas effectively.

The contents of the joint research program are as follows:

1) Joint research on the enhancement of remote sensing technology, including a method to extract and process basic information (with Thailand) and joint research on low-cost, high-performance processing systems and software.

2) Joint research on the explication of environmental characteristics of tropical areas, including environmental characteristics of the forest (with Malaysia), environmental characteristics of agriculture (with Thailand), and investigation of nonrenewable resources (with Indonesia).

3) Physical and biological oceanography in tropical areas (with Thailand and Indonesia, respectively).

VI. Conclusion

Concerning the commercialization of remote sensing in Japan, some application fields are operational or preoperational, but basically they are still in the research stage. Users of remote sensing data currently prefer or welcome high-resolution data such as Landsat TM or SPOT multi spectral band (XS).

However, in order to achieve the commercialization of remote sensing in Japan, the continuity of satellite remote sensing data in the future is essential. Therefore, the development and future planning of remote sensing satellites should be continued.

Communications and Broadcasting Satellites in Japan

H. Dobashi,* E. Kimura,* and K. Aikyo†
Telecommunications Satellite Corporation of Japan, Tokyo, Japan

Abstract

This paper outlines domestic communications and broadcasting satellite systems in Japan. The development of these systems was initiated by the Japanese government in 1972. For the purpose of various experiments in Japan, including reaching remote islands, a communications satellite (CS) and a broadcasting satellite (BS) were launched in 1977 and 1978, respectively. Succeeding the CS and BS, the first operational satellite services, CS-2s and BS-2s, started their operations in 1983 and 1984, respectively. These four satellites have been successfully operated by the Telecommunications Satellite Corporation of Japan (TSCJ). The next generation communications Satellites, CS-3a and CS-3b, were launched in February and September 1988 and have been used for the increased demands of telecommunication services. The next generation broadcasting satellites, BS-3a and BS-3b, are expected to be launched in summer of 1990 and 1991, respectively, and intend to broadcast three television programs, including newly developed high definition television. According to the new regulation policy in the communication field, two new commercial domestic communication service companies were founded in 1985, and each planned to launch two communications satellites in 1989.

Introduction

To the casual observer, it might seem difficult to find a necessity for communications and broadcasting satellites in Japan, where a terrestrial telephone network as well as television broadcasting are well established as fundamental infrastructures in society. There are, however, at least two factors that provide good reasons for the use of these

---

Copyright © 1990 by the American Institute of Aeronautics and Astronautics, Inc. All rights reserved.
*Vice President.
†Director, Kimitsu Satellite Control Center.

satellites even in Japan. They are mainly reflected in Japanese geographical conditions. One is natural disasters, which are unavoidable in Japan. There are earthquakes and typhoons that may paralyze almost all terrestrial networks.

The other is mountainous areas and remote islands, such as the Ogasawara islands, where no microwave or cable service can be economically feasible. Accordingly, development of satellites was started as a national project. Potential advantages of communications and broadcasting satellites in Japan have been recognized over the past decade. This paper traces the developments in the field of communications satellites and broadcasting satellites.

Development of Communication and Broadcasting Satellites

The development of communications and broadcasting satellites in Japan was initiated by the government in 1972.

As the first stages, medium size satellites of about 350 kg of on-orbit weight for experimental purposes (CS and BS) were built by the National Space Development Agency of Japan (NASDA) and successfully launched in 1977 and 1978, respectively, by Delta 2914 of NASA. The experiments with CS and BS were mainly conducted by Ministry of Posts and Telecommunications (MPT) using its Kashima Earth Station of Communications Research Laboratories for several years with the participation of Nippon Telegraph and Telephone Corporation (NTT) and Japan Broadcasting Corporation (NHK).

After the experiments with CS and BS, it was decided, as the next stage, to build operational systems. Two communications satellites (CS-2a and -2b) and two broadcasting satellites (BS-2a and -2b) were built and launched by NASDA using its N-II launch vehicle in the years 1983-1986.

The design of CS-2 and BS-2 spacecrafts has mostly inherited that of CS and BS, respectively. However, the design life was extended from three yr to five yr, and some changes have been implemented to reflect the flight experiences obtained with the experimental satellites. Figures 1 and 2 show the configuration and the major characteristics of CS-2 and BS-2.

CS-2s were mainly used by NTT as a supplement to its nationwide network. Two out of six Ka-band transponders aboard each CS-2, however, are shared by several small capacity links of users such as government authorities, railway companies, and electric power companies. CS-2s were the first operational communications satellites in the world to be equipped with Ka-band transponders.

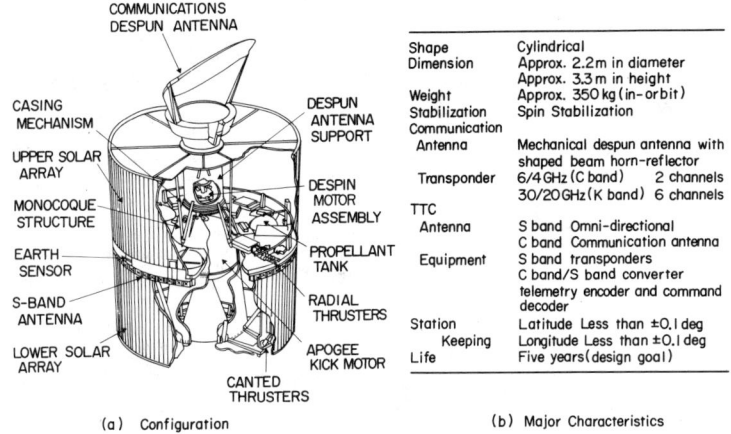

Fig. 1 Summary of communications satellites CS-2.

Two Ku-band 100-W transmitters aboard the BS-2 are used for NHK's television service according to the plan of WARC--BS-77. One channel is for conventional service directed at mountainous areas and remote islands, and another one is for a new service called "around the clock broadcasting".[2] The number of BS-2 television receivers was expected to exceed one million in 1988. The transmitted signal is in the NTSC (National Television System Committee) mode for video and PCM (Pulse Code Modulation) mode for audio using the 12-GHz band frequency. The target for the receiving antenna size is less than 75 cm in diameter in the central part of Japan, but a highly efficient antenna 45 cm in diameter can be now used satisfactorily in urban areas.

Orbit Control System of the CS-2/3, and BS-2 in Japan

In 1979, the Ministry of Posts and Telecommunications established a new organization called the Telecommunications Satellite Corporation of Japan (TSCJ) in cooperation with NTT, Kokusai Denshin Denwa Co., Ltd (KDD) and NHK. The main tasks at TSCJ are stationkeeping and housekeeping of both communications and broadcasting satellites through its Kimitsu Satellite Control Center located some 50 miles SE of Tokyo. The center was completed in August 1982. At present, four satellites (CS-3s and BS-2s) are under control of the center. The ground facilities of the center were designed in such a way that satellites can be monitored and controlled by a single station. As BS-2a and BS-2b have to be

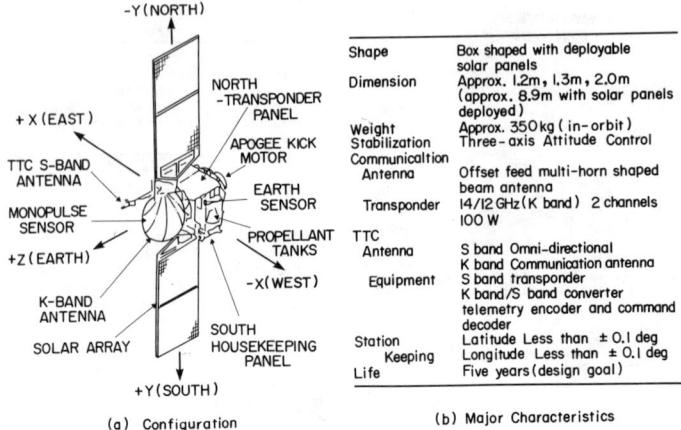

Fig. 2 Summary of broadcasting satellites BS-2.

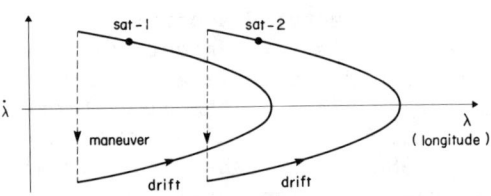

Fig. 3 Longitudinal station keeping by synchronized method.

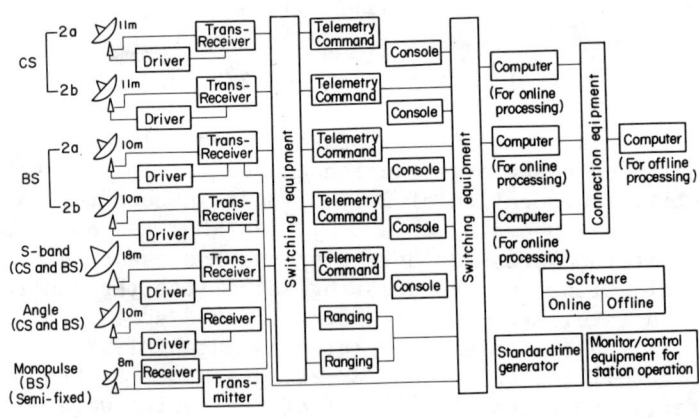

Fig. 4 Diagram of satellite control system at the center.

Table 1 Major characteristics of CS-3

(a) Space major characteristics

Shape	Cylindrical, diameter: 218 cm, height: 356 cm (including antenna)
Weight	Approx. 1099 kg (at launch)
	Approx. 550 kg (beginning of life)
Attitude control	Spin-stabilized
Life	≥ 7 yr (design life)
Launch site	Tanegashima Space Center
Launch date	CS-3a February of JFY[a] 1987
	CS-3b September of JFY[a] 1988
Satellite location	CS-3a 132° E long
	CS-3b 136° E long

(b) Mission major characteristics

	Ka-band	C-band
Frequency Uplink	27.5 ~ 29.25 GHz	5.925 ~ 6.425 GHz
Downlink	17.7 ~ 19.45 GHz	3.7 ~ 4.2 GHz
XPDR power amp.	TWTA	SSPA
Channels	10(+5 backups)	2(+1 backup)
Operating channels in	10	2
Antenna coverage area	Major Japanese islands, including Okinawa	All Japan

[a] JFY is Japanese fiscal year.

operated at the same longitude within ±0.1° E-W and N-S, they are controlled by the synchronized method shown in Fig. 3 to avoid collisions between them.[3] The total diagram of the control system is shown in Fig. 4. For economic and operational availability, the ground facilities for CS-3 and BS-2 have been kept compatible with each other and facilities have been shared as much as possible. The center has eight antennas.

Feature of CS-3 and BS-3 System Programs

CS-3 System

The new generation spacecrafts of communications satellites, CS-3a and CS-3b, were successfully launched by NASDA using its H-1 launch vehicles[4] in February and September 1988 and were located in geostationary orbits at 132° and 136° E long, respectively. The weight of CS-3s at the beginning of life (BOL) was about 550 kg in geostationary orbit, and the satellite's life is designed at more than seven yrs.

The major characteristics of CS-3 are summarized in Table 1. The Japanese archipelago including the Okinawa islands, is covered by the Ka-band, while remote islands are covered by the C-band. Figure 5 shows the CS-3 satellite coverage area. CS-3 and CS-2 development program and launch schedules are shown in Fig. 6.

The CS-3's spacecraft has a spin-stabilized body with a mechanically despun communication antenna. The cylindrical part of the body is approximately 218 cm in diameter and 243

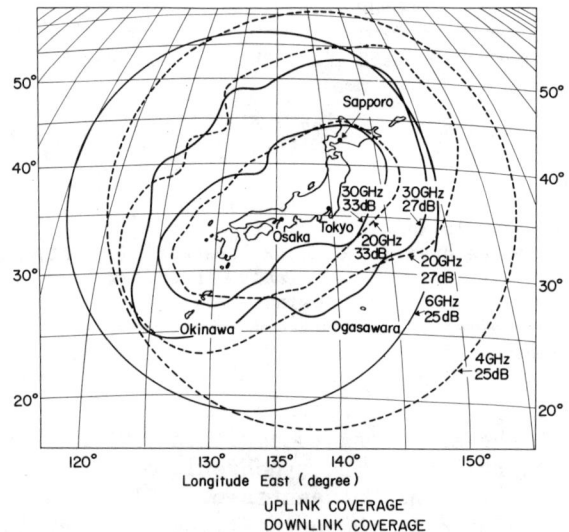

Fig. 5　CS-3 satellite coverage area.

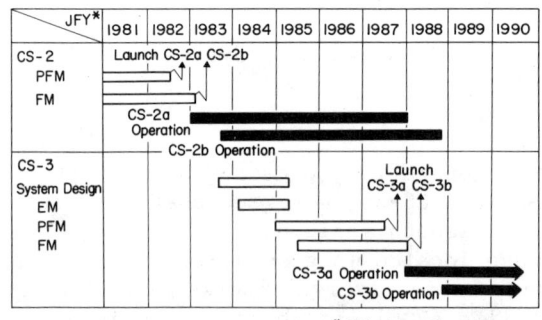

Fig. 6　CS development and operation schedule.

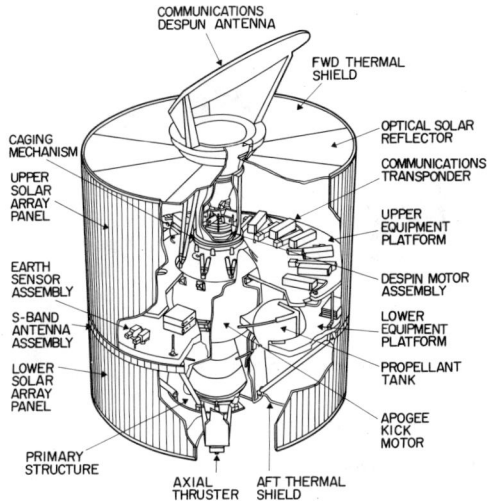

Fig. 7 Configuration of CS-3 spacecraft.

cm in height. These satellites were developed to meet increasing and diversified communication needs. The number of Ka-band transponders is increased to 15, including five spares, as compared with six with no spare for the CS-2. The C-band transponders are composed similar to those of the CS-2.
For the CS-3, however, solid state power amplifiers replace travelling wave tube amplifiers (TWTA) to improve the reliability of onboard transponders. Full eclipse operation of Ka-band and C-band transponders is possible. Newly developed gallium arsenide (GaAs) solar cells are adopted to CS-3 as its main power source for the first time in space application. Figure 7 shows the configuration of the CS-3 spacecraft.

Having achieved the objective set for the initial stages with the CS-2 and -3 series, we now envision that a mobile communications system will become an important target up to the late 1990s because of the Japanese geographic condition (surrounded by the sea). NASDA has already launched the ETS-5 (Engineering Test Satellite-5) for this purpose.

BS-3 System

The follow-on broadcasting satellites BS-3s are now in the fabrication phase, aiming at launch by H-1 vehicles in summer of 1990 and 1991, respectively.
These satellites are to succeed the broadcasting service by the BS-2 satellites in response to the increasing and diversified demands for such service and are designed for seven yrs. mission life. The BS-3 spacecraft weighs ap-

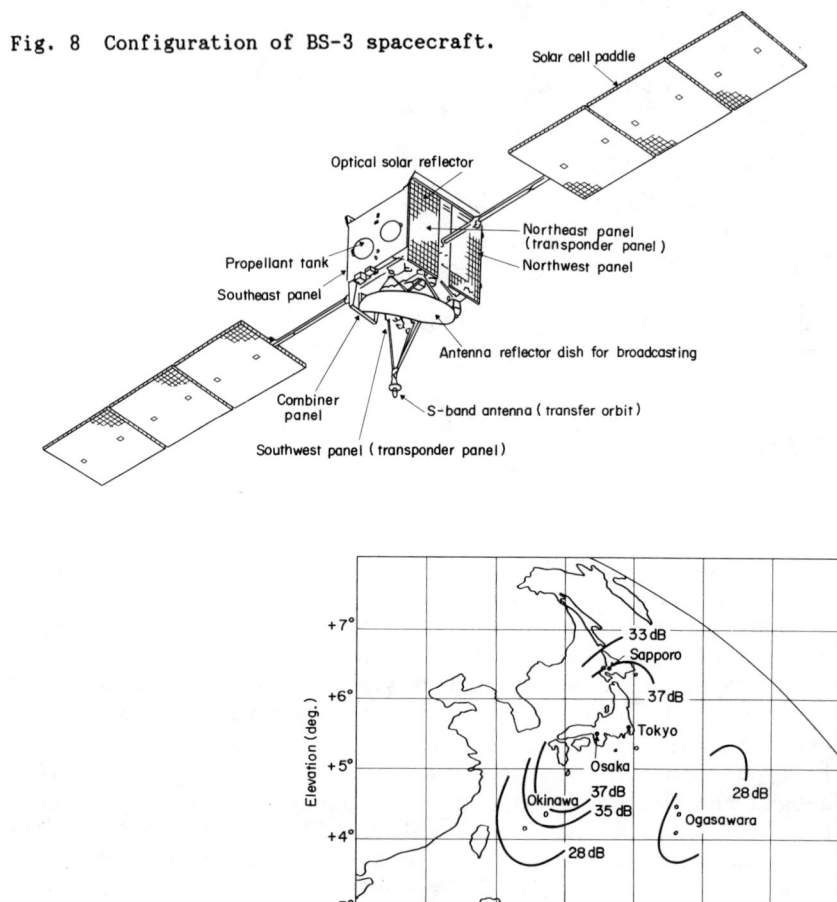

Fig. 8 Configuration of BS-3 spacecraft.

Fig. 9 BS-3 antenna gain.

proximately 550 kg at BOL and will be placed at the longitude of 110° E. It is based on a three-axis stabilized biased momentum system and has a box-type body. Its top panel during the launch phase becomes the Earth-facing panel on geostationary orbit.[5] The top panel supports the mission antenna, attitude sensors, TT&C (Tracking, Telemetry and Command) antenna, and some elements of the transponder. Two side panels, set to face N and S on geostationary orbit, are equipped with the transponders and spacecraft control components. Two three-section solar panels are expanded at the opposite sides of the spacecraft. Figure 8 shows the configuration of the BS-3 spacecraft.

Table 2 Major characteristics of BS-3

(a) Spacecraft Major characteristics

Shape	Box with deployable solar cell paddles
Weight	Approx. 1100 kg (at launch)
	Approx. 550 kg (beginning of life)
Attitude control	3-axis stabilized
Design life	7 yr
Launch vehicle	H-1
Launch site	Tanegashima Space Center
Launch date	BS-3a: summer of JFY 1990
Orbit	Geostationary orbit, 110° E long (BS-3a, 3b)

(b) Mission characteristics of BS-3

Channel	3 (broadcasting)
	1 (wideband relay)
TWT output (W)	$\geq$ 120 (broadcasting)
	$\geq$ 20 (wideband relay)
EIRP (dBW) (in main islands)	$\geq$ 55.5 (broadcasting)

The BS-3 antenna is an elliptical offset reflector with a new elliptical-type corrugated horn and a rectangular one-.[6] The antenna gain of BS-3 is shown in Fig. 9. Each BS-3 has six 120-W 12-GHz TWTAs implemented in a two-for-one redundancy configuration and a single wide bandwidth (60MHz) 20-W 12-GHz TWTA. The electric performance tests of the engineering model of TWTA for BS-3 have been already performed successfully in simulated solar ecliptic conditions.[7]

The broadcasting channels with 27 MHz bandwidth will be used by NHK and Japan Broadcasting, Inc. (JSB) which is a newly organized private broadcasting company. The major characteristics of BS-3 spacecraft are summarized in Table 2.

For the new broadcasting service by BS-3, high definition television (Hi-Vision) is now drawing much attention as a prospective application, and the service will be available within 27 MHz bandwidth by the development of a transmission system called MUSE (Multiple Sub-Nyquist Sampling Encoding), which utilizes a bandwidth compression.[9] Recently, TSCJ has been entrusted to use one 120-W transponder aboard BS-3 for experiments with Hi-Vision. BS-3 development schedules are shown in Fig. 10.

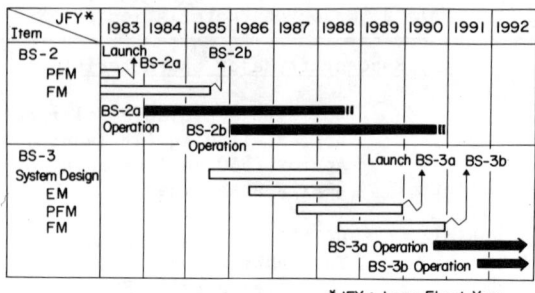

Fig. 10 BS development and operation schedule.

Activity of Commercial Domestic Communications Services in Japan

In view of the results so far successfully achieved by CS-2 and other communication satellites, the advantages of the domestic communication services in Japan have been well recognized. With the enactment of the Telecommunication Business Law, two new satellites communication service companies, Japan Communication Satellite Company, Inc. (JCSAT) and Space Communication Corporation (SCC), were both established in 1985. It was expected that JCSAT and SCC each launched two communications satellites in 1989 and provided network services such as satellite news gathering, television program and data transmission, and very small aperture terminal (VSAT), etc. These satellites will be controlled in stationkeeping and housekeeping by their own ground control facilities.

According to the companies' announcements, these satellites systems are as follows:

The JCSAT space segment is being designed and manufactured by the Hughes Aircraft Company and will consist of two HS-393 series spacecrafts with a 10-yr mission life. The first spacecraft (JCSAT-1) will be positioned at the longitude of 150° E by the Arian lV vehicle, and the second spacecraft (JCSAT-2) will be positioned at the longitude of 154° E by Titan lll vehicle. The spacecraft has a spin stabilized platform and has the 40 Ku-band TWTAs for the communication payload.[9]

The SCC three-axis stabilized spacecrafts, named Superbird-A and B and designed and manufactured by Ford Aerospace and Communications Company, will be launched by Arian lV into geostationary orbit at longitudes of 158° and 162° E, respectively, 10 yr mission life. The spacecraft communication payload contains the 22 Ku-band and 21 Ka-band TWTAs.[10]

Conclusion

Communications and Broadcasting satellites were launched for experimental purposes in the late 1970s. Taking into account the geographic condition of Japan, the potential advantages of communication and broadcasting satellite services were recognized at these earlier stages. The follow-on first operational satellite systems with CS-2s and BS-2s were successfully promoted to meet the demand for domestic telecommunications and direct broadcasting services.

JCSAT and SCC planned to launch commercial-based communication satellites in 1989 and intend to start new domestic telecommunications services.

BS-3s, the second generation broadcasting satellites to be launched in 1990 and 1991, will be able to broadcast Hi-Vision, which will present a high-definition television picture with amazing clarity on a large-screen display.

The "Information Oriented Society" is a key term in understanding the society and economy of Japan in the 1990s. The authors believe that the previously described satellites will be very important facilities for the Information Oriented Society.

Acknowledgment

The authors would like to thank H. Hirose, President of TSCJ, and all of the members of MPT, NASDA, NTT, NHK, JSB, and other companies who led and participated in the development of communications and broadcasting satellites.

References

[1] "Japan's CS (Sakura) Communications Satellite Experiments," Institute of Electrical and Electronics Engineers Transaction Vol. AES-22, No. 3, May, 1986.

[2] Hasegawa, T., "An Overview of the Japanese Satellite Broadcasting Program," AIAA 12th International Communication Satellite Systems Conference, Paper 88-0807, 1988.

[3] Dobashi, H., Nonobe, Y., Shiro, I., and Ikeda, K., "Some Control Experiences with CS-2s and BS-2s," AIAA 12th International Communication Satellite Systems Conference, Paper 88-0879, 1988.

[4] Mochizuki, M., Sogame, E., Shibata, Y., "Status Report of the H-1 and H-11 vehicles," AIAA 12th International Communication Satellite Systems Conference, Paper 88-0853, 1988.

[5] Frohbieter, J.A., and Duffy, T.J. "Japanese Broadcasting Satellite (BS-3)," 16th International Symposium on Space Technology and Science, Vol. 2, Sapporo, Japan, 1988, p. 2037.

⁶Miura, S., Toyama, N., Ohmaru, K., Shogen, K., Obuchi, T., and Miyata, Y., "Electrical Performance of BS-3 Shaped Antenna," AIAA, 12th International Communication Satellite Systems Conference, Paper 88-0876, 1988.

⁷Akanuma, T., et al., "Development Activity of 12 GHz 120 W TWTA for Japanese Broadcasting Satellites of BS-3," AIAA 12th International Communication Satellite Systems Conference, Paper 88-0833, 1988.

⁸Ninomiya, Y., Ohtsuka, Y., and Izumi, Y., "A Single Channel HDTV Broadcast System: The MUSE," NHK Lab. Note, No. 304, 1984.

⁹Takamatsu, H., and Nagai, Y., "JCSAT Space Segment," 16th International Symposium on Space Technology and Science, Vol. 2, Sapporo, Japan, 1988, p.2065.

¹⁰Ogata, M., and Louie, M., "Optimization of Frequency Utilization for Communications Satellite System in Japan," AIAA 11th International Communication Satellite Systems Conference, Paper 86-0725, 1986.

Space Research Satellite Program of Japan

Tomonao Hayashi*
Institute of Space and Astronautical Science,
Sagamihara, Kanagawa, Japan

Abstract

The space research program in Japan was started by the leadership of the Institute of Space and Astronautical Science(ISAS). From the first orbiting of the Japanese satellite named "Ohsumi" in 1970, 17 scientific satellites for various mission objectives have been launched in 18 years. Their scales are not so large due to the limitation of the capability of launchers, which ISAS has developed, but owing to the steady pace of the launching, scientists have been able to proceed their research works along their long term prospects, and have been contributing to space exploration with international appreciation. In this paper the management for developing scientific satellites in ISAS is reviewed. After some explanations on the historical background, the way of selection of the mission, system design, program control, and tests for scientific satellites are described.

Introduction

In Japan there are two organizations which are responsible to space development. The one is National Space Development Agency(NASDA), and the other is ISAS. NASDA is in charge of the development of application satellites, and ISAS is in charge of scientific exploration of space.

The administrative structure on the space development in Japan is shown in Fig.1.

Space Activities Commission is responsible to coordinate the national space programs, and makes the review of the future plan once a year.

Copyright © 1990 by the American Institute of Aeronautics and Astronautics, Inc., All rights reserved.

*Professor.

Historical Background of ISAS

Space research activities in Japan originated from the University of Tokyo in 1955 for participation in the International Geophysical Year(IGY) program held in 1957-1958. Team members of the Institute of Industrial Science, University of Tokyo, succeeded in observing the upper atmosphere in June 1958 by utilizing Kappa-6 rocket, which they had developed.

In 1965, as one of the attached research institutes of the University of Tokyo, ISAS was established, and was reorganized in 1981 as to belong directly to the Ministry of Education, Science, and Culture. This is one of the National Interuniversity Research Institutes and is in charge of the research on space science and technology by utilizing satellite, sounding rocket, and balloon.

Professors, research assistants, and graduate students from scientific and technological fields have thus been able to participate in space development in the structure of a university. Various kinds of sounding rockets have been developed and applied to space exploration. Sounding rockets dedicated now for space research are shown in Fig.2. They are all solid propellant rockets, and more than 350 rockets were launched up to the present from Akita and Kagoshima space centers for the use of space science and technology. By making use of the simplicity of operation more than 30 single stage rockets were launched at Showa Base, Antarctica.

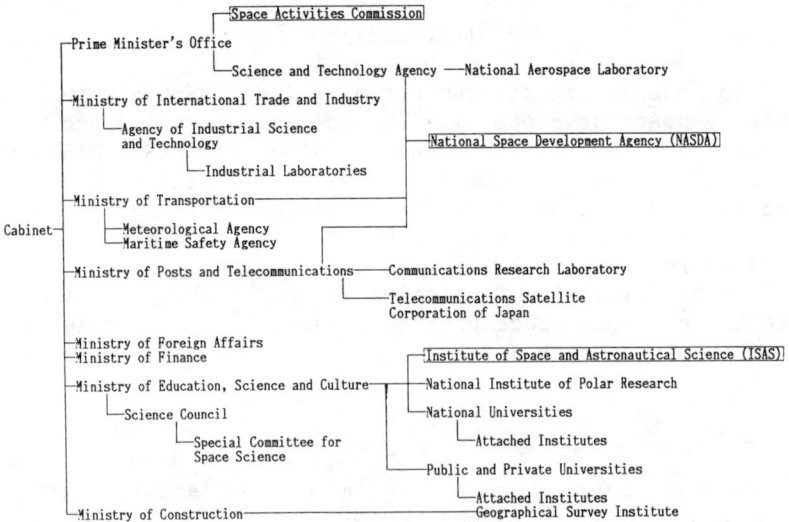

Fig. 1 Administrative structure on the space development in Japan.

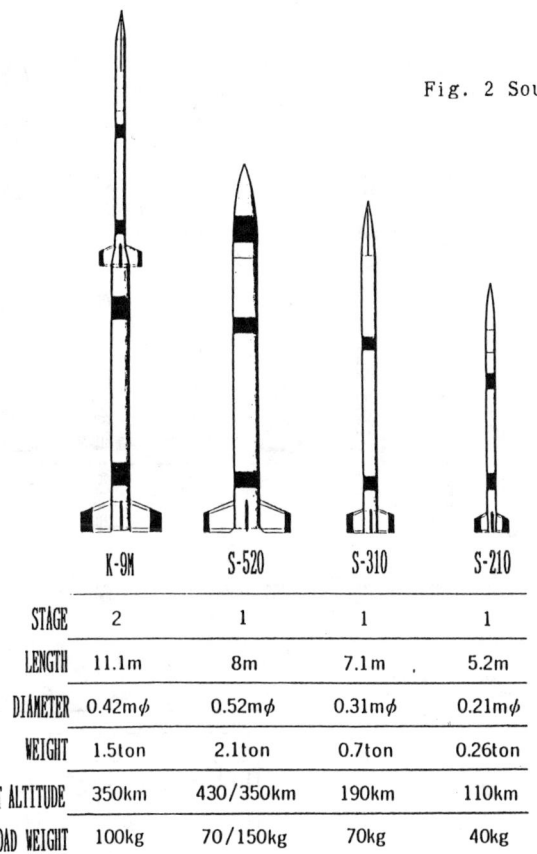

Fig. 2 Sounding rockets.

	K-9M	S-520	S-310	S-210
STAGE	2	1	1	1
LENGTH	11.1m	8m	7.1m	5.2m
DIAMETER	0.42mϕ	0.52mϕ	0.31mϕ	0.21mϕ
WEIGHT	1.5ton	2.1ton	0.7ton	0.26ton
SUMMIT ALTITUDE	350km	430/350km	190km	110km
PAYLOAD WEIGHT	100kg	70/150kg	70kg	40kg

Many of the scientific payloads, which were tested by sounding rockets, have been installed on scientific satellites, and good results have been obtained.
A series of satellite launchers(Fig.3) also were developed by ISAS. A three stage solid motor M-3SII is now dedicated to the launching of scientific satellites.

Scientific Satellites

Since the launch of the first satellite "Ohsumi" by using Lambda rocket, ISAS has launched 17 spacecraft into orbit in 18 years by using various versions of Mu launch vehicle. The history of the scientific satellites launched by ISAS, including the future plan, is shown in Fig.4.
The missions of satellites are categorized into five groups corresponding to the study groups of ISAS. It is evident that the launchings have been made at the pace of one satel-

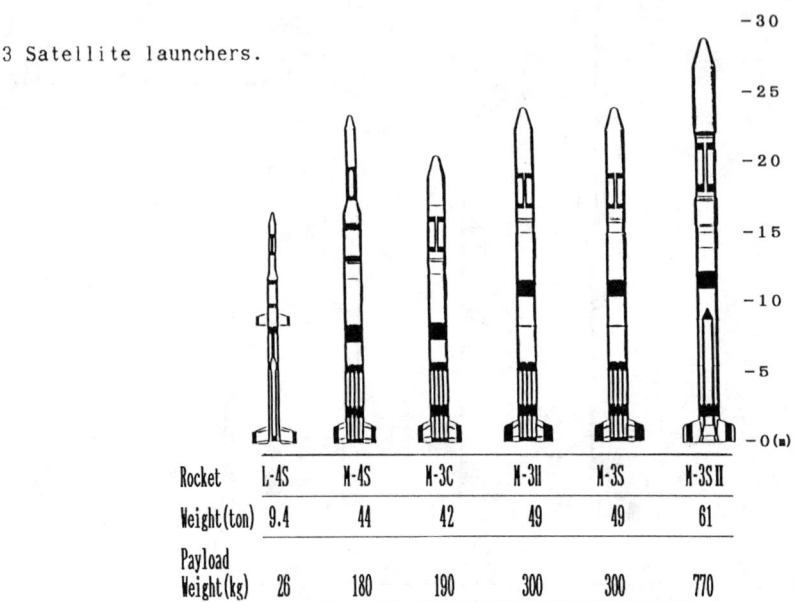

Fig. 3 Satellite launchers.

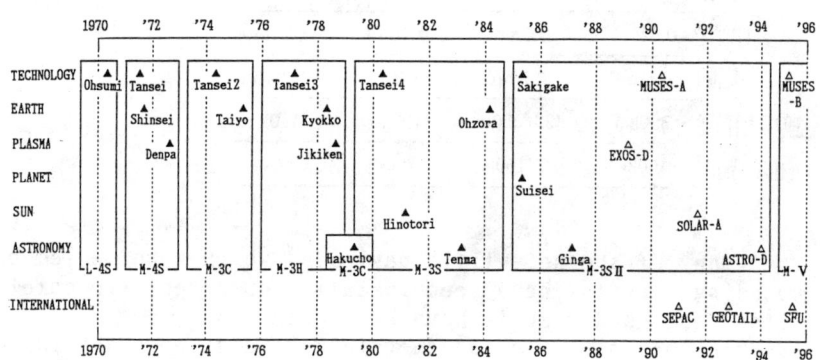

Fig. 4 History of scientific satellites launched by ISAS.

lite per year. The mission objectives and other comments including future plans are listed in Table 1a-c.

Management for Developing Scientific Satellite in ISAS

Selection of the Mission

Selection of the mission is made at a board committee of ISAS, where they discuss the candidates for future missions. The selection of the candidate is made every year

Table 1a Scientific satellites in Japan

Scientific satellite	Mission	Weight kg	Orbit Altitude km	Inclination deg	Launch year	Launch vehicle
OHSUMI	Confirmation of gravity-turn launch technology	23.8	Perigee 337 Apogee 5,151	31.1	1970	L-4S
TANSEI (MS-T1)	Confirmation of M-4S launch capability and test of on-board sub-systems	63	Perigee 990 Apogee 1,110	29.6	1971	M-4S
No.1 SHINSEI (MS-F1)	Observation of ionosphere, cosmic rays, HF-band solar radio emissions	66	Perigee 870 Apogee 1,870	32.1	1971	M-4S
No.2 (REXS)	Observation of plasma electromagnetic waves, geomagnetism	75	Perigee 250 Apogee 6,560	31.0	1972	M-4S
TANSEI-2 (MS-TZ)	Confirmation of M-3C launch capability and test of attitude control by means of magnetic torquing	56	Perigee 288 Apogee 3,236	31.2	1974	M-3C
No.3 TAIYO (SRATS)	Observation of solar soft X-rays, solar UV radiations, UV terrestrial corona lines, ionospheric parameters	86	Perigee 260 Apogee 3,140	31.6	1975	M-3C
TANSEI-3 (MS-T3)	Confirmation of M-3H launch capability and tests of magnetic attitude stabilization and UV photometry	130	Perigee 791 Apogee 3,813	65.7	1977	M-3H
No.5 KYOKKO (EXOS-A)	Observation of electron density and temperature, auroral particles and UV imaging	126	Perigee 636 Apogee 3,977	65.4	1978	M-3H
No.6 JIKIKEN (EXOS-B)	Observation of electron density of energetic particles, plasma waves, electric field	90	Perigee 227 Apogee 30,051	31.1	1978	M-3H

Continued

Table 1b Scientific satellites in Japan

Scientific satellite	Mission	Weight kg	Orbit Altitude km	Inclination deg	Launch year	Launch vehicle
No.4 HAKUCHO (CORSA-b)	Observation of cosmic X- and gamma-ray radiations	90	Perigee 545 Apogee 577	29.9	1979	M-3C
TANSEI-4 (MS-T4)	Confirmation of M-3S launch capability and tests of attitude control systems of momentum wheel, magnetic torquing and MPD arc-jet thruster and X-ray Bragg spectrometer, star mapper	185	Perigee 521 Apogee 606	38.7	1980	M-3S
No.7 HINOTORI (ASTRO-A)	Observation of solar flare radiations	120	Perigee 350 Apogee 600	31.0	1981	M-3S
No.8 TENMA (ASTRO-B)	Observation of cosmic X- and gamma-ray radiations	216	Perigee 497 Apogee 503	31.5	1983	M-3S
No.9 OHZORA (EXOS-C)	Remote sensing of the middle atmosphere and ionospheric plasma	210	Perigee 354 Apogee 865	75.0	1984	M-3S
SAKIGAKE (MS-T5)	Preliminary test for the launching of deep space probe	138	Heliocentric Perihelion 1.22×10^8 Apohelion 1.50×10^8		1985	M-3S II
No.10 SUISEI (PLANET-A)	Observation of comet Halley and interplanetary plasma	140	Heliocentric Perihelion 1.01×10^8 Apohelion 1.51×10^8		1985	M-3S II
No.11 GINGA (ASTRO-C)	Observation of the nuclei of active galaxies in X-rays, and various X-ray sources	420	Perigee 504 Apogee 674	31.0	1987	M-3S II

Table 1c Scientific satellites planned in Japan

Scientific satellite	Mission	Weight kg	Orbit Altitude km	Inclination deg	Launch year	Launch vehicle
No.12 AKEBONO	Observation of particles in the magnetosphere to study the mechanism of acceleration of auroral particles	300	Perigee 300 Apogee 8,000	75.0	1989	M-3SII
No.13 MUSES-A	Confirmation of swingby technology and tests of lunar orbiter and dust counter	195	Double lunar swingby		1990	M-3SII
No.14 SOLAR-A	Observation of high energy phenomena on the sun through X-ray and gamma-ray in time of maximum solar activity	420	Perigee 500 Apogee 800	31.0	1991	M-3SII
Magnetosphere Observation Satellite GEOTAIL	Observation of energy flow and transformation in the earth's magnetotail	930	Double lunar swingby		1992	NASA VEHICLE
No.15 ASTRO-D	Observation of X-ray sources with X-ray reflective mirrors	420	Perigee 500 Apogee 800	31.0	1993	M-3SII
No.16 MUSES-B	Observation of quasars and other distant celestial objects by use of space VLBI technique	300	Perigee 1000 Apogee 10000	31.0	1994	M-3SII

based on a lot of proposals submitted to Scientific Satellite Symposium, Space Observation Symposium, and other meetings for specific study groups, which are held by ISAS.

A working group is organized for the specific theme of the candidate and conceptual design of the satellite is started. The final plan settled at the committee of project coordination of ISAS is proposed to the review meeting of Space Activities Commission, and after getting its approval a project team for the specific mission is organized. The budget for its development usually comes in the following five years from the Ministry of Finance through the Ministry of Education.

System Design of the Spacecraft

The project team consists of science and technology specialists from ISAS and engineers from several manufacturing companies concerned. The team usually handles the mission planning, development of spacecraft, overall test, launch operation, satellite operation, data acquisition, and analysis in ISAS.

Most of the scientific satellite needs five years from fixing its conceptual design up to the launch. Sequential phases concerning prototype model(PM) and flight model(FM) in this period are shown in Fig. 5.

For effective use of the budget prototype model(PM) is not usually made in the correct sense of the word, but is made of engineering models for newly developed units, and also a structural model for both thermal and mechanical preliminary tests. Design meetings, held about once a month consist of all the members of the project team and are chaired by a project manager. The meetings play a very important role to proceed the development of spacecraft because detailed design of the instruments and scheduling are reported, discussed, and reviewed. Some of the specific items are sent to subgroups for investigation or discussion, and the results obtained are requested to be reported at the next design meeting.

As for an example of the weight control of a spacecraft, the mission group would usually like to mount more payload even if it were one gram. On the other hand, the technology group being responsible to make its system design, based on the capability study of the vehicle developed by themselves, would like to reduce the payload weight even if it were one gram for getting the system margin to take care for probable cotingency. In most cases "gentlemanlike" adjustment is performed between science and technology teams, but when the weight increase for an instrument is inevitable, overall review for the spacecraft to reduce the weight is made, and at the same time possible enhancement of the vehicle capability and optimization of the launching orbit also are considered.

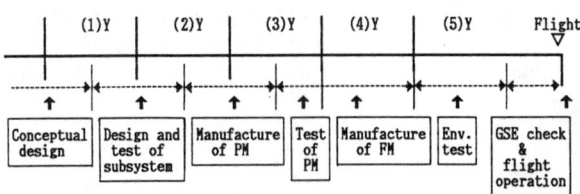

Fig. 5 Sequential phases for developing satellite.

The process for fixing design through the design meeting is shown in Fig.6.

The fact that the ISAS is responsible to perform space exploration based on its research groups on both science and technology, which cover the fields from launch vehicle to space observation, allows of tight cooperation of the teams, flexible management, and technology transfer to the follow-on missions.

Program Control

For designing spacecraft by adopting up-to-date technologies within the limit of budgetary appropriation, it is difficult for ISAS to have a contractor system to ask the manufacturing based on the whole specification of the spacecraft at the beginning phase of the plan. The importance of the design meeting above mentioned is based on this reason.

The program management of scientific spacecraft contains three control items: 1) configuration and data; 2) reliability and quality; and 3) schedule, safety, etc.

The management is self-imposedly controlled by each manufacturing company along guidelines defined at the design meeting.

The most important documentation for configuration and data control is a document named System Confirmed, in which mission definition, orbit, attitude, structure, weight, arrangement of onboard instruments, function and characteristics for each subsystem, command items and corresponding codes, telemetry format, and assignment for connector pins, etc. are described. The contents are continuously revised

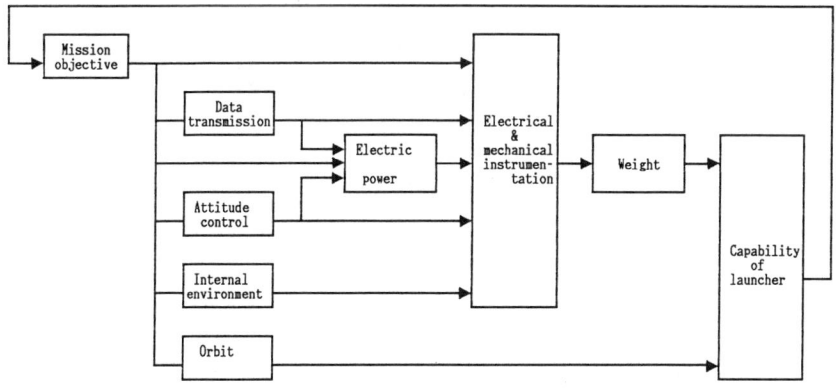

Fig. 6 Process for fixing design.

along the progress of the design, and up-to-date information is always provided to the members concerned. Selection of parts is discussed at the parts meeting held by ISAS. The reliability is maintained by selecting parts from those which belong to MIL-S class and those having space flight evidence. Reliability and quality are maintained by self-imposed control of each manufacturing company being responsible to the manufacturing of subsystems.

In order to limit the expense, redundancy is usually not considered in component level, but is solely applied to the important circuit level.

Several changes in the design are usually inevitable but actual changes are mostly made at the transfer phase of PM to FM, based on the test results of PM.

Test of Scientific Spacecraft

The hardware of the spacecraft is completed one year before the launch. All subsystems are brought together at ISAS, and after the subsystem test, they are integrated in an assembly room to check their mechanical and electrical interfaces.

Overall test is then conducted for various kinds of items, which are listed in Table 2. The order of test items for overall test is arranged for each satellite so as to

Table 2 Test items

Test items	Measurement or checkout	Adjustment*	Test
Center of gravity	○		
Weight	○	△	
Dynamic balance	○	△	
Shock			○
Vibration			○
Mechanical function	○	△	
Moment of inertia	○	△	
Sensor alignment	○	△	
Electro-magnetic compatibility	○	△	
Temperature			○
Thermal vacuum			○
Solar cell power	○		
Sensor function	○		
Magnetic moment	○	△	

* : Adjust if necessary

minimize the time required. For example, temperature test under atmospheric pressure is usually followed by thermal vacuum test. This is very effective for saving time to replace the parts which have happened to fail for the sole reason of their temperatures.

Conclusion

Scientific satellites in Japan have been launched at the pace of one satellite per year. The size of the satellite is not so big due to the limitation of the capability of the launch vehicle, but owing to the steady pace of the launching, Japanese scientists have been able to proceed their research program on a long term prospect for each field of research, and have been getting international appreciation. This is because the scientific satellite program being conducted in ISAS originated in the framework of a university. New technologies are sometimes provided from ISAS in the way of the development of scientific satellites to manufacturing companies. Provided that those technologies seem to be applicable to other bigger fields such as to application satellites or to other transportation systems, companies would like to challenge them taking a longer view of matter at the cost of their present expense, and inexpensive scientific satellites have been realized. For keeping up such a status, all personnel concerned in ISAS are requested to make their constant efforts in academic fields.

The procedures for the system design and management have been built up as the result of trial and error, and better ones to adapt to the change of technological environment are continuously sought.

In recent years several international programs have been started in ISAS. Most of the international cooperative works have started from discussions based on person-to-person contacts. On the way of maturing their ideas, interorganization arrangements are discussed, and finally international cooperations are realized. Although explanations on our standpoint for the development are sometimes necessary, tight cooperations are being carried out owing to the flexible response based on mutual understanding.

Mobile Satellite Communications: Applications for Developing Countries

Wolf D. von Noorden*
International Maritime Satellite Organization, London, United Kingdom

Abstract

Since the start of its communications services in February 1982, the International Maritime Satellite Organization's satellite system has mainly benefited the fleets of the major maritime nations. But it would be misleading to conclude from this that Inmarsat will remain primarily at the service of the developed world. This paper describes how Inmarsat's institutional commitments, the evolution of space technology, and the introduction of new service applications will enable Inmarsat to contribute to economic and social developmental goals. In particular, this paper will show how a mobile-satellite system can offer a range of applications and reach user communities that cannot be served by fixed systems and will show how economies of scale achievable through a global network can put highly reliable communications within financial reach of most potential users.

Needs Identified

In the 1960s satellite technology was seen as a possible boon to education through television transmissions to widely scattered communities, but little attention was paid to the capabilities of satellites for telephone, data, and other services. The 1970s saw increasing use of satellites by developing countries, either through the International Telecommunications Satellite Organization (INTELSAT) or on domestic or regional systems, such as Indonesian Communications Satellite System (PALAPA), India's Satellite System (INSAT), Arab Satellite Communications

---

Copyright © 1990 by Dr. Wolf D. von Noorden. Published by the American Institute of Aeronautics and Astronautics, Inc., with permission.
*General Counsel, Office of the General Counsel, Inmarsat.

Organization (ARABSAT), and Brazil's Satellite System (BRAZILSAT).[1]

In the 1980s the United Nations and the International Telecommunication Union (ITU) have exposed the disparity in the distribution of telecommunications facilities between the developed and the developing world. The Second United Nations Conference on the Exploration and Peaceful Uses of Outer Space (UNISPACE 82) recognized the key role that international satellite communications organizations, including Inmarsat, played in development, and it also acknowledged the need for low-cost satellite terminals.[2]

The ITU's Maitland Commission Report in 1984 revealed that in many developing countries telecommunications systems were inadequate to sustain even the barest of essential services and that deficient communications could be a major hindrance to a country's ability to maintain essential services in the fields of health, education, disaster relief, transportation, economic development, and other social endeavors. It highlighted Inmarsat's role and recognized that small satellite terminals and portable equipment were adaptable to remote and rural areas.[3]

Other reports attest to the striking lack of communications services in isolated areas and conclude that where there is no existing infrastructure, or where other forms of communication are not economically viable, satellite communications may be the only answer. The Indian Space Research Organization (ISRO) reported in 1987 that it is in remote or rural areas, which are prohibitively expensive to reach by conventional terrestrial means, where satellites come into their own. An important requirement for communications in rural or remote areas is their connectivity to the leading cities of the nation. Such connectivity using traditional networks is a very lengthy and expensive process. Satellite systems are preferable. This is true for all locations without infrastructures, i.e., where no telecommunication services exist.[4] Other studies in countries such as Kenya[5] and Thailand[6] have shown the desperate lack of telephones in the countryside compared with the urban areas.

In urban areas the use of cellular radio technology for mobile communications may initially be cost effective. Its usefulness, however, depends directly on the number and dispersion of the base stations included in a network. The fewer the base stations, the smaller the range of the mobile communications terminal. To be economically viable, a large number of users would be required in order to justify the investment in base stations. Satellites, however, are insensitive to distance and are therefore preferable for communications to wide areas and to thin routes, where the market density is too small to sustain terrestrial systems.

Despite the apparent advantages of satellite communications, there are still serious obstacles to the implementation of such communications. The lack of fully-fledged demand in the rural districts of many developing countries may make it difficult for the telecommunications business to be profitable. Harsh climatic and geographical conditions and the absence of sufficient electrical power supplies could make matters worse.[7] A key question for any country contemplating the expansion of telecommunications infrastructure and services is whether there is a viable and sustainable market.

Doubts about viable markets in the past have meant that there has been little effort to address thin-route telecommunications as an economically justifiable development and investment activity.[8] But things are changing. A model developed some years ago and used to predict the impact of investment in satellite earth stations for thin-route telephony showed a small but significant contribution of telecommunications investment to national economic growth, as measured by gross domestic product (GDP).[1]

In 1988 an ITU study worked out a generally applicable methodology on how to assess the value of information and to quantify the effects of the availability of telecommunications on transport systems. It was found that the costs of transporting goods could be substantially reduced.[9] In its study ISRO concluded that the establishment of inexpensive, thin-route communication services would have a significant impact on the local economy. Using the new Inmarsat-C, a small, inexpensive data terminal, as an example, ISRO estimated that in India the potential demand for this service could be around 3000 terminals, at a terminal cost of $5000 and a per message cost of $1. At a per terminal cost of $1000 and a per message cost of $0.30, it is estimated that the potential demand could be as high as 50,000 terminals.[4]

Inmarsat's own research into the worldwide fishing industry is very revealing. Among the world's top fishing nations, some of the leaders such as Japan, Spain, and the U.S. use the Inmarsat system extensively. But others such as China, Chile, and Peru, which has more than 500 ocean-going fishing vessels, have not yet taken to satellite communications and constitute a big potential market for both large Inmarsat-A terminals and the new, small Inmarsat-C terminals.

A number of other important fishing nations, such as Thailand, Cuba, Indonesia, and the Philippines, have large numbers of vessels that are well suited to the Inmarsat-C system, and some have already expressed great interest in it.

In responding to the needs of developing countries, the emphasis in the past has been on the expansion of telecommunications for fixed services for uses such as telephone, broadcasting, and education. But mobile-satellite services can offer greater flexibility through a variety of applications and by complementing terrestrial and fixed-satellite systems without the need for expensive, supporting infrastructures. This opens up new horizons for the shipping and maritime industries, airlines, road and rail networks, and exploration and resource industries. The availability of small, transportable earth stations at a declining cost enables remote or inaccessible areas to be linked by communications in a way not possible through fixed systems.

Inmarsat's Institutional Commitment to Development

Inmarsat's charter is based on the principles established by the United Nations that satellite communications should be available to all nations on a global and nondiscriminatory basis and that outer space should be used for the benefit of all countries.[10] The organization's initial mandate to serve the maritime community has been extended to cover aeronautical services and is being further extended to cover land mobile services in recognition of the worldwide demand for all types of mobile-satellite communications.

The Inmarsat space segment is open for peaceful, nondiscriminatory purposes by all nations, whether members of Inmarsat or not. The organization is required to provide services in all geographical areas where there is a need for mobile communications. Its current membership of 59 states represents developed and developing regions. Its managing Council includes four members elected with due regard to the interests of the developing countries. All member states are represented equally at the Assembly of Parties. Thus developing countries participate in the management and financing of the organization and in the formulation of its general policy and long-term objectives.

The commercial principles underlying Inmarsat's operations can work to the advantage of developing countries. International ownership and control of the organization ensures that in the planning of the space segment and of new mobile services the interests of all nations will be considered. While enjoined to provide revenue-earning services without the benefit of any subsidies, Inmarsat is not a profit-making body in the ordinary sense, but it uses excess revenues to stabilize or reduce charges. The procurement policy of Inmarsat encourages worldwide competition in the supply of goods and services. Neither national quotas nor preferences for the industries of member states are permitted, thereby providing an opportunity for developing countries to participate in the procurement process.

An important institutional obligation of Inmarsat is to cooperate with the United Nations and to take account of the regulatory requirements of the International Maritime Organization (IMO), the International Civil Aviation Organization (ICAO), and the ITU. This obligation underlies Inmarsat's activities, which conform to the recommendations of UNISPACE 82 and the Maitland Commission, as subsequently described.

Existing Services to Developing Countries

The high cost of Inmarsat's initial mobile terminal, the Inmarsat-A ship earth station (SES), was a factor that impeded the rapid adoption of maritime satellite communications in developing countries. But the maritime satellite technology proved at an early stage to be particularly adaptable for many uses on land in developing countries in areas where fixed-satellite or terrestrial services were uneconomical to install or unavailable for any reason. These uses were facilitated by the marketing of small, portable, Inmarsat-A terminals.

Inmarsat responded to this demand by authorizing an exception for the use of its maritime system for land-based services, conditional upon the nonavailability of other telecommunications facilities. The use of the system for humanitarian purposes, i.e., for emergency relief operations on land, was also sanctioned, pursuant to the requirement of the ITU that the international telecommunication services give absolute priority to safety of life communications.[11]

Thus, since 1982 the Inmarsat system has been used for many hundreds of applications in Africa, Asia, and Central and South America, in remote areas, or where adequate facilities either were unavailable or had broken down. The applications include transportation, health care, emergency surgery, agriculture, seismology, oil exploration, and coordination of food and relief supplies. It has also been used for disaster relief, e.g., following the Mexican and Armenian earthquakes, the volanic eruption in Colombia, and hurricane damage in the Caribbean. Today, about 10% of Inmarsat Standard-A terminals are being used on land.

While these land-based services have fulfilled many needs in developing countries, cost factors as well as institutional barriers have been a deterrent to widespread use of mobile-satellite communications by the maritime, aeronautical, and land transport communities.

Towards Cheaper Communications

It has always been realized that cheaper communications were the key to more extensive use of mobile satellite services. Already, economies of scale have had an impact on cost. Between 1982 and 1990, when users of Inmarsat services rose from 1000 to more than 10000, Inmarsat's utilization charges have decreased in real terms by up to 50%. The cost of the Inmarsat-A SES has declined from $75,000 to as low as $25,000. This lower price is competitive with traditional High Frequency/Medium Frequency radio equipment on a sea-going merchant vessel.

But that was only a beginning. In line with the UNISPACE 82 and Maitland Commission recommendations, Inmarsat has developed a new, low-cost mobile terminal, the Inmarsat-C, introduced into service in 1989. It will open up the Inmarsat system to more categories of user communities and a wider range of applications.

Inmarsat-C SESs provide data message services from small, portable, lightweight equipment. It weighs about 6 kg and can be battery powered. Its cost in volume production is likely be less than $5,000 per unit. The new terminal uses the radio frequency spectrum very efficiently, thereby eliminating the need for bigger and more expensive satellites to accommodate demand. A number of companies are already manufacturing Inmarsat-C terminals, and coast earth station facilities are being upgraded to handle Inmarsat-C communications traffic. Work is also progressing on the development of low-cost, mobile telephony terminals.

Other measures to provide cheaper communications include work on the development of a low-cost, mobile, telephony terminal. Inmarsat has also recently introduced lower charges during off-peak hours, typically involving reduction in charges of around 30%. With the introduction of a new generation of terminal called Inmarsat-B from 1991 onwards, charges will probably be reduced substantially at all times of day.

Potential Applications of Small Mobile-Satellite Terminals in Developing Countries

The advent of low-cost, portable, satellite terminals opens up new opportunities for communications and should have a great impact on development. Some potential applications are:

1) <u>Improved maritime communications</u> -- Inmarsat-C will extend the benefits of satellite communications to the smallest of ships, fishing boats, and even life rafts. Maritime operations in the developing world, particularly off the coasts of Africa and Southern Asia, still suffer from inadequate communications and the lack of navigational aids. This paper has already referred to the potentially widespread use of satellite communications by the fishing fleets of some developing countries. Amendments adopted late in 1988 to the International Convention on the Safety of Life at Sea (SOLAS), to incorporate the Global Maritime Distress and Safety System, will provide for the Inmarsat system to be used as part of the mandatory carriage requirements for ocean-going vessels. An example is the SafetyNet service of Inmarsat's Enhanced Group Call (EGC) system, whereby maritime safety messages will be broadcast to all ships by area. This will offer a safety communications system to countless small

fishing vessels. Inmarsat is also working on the development of a worldwide satellite navigation capability, which is expected to benefit all nations in the years to come.

2) **Air traffic services** -- In many developing countries, e.g., over some of Africa and parts of south and southeast Asia, communications for the provision of air traffic control services are inadequate. Pilots are often forced to perform their own air traffic control in mid-air using vhf radio. Inmarsat's aeronautical communications services, which were introduced on a regular basis in 1990, will offer high quality communications services to the aviation community worldwide, although the initial capital cost of airborne mobile equipment will not be cheap. Africa and, indeed, most developing nations would need reliable low-cost aeronautical fixed networks capable of interconnecting with other local and international airports and cities. One method of satisfying this need may be to establish a rationalized aeronautical fixed telegraph network (AFTN) using inexpensive Inmarsat-C terminals to exchange brief telex messages relating to aircraft arrival/departures, flight plans, position reports, and meterological information.

3) **Land transport** -- Communications are becoming increasingly important to land transport systems, helping to coordinate the routing of rail and road traffic, to locate goods in shipment, and to improve efficiency. Mobile-satellite systems, while not cost-effective in urban areas or in places where alternative telecommunications links exist, can help in the management of long-distance transport networks, trucking, and rail operations through rural and remote areas. During 1988 an Inmarsat vehicle, equipped with an Inmarsat-C terminal, gave demonstrations in many countries in Europe and North America of two-way data or text communications and other specialized applications, such as automatic position reporting. For example, in Europe the vehicle was tracked from London by a series of some 300 position reports sent automatically via an onboard Inmarsat-C terminal to Inmarsat headquarters.

4) **Rural and health services** -- Mention has already been made of Inmarsat-C's potential use in remote and rural areas. The use of mobile-satellite terminals can help officials at remote field locations keep in touch

with central authorities, thus improving the efficiency of administrative and government services, health care, agriculture, resource management, forestry, and many other activities. An example of a specialized application is the use of slow scan television for medical care. This enables, for example, images of a delicate operation being performed by a doctor in a remote area to be relayed to a consultant in the capital city or another country for immediate specialized advice.

5) **Early warning systems** -- Mobile-satellite terminals can be installed at remote locations, such as unmanned weather or environmental stations, providing a constant flow of information on changing conditions and an early warning capability. This type of use could have enormous implications for meteorological systems, forecasts of natural hazards, and other services.

6) **Emergency relief** -- In times of natural or man-made disasters there is usually an immediate need for communications to assess damage, to report conditions at the disaster site, and to coordinate relief activities. Mobile-satellite terminals can be quickly transported to the site by road or by air to reestablish broken or nonexistent communication lines and to help organize the relief efforts of government organizations and international relief agencies.

Recent Activities

The evolution of new low-cost technology has been complemented by recent policy decisions and activities by Inmarsat directly related to the use of mobile-satellite telecommunications in developing countries.

At its Fifth Session in October 1987, the Inmarsat Assembly of Parties called for studies on the development of low-cost coast earth stations (CESs). Some countries have found the cost of procuring full specification Inmarsat CESs prohibitive, and the land line costs between such countries and existing CESs can make the total cost of Inmarsat services unattractive to some potential users. These factors may thus be an impediment to the availability of Inmarsat services in all countries where there is a need. Consequently, Inmarsat has been studying the benefits and applications of low-cost CESs, possibly having a more limited range of facilities, in order that more countries may benefit directly from access to Inmarsat services.

One such cost-effective solution has been use of a conventional mobile earth station as an L-band CES, offering telex only and which works in conjunction with an existing CES. The Inmarsat Council has authorized the use of L-band CESs in China and Nigeria in recognition of the inadequacy of other facilities.

With the advent of Inmarsat-C, Inmarsat has consulted with development agencies and banks and with national, regional, and international organizations to identify potential applications in developing countries. The results are encouraging. For example, the use of Inmarsat-C was seen by the United Nations Educational, Scientific and Cultural Organization (UNESCO) as a possible link between its regional offices and field personnel engaged in rural projects, by United Nations International Children's Emergency Fund (UNICEF) in support of relief and rehabilitation projects, and by United Nations Disaster Relief Coordinator (UNDRO) for use in emergency situations. The African Maritime Radiocommunications Development Project, undertaken by the ITU and aimed at improving the inadequate maritime communications infrastructure in Africa, is another possible candidate. The Permanent Interstate Committee for Drought Control in the Sahel has also expressed interest in the use of Inmarsat-C.

In 1988 the Inmarsat Council approved a Programme of Cooperation with Developing Countries containing the following elements:

1) Professional assignees -- The first element will enable professional assignees from developing countries to work at Inmarsat. Whether the assignee comes for training or to contribute his or her experience to the organization, Inmarsat is applying a "learn-by-doing" philosophy to this element of the program.

2) Technical assistance -- The second element allows developing countries to apply to Inmarsat for technical assistance in the establishment or use of mobile-satellite communications facilities.

3) Demonstration program - The third element is a demonstration program, by means of which developing countries can apply to Inmarsat for use of Inmarsat-C terminals free of charge in new and innovative applications. Several criteria have been established

to evaluate applications, notably that the application has the support of the local telecommunications authority and that it should be demonstrably cost effective.

Though the resources initially committed to the Programme of Cooperation are limited, it will be implemented in cooperation and consultation with international agencies concerned with development programs and issues. The program is expected to increase awareness of the benefits of mobile-satellite communications, which could be of particular value to developing countries.

In a further attempt to coordinate its activities with the developing world, Inmarsat, on the authority of the Assembly and Council, is establishing formal relations with intergovernmental, regional, telecommunications organizations in Africa and Asia, i.e., the Arab Telecommunication Union (ATU), ARABSAT, the Pan-African Telecommunications Union (PATU), and the Asia-Pacific Telecommunity (APT). Regional seminars in Africa and Asia are also being planned.

Conclusions

This paper mentions some of the needs of the developing countries for improved communications and outlines various ways in which mobile-satellite services could provide a way forward. But formidable problems remain.

Perhaps the high cost of such systems heads the list of problem areas. The ITU's Maitland Commission encouraged developing countries to run telecommunications as separate, self-sustaining enterprises operated on business lines and separate from postal services and similar undertakings.[3] It also concluded that no development program of any country should be regarded as balanced, properly integrated, or likely to be effective unless it accords a corresponding priority to the improvement and expansion of telecommunications. But these objectives involve structural changes which are often impeded by serious lack of capital. Furthermore, developing countries are often faced with a dilemma as to how to cope with the many competing priorities for scarce resources.[12]

The various steps being taken to reduce the costs of access to mobile-satellite services described previously may help to solve some of these problems. And the availability of a global system, with the economies of scale that it can achieve, provides advantage for countries that cannot afford their own dedicated facilities.

The question of cost, and the associated policy decisions on implementation or restructuring that need to be taken, are partly related to the issue of awareness. This means awareness of how mobile communications can meet many communications needs at acceptable cost levels; awareness of how efficient communications can contribute to economic development; and awareness of what international assistance is available. Work being done by the United Nations and other international bodies and regional seminars attended by government financial and technical planners and policy decision makers will undoubtedly help to meet these ends. Another effective means is to provide direct evidence, on the spot, of the utility of potential services. This underlines the Inmarsat-C demonstration element of our Programme of Cooperation.

A further serious impediment to the wider use of mobile-satellite communications is the regulatory restrictions that apply to the use of mobile-satellite terminals. Concerns about bypass of local networks, loss of revenue, and other concerns, though understandable, have already proved to be a serious problem in some disaster situations. Regulatory restrictions even exist in some areas where there are no alternative facilities of any kind.

Inmarsat has sought to cope with this problem in a number of ways. In the maritime field, the Inmarsat Assembly in 1985 adopted an International Agreement on the Use of Inmarsat Ship Earth Stations in the Territorial Sea and Ports. This agreement, which is open for signature by all states, is designed to achieve a worldwide regime enabling ships to use their satellite terminals in all national waters so as to improve maritime safety and contribute to more efficient ship management.

In the aeronautical field, Inmarsat is consulting with its members and with ICAO about ways in which all aircraft flying over national airspace can be authorized to communicate via satellite for international public correspondence as well as for air traffic services.

In the land mobile field, some recognition of the problem is emerging. For example, the Pan American Health Organization (PAHO) has been of assistance in obtaining regulatory approval for the use of Inmarsat mobile terminals in many Latin American countries in the event of a disaster. Within Europe efforts are being made to secure a common operating licence for mobile terminals on long-distance transport. Inmarsat is undertaking a survey of the domestic regulatory regimes in all countries in order to facilitate the future operation of land mobile services across frontiers and within national boundaries. However, the support of all developing countries, unilaterally, regionally, or on an international basis, will be needed if these obstacles are to be overcome.

The decision taken by the Inmarsat Assembly of Parties at an extraordinary session in January 1989 to extend Inmarsat's mandate for land mobile communications is a major landmark for the organization. Inmarsat can now plan to provide access for all developing countries to highly reliable mobile-satellite communications to meet the many urgently needed services in the maritime and aeronautical areas and on land, as identified previously. This is a challenge and a responsibility that will guide Inmarsat's efforts in the years ahead.

References

[1] Hudson, H.E., <u>Satellite Communications for Developing Countries: From Conjecture to Reality</u>, Space Communication and Broadcasting 3, North-Holland, Amsterdam, Netherlands, 1985.

[2] <u>Report of the Second United Nations Conference on the Exploration and Peaceful Uses of Outer Space</u>, A/CONF.101/10, Vienna, Austria, Aug. 1982, paragraphs 62-77 and 145-161.

[3] <u>Report of the Independent Commission for World-Wide Telecommunications Development</u>, International Telecommunication Union, Dec. 1984.

[4] <u>Study on the Use of INMARSAT Standard-C Terminals for Mobile Communications in Remote and Rural Areas</u>, Indian Space Research Organization, Oct. 1987.

[5] <u>Contribution of Telecommunications to the Earnings/Savings of Foreign Exchange in Developing Countries</u>: International Telecommunication Union, Geneva, Switzerland, Apr. 1988, p.15.

[6] Chu, G. C., and Srivasal, C., <u>Cost Benefit Analysis as a Guide for Investment in Telecommunications: a Study in Rural Thailand</u>, International Telecommunication Union World Telecommunication Forum, Africa Telecom, Nairobi, Kenya, Sept. 1986.

[7]Sugiyama, Elsuka, *Japan's Current International Cooperative Efforts in the Telecommunications Field*, Asia Pacific Telecommunity Seminar on Telecommunications Development, Tokyo, Japan, Sept. 1987.

[8]Lauffer, S., and Robertson, T., *A Study of The Impact of US Separate Satellite Systems on Developing Countries*, Prepared for the U.S. Department of State, Bureau of International Communications and Information Policy, Washington, D.C., U.S., 1987, p. 5.

[9]*Benefits of Telecommunications to the Transportation Sector of Developing Countries*, International Telecommunication Union, Geneva, Switzerland, Mar. 1988, p. iv.

[10]*Convention on the International Maritime Satellite Organization (INMARSAT)*, London, England, 1976.

[11]*International Telecommunication Convention*, Nairobi, Kenya, 1982, Article 25.

[12]Kouaja, H.E.B.,*Telecommunication Development and the New World Economy*, International Telecommunication Union Forum, Geneva, Switzerland, Oct 1987, Part I, p. 325.

Remote Sensing Program of the Federal Republic of Germany

A. Langner*
DFVLR, Koeln, Federal Republic of Germany
and
H. Schüssler†
Dornier GmbH, Friedrichshafen, Federal Republic of Germany

Abstract

The national remote sensing activities in the Federal Republic of Germany (FRG) are complementary to the European earth observation program of the European Space Agency (ESA) in which the FRG takes substantial part. The activities sponsored by the Federal Ministry for Research and Technology cover preparatory scientific and technological studies, development of new space-borne instruments, and demonstration projects to promote use of remotely sensed data for geoscientific application. Main emphasis is given to instrument development, in particular to active microwave sensors.

This paper presents the development line for microwave sensors covering the X-band imaging radar (X-SAR) and the millimeterwave atmospheric sounder (MAS). The activities on an optical imaging instrument like the modular optoelectronic multispectral scanner (MOMS) and the precise range and range rate equipment (PRARE) that will be flown on ERS-1 in the 1990s are described.

I. Introduction

The past activities of the FRG in the field of remote sensing of the Earth's surface from space started with the aim to gain experience in the evaluation and interpretation of remotely sensed data available from meteorological and land observation satellites. In the middle of the 1970s an aircraft measurement research program was organized and the core of a user community was established.

At the same time, a mid-term Earth observation program was defined that placed emphasis on access to data from Earth observation satellites operated by NASA, the National Oceanic and Atmospheric Administration (NOAA) and ESA and on

---

Copyright © 1990 by the American Institute of Aeronautics and Astronautics, Inc. All rights reserved.
*Dept. PT WRF/WRT
†Dept. RGK

support of national instrument development. The initial intent was not to develop and operate national satellites but to take substantial part in programs of ESA like the METEOSAT, the EARTHNET, and the remote sensing satellite (ERS-1) programs and use flight opportunities for research instruments offered mainly by NASA and ESA for Spacelab and Space Shuttle missions.

The latter resulted in successful experiments performed with the metric camera on the first Spacelab mission in 1983 and with the MOMS on STS 7 and 11 in 1983 and 1984. The development of a microwave remote sensing experiment (MRSE) operating in the X-band started in the late 1970s, but it failed in the first Spacelab mission because of a high-power amplifier failure. The subsequent development of MRSE led to the definition of the X-SAR project, which is performed in cooperation with Italy and the United States of America. Another field of instrument development covered atmospheric physics and the measurement of radiation balance parameters.

The national Earth observation activities have been incorporated into the overall German space program. Its fifth mediumterm program is presently under preparation. It will consider decisions taken recently by the ESA Council on the ministerial level, along with relevant long-term strategy for their implementation in the ESA Earth observation program and recommendations from international meetings of coordinating bodies like the Committee on Earth Observation Satellites (CEOS), the International Polar-Orbiting Meteorological Satellite Group (IPOMS), etc.

In summary, it can be stated that the FRG will contribute significantly to future remote sensing activities to observe the system "Earth" from space.

II. National Activities and Programmatic Structure

The national program is determined by the envisaged needs for Earth observation data of the German user community. Such needs are seen both for applications and for basic research.

Fields of interest are:
1) Weather forecasting, monitoring of environment, and climate research.
2) Sea state information and forecasting.
3) Regional planning and land use information.
4) Basic geodetic data.

Meteorological data are not only required from geostationary orbit but also from polar missions. Special emphasis will be given in addition to cloud image missions, vertical sounding of temperature and humidity, and measurements of the wind field at sea surface level. Of increasing importance will be the acquisition of data that provide information on transport processes in the atmosphere and oceans, sea surface temperature, ocean color, and trace gases and the chemistry of the atmosphere. Climate research requires provisions for maintaining long-term archives, measurements of radiation balance parameters, and long-term climate studies on the basis of global data.

Sea state information is required for applications such as ship routing, offshore activities, and meteorological, oceanographic, and climate research. Land use information will be needed for statistical and planning purposes and will support German activities for development aid. The national program will concentrate on securing long-term access to Earth observation data and will promote the distribution and use of data by the private sector.

In Germany the need for geodetic data results from topographic survey requirements and from improvement of knowledge of the tectonics and crustal dynamics of the Earth, the poles' motion, and the Earth's rotation. Other users are requesting a better knowledge of the Earth's gravity and magnetic field.

In order to meet the objectives of the German space program in Earth observation, the following measures will be taken:

First, FRG will participate in:

1) Operational geostationary and polar-orbiting meteorological satellite systems of the next generation.

2) Maintaining a global observation system for ocean and environment - especially atmosphere - monitoring.

3) Performing a measurement program for basic geodetic data, to be planned and implemented in close international cooperation and coordination.

Second, the FRG will continue to support:

1) Preparatory studies and technological predevelopment.

2) Instrument development with special emphasis on microwave remote sensing sensors and instruments for atmospheric and climate research.

3) Data handling and data evaluation projects with the goals to promote the use of Earth observation data for improved applications, to satisfy user requests, and to increase the involvement of the private sector.

The German activities in the field of Earth observation from space are sponsored by several federal ministries (Ministry for Research and Technology, Ministry for Transport, Ministry of the Interior and others), the German Research Foundation, and ministries of the federal states.

The **Ministry for Research and Technology (BMFT)** provides funds for the development of space-borne research instruments and the promotion of relevant technologies. BMFT contributes to the Earth observation program of the ESA and finances the budget of the German Aerospace Research Establishment (DLR) to a large extent. The execution of the BMFT-sponsored part of the national program is conducted by the DLR Project Management Department (Technology and Utilization Program Division). This includes payload project management and program planning support to the German delegates in ESA program boards.

Other DLR institutions are involved in development and research projects, data acquisition, and data evaluation and distribution as well as archiving. For instance, DLR is acting as the national point of contact (NPOC) for ESA-EARTHNET and as a processing and archiving facility (PAF) in the ERS-1 exploitation phase.

The **Ministry for Transport (BMV)** is supporting all activities connected with operational space meteorology and, therefore, contributes to the operational METEOSAT program that is under the responsibility of the European EUMETSAT organization.

The **Ministry of the Interior (BMI)** is supporting, for example, activities in the field of applied space geodesy that are performed at several institutes and establishments.

The **Ministry for Economic Cooperation (BMZ)** is responsible for sponsoring development aid projects that also include in some cases training and user assistance to apply remote-sensing techniques in developing countries. In the preparation and execution of such activities the ministry is supported by the German Foundation for International Development, the German Aerospace Research Establishment, the Federal Establishment for Geoscience and Natural Resources, and other institutions.

The **German Research Foundation (DFG)** sponsors special sectors of interdisciplinary research at universities. It also provides funds to numerous individual scientists and groups of researchers at universities for basic research and applied science projects.

In summary, it can be stated that Earth observation activities are well established and funds are provided from several governmental and private sources. To the extent necessary those activities are coordinated at the governmental level.

It can be expected that the organizational and programmatic structure of all space activities will be improved in the near future if a German space agency is founded.

III. Space-borne Instruments under Development

A. X-SAR/SIR-C

The X-SAR is a synthetic aperture X-band radar (9.6 GHz). It will be part of the "Space Radar Lab" (SRL) being developed jointly by NASA and the German Federal Ministry for Research and Technology/the German Aerospace Research Establishment in cooperation with Agenzia Spaziale Italiana (ASI) in Italy. Dornier (Germany) is responsible as the prime contractor with Selenia Spazio (Italy) acting as co-contractor. NASA provides an L/C-band/multipolarization SAR developed by the Jet Propulsion Laboratory of the California Institute of Technology. Space Radar Lab-1 (SRL-1) will be the first multispectral/multipolarized SAR flown in space and will provide images complementary to those of optical sensors.

The X-SAR is one of the most ambitious joint ventures between Germany and Italy in the field of Earth observation from space.[1] It will essentially contribute to the development of high-tech standards in German and Italian space industries. Integrated together with the NASA Shuttle imaging radar SIR-C, this SRL will be operated for approximately 50 hours during each of the envisaged three flights. The first one is scheduled for 1991, and reflights are planned at 6- or 18-month time intervals to allow monitoring of seasonal changes.

This radar lab will extend our knowledge of the conditions and dynamic processes on the Earth surface and contribute to more precise and more reliable recording and monitoring of renewable and nonrenewable resources on our planet. Furthermore, the radar lab is considered a forerunner to an "SAR facility" to be flown on NASA's Earth observing system (EOS).

X-SAR/SIR-C realizes three new concepts:
1) Multifrequency (X-, C-, and L-bands).
2) Multipolarization (horizontal and vertical in any combination).
3) Variable incidence (15-60°).

The mechanical implementation of X-SAR in the SRL pallet is shown in Fig. 1. Its four major electronics boxes are located on cold plates on the pallet close to the SAR antenna rotation axis. High-power radar signals are routed from the radio frequency electronics box (RFE) to the X-SAR antenna, and their echoes are sent back via the same flexible waveguides and rotary joints. The X-SAR antenna itself is a planar slotted waveguide array made of metallized carbon fiber reinforced plastics in order to ensure low mass, high stiffness, and good electrical performance.

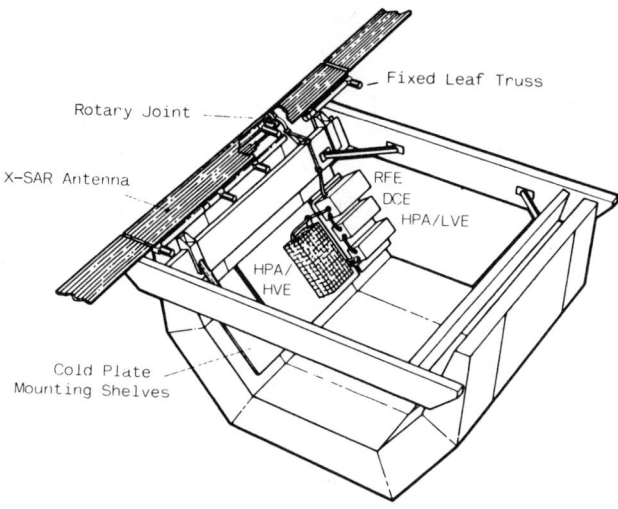

Fig. 1 Mechanical implementation of X-SAR in the Space Radar Lab pallet.

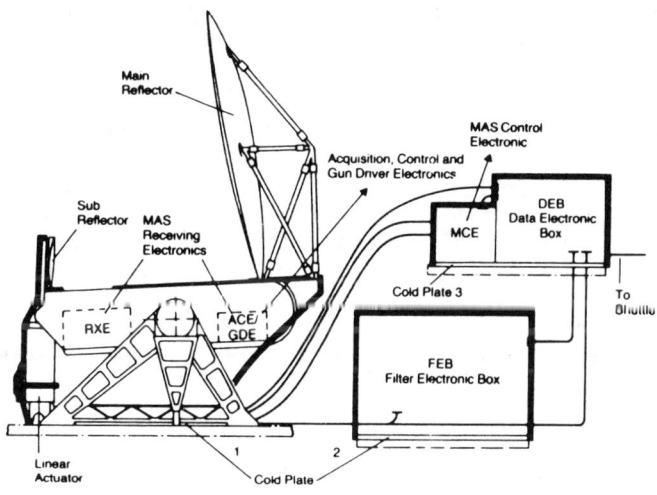

Fig. 2 Schematic configuration of MAS.

The pointing of the three antennas is provided by an antenna support structure carrying all antennas. The total SAR antenna consists of three mechanical panels, each of 4 m x 0.4 m size. During launch and landing phase the outer panels are locked in front of the center panel.

For operation they are unfolded in order to provide a total antenna surface of 12 m x 0.4 m. RF connections be-

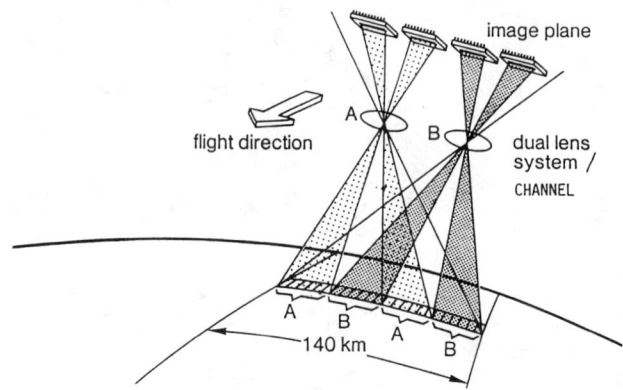

Fig. 3 MOMS-01 dual optics principle for scan line extension beyond one CCD array.

tween the panels are provided by choke flanges. The X-SAR antenna panels are mechanically integrated onto the SIR-C antenna structure. The total X-SAR/SIR-C antenna assembly is 4 m wide and 12 m long.

The high-power amplifier (HPA) amplifies low-power transmitting signals to the required high-power level for transmission via the X-SAR antenna. The amplifying element is a travelling-wave tube being manufactured by AEG, Germany. After experience gained with the MRSE on the Spacelab 1 mission, the high-power amplifier is designed such that problems of high-voltage arcing and multipaction discharge in the Shuttle environment can be avoided. Therefore, the X-SAR high-power amplifier will be put into a sealed container, providing enough air pressure to overcome problems of critical environmental conditions in the cargo bay.

The third critical field of technology is the real-time SAR ground processor. This processor will be used for quick-look SAR processing during the X-SAR/SIR-C mission. It will provide high spatial resolution (35 m), but spatial and radiometric corrections will be limited for the benefit of a quick throughput. It is a high-speed pipeline processor dealing with a data input rate of about 45 Mbytes/s. It continuously produces an SAR image stripe in accordance with an overflight speed of about 7 km/h.

B. Millimeter-Wave Atmospheric Sounder (MAS)

The MAS is a remote-sensing instrument to be used for the passive sounding of the Earth's atmosphere onboard the

Table 1 MOMS-02 specification

Spectral bands	440-505 nm
	530-575 nm
	645-680 nm
	770-810 nm
	15 m resolution
	(ground pixel size)
3 stereo channels for in-track stereo data acquisition	
Spectral range	530-810 nm
	10 m resolution for both tilted modules
	5 m resolution for nadir looking module
Stereo convergence angle	24.5° off nadir
Radiometric resolution	8 bits selected out of 12 bits
	onboard compression of panchromatic 6 bits
Data storage	recorder DDR 100, max. input rate 100 Mbit/s
	total storage capacity 5 1/2
Conceptual lifetime	3 Shuttle missions
MASS	208 kg
Power	350 W
Swath width	
- multispectral imaging	50 km
- stereo imaging	36 km

Space Shuttle in three Atmospheric Laboratory for Application and Science (ATLAS) missions. The instrument is designed to measure the emitted microwave radiation of the atmosphere at altitudes between 20-100 km in the 60-204 GHz range with an altitude resolution of better than 4 km. MAS data will permit determination of the composition, pressure, and temperature of the Earth's atmosphere in that range above the surface. In April 1985, Dornier was entrusted with a contract by the Federal Ministry for Research and

Technology, represented by the German Aerospace Research Establishment, for the development of the experiment facility and the implementation of systems integration and acceptance testing.

The Earth's atmosphere not only determines the current weather and climate but is one of the important elements of the biosphere, together with the pedosphere (soil and weathering rock) and hydrosphere. It is a very sensible and interlinked system with many different interactions varying with location and time.

Its main components are nitrogen und oxygen, but there are also several other gases that, in spite of their small amount, are of extraordinary importance. These latter gases, for example, hydrogen, carbon dioxide, and ozone, follow different distribution patterns in space and time and are partly subject to considerable modifications.

The following objectives can be achieved with MAS[2]:

1) Monitoring of changes in the ozone altitude profile and the total ozone volume throughout a sunspot cycle on several ATLAS missions.

2) Observation of time-related changes of water vapor and chlorineoxide during the same period.

3) Establishment of an experimental data base on the spatial distribution of the constituents ozone, water vapor and chlorineoxide as well as temperature, which can be used to test photochemical and dynamical models of the middle atmosphere.

The main task of the MAS is measuring the composition of the middle atmosphere at altitudes between 20-100 kilometres. The MAS uses the Earth limb sounding method for its measurements. This method differs from the vertical sounding method by high sensitivity and good altitude resolution. With a 1-m antenna on a 250-km orbit, a height resolution of 10 km is possible at 60 GHz and of better than 4 km at 180 GHz. The MAS experiment is designed to receive and measure Earth-derived radiation in the microwave range with two operational modes, the scan mode and the pointing mode. In the scan mode the antenna is moved within a given angular range in elevation with a constant speed of 1.25°/s (at an orbit altitude of 250 km). In the pointing mode the antenna is aligned to a given angle within a predetermined angular range and remains fixed during the measuring phases. The azimuth angle remains fixed in both modes. In flight the equipment is calibrated periodically by hot and cold radiation sources. The MAS is shown in Fig. 2.

The antenna picks up microwave radiation and transmits it to the MAS receiving electronics. There the high-frequency antenna signals in the 60-204 GHz range are converted to five intermediate frequencies.

The MAS control electronics (MCE) receives the start and stop commands and velocity control commands from the MAS data electronics (MDE) and controls the antenna motion via the scan and pointing mechanism. The data and commands between MAS and CDMS are routed via the remote acquisition unit interface. Scientific data are passed to the fast data channel via the high-rate multiplexer.

In-flight testing of the MAS by the research institutes involved Lindau/Germany (MPAe), Institute for Applied Physics, Bern University/Switzerland (IAP), Naval Research Laboratory, Washington D.C./U.S. (NRL) and Pennsylvania State University/U.S. (PSU) - is planned on the ATLAS-1 mission. Further flights of MAS are planned on ATLAS-2 and -3.

C. The Modular Optoelectronic Multispectral Scanner

The MOMS is an instrument for optical remote sensing of the Earth surface. It was developed by Messerschmitt-Bölkow-Blohm by order of the German Federal Minister for Research and Technology under contract of the German Aerospace Research Establishment.

The basic idea behind the MOMS concept is to construct a flexible-modular instrument on the basis of CCD "pushbroom array" sensor technology, which can be adapted to different mission requirements or platform types as a complementary instrument to the existing Landsat and SPOT (Systéme Probatoire d'Observation de la Terre) systems.

The MOMS instrument development started in 1979 with EOS, a single optics instrument with one array in the focal plane to demonstrate the feasibility of the concept.

For space verification of hardware and demonstration of the capabilities of optoelectronic scanner data for Earth resources mapping, a two-channel instrument MOMS-01 with 20-m resolution (ground pixel size) was flown twice in space onboard Space Shuttle STS-7 (June 1983) and STS-11/41-B (February 1984) flights.[5]

The main objectives of the experiments were to obtain experience concerning the adjustment and calibration of the novel type of instrument, featuring:

1) Modular assembly of the sensor unit and its peripherals, i.e., optical system per channel; focal plane assembly; supply systems (power, and thermal control); signal processing and control electronics box; and data storage with high density digital tape recorder.

2) Dual lens principle for scan line extension to four CCD arrays. (see Fig. 3).

3) Precise pixel correlation between spectral channels.

In terms of geoscientific application the missions yielded approximately 300 individual scenes with each one

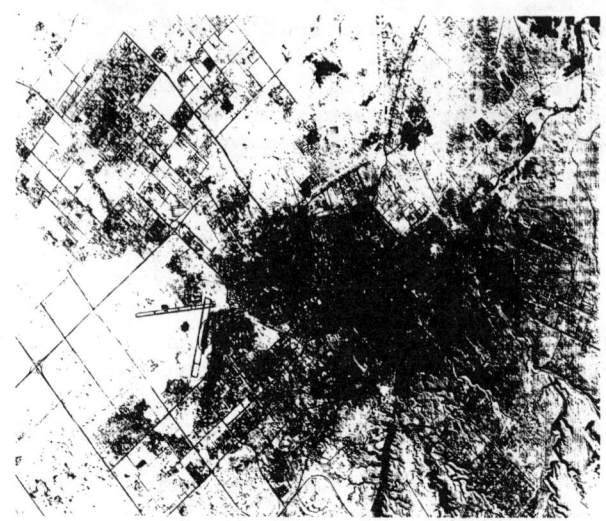

Fig. 4 MOMS-01 scene of Riyadh in the visible range (575-623 nm) at a ground resolution of 20 m.

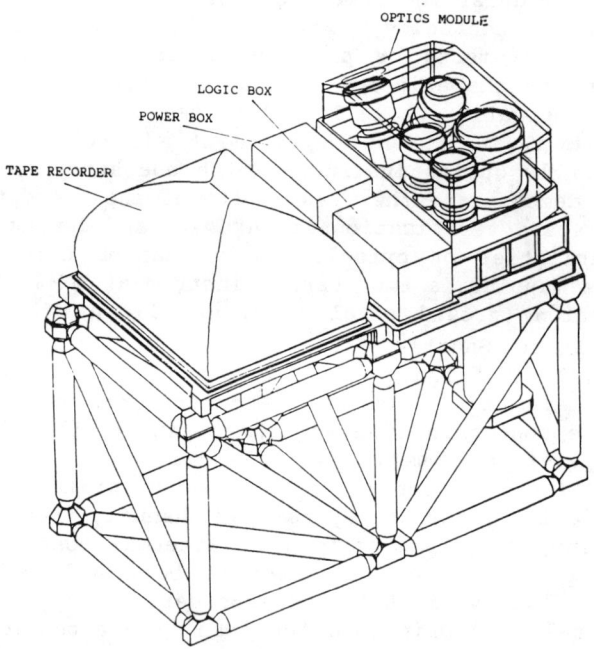

Fig. 5 Conceptual design of MOMS-02.

covering an area of 90-140 km. The evaluation was conducted within the framework of an announcement of opportunity and a cooperation agreement between NASA and BMFT. As an example see Fig. 4.

Based on the successes obtained with MOMS-01, the development of a more advanced MOMS-02 scanner was initiated. It is scheduled for spacetest within the German Spacelab D-2 mission in December 1991.[3,4]

MOMS-02 for the first time in civil Earth observation meets the requirement of simultaneous multispectral and stereoscopic data collection. To achieve this, MOMS-02 combines two units (see Fig. 5):

1) A four-channel nadir-looking multispectral scanner with two optics, under which the spectral bands are arranged pair-wise with filter plates mounted directly to the arrays.

2) A so-called "three-line scanner" responsible for acquiring geometric surface data, realized by grouping two optic modules tilted by 24.5° - one looking forward in flight direction, the other backward - vs the nadir-looking, high resolution module; the spectral range of all three modules is panchromatic (Table 1).

As basic conditions in the layout of the MOMS-02 instrument parameters, the following application requirements were considered:

1) Spatial resolution to comply with the accuracy standard of maps at a scale 1:50,000 and larger.

2) Simultaneous combination of stereoscopic and multispectral surface measurements.

3) Selection of band position and bandwidth with a focus on vegetational investigations.

These user-relevant parameters (see Table 1) have been elaborated by scientists of photogrammetric and thematic disciplines during the preparatory phase of the MOMS-02 experiment.[6,7] To meet the different scientific objectives in photogrammetry and/or thematic mapping, three different data acquisition modes have been foreseen, switchable by ground command. For photogrammetric applications, the mode is all three panchromatic stereo channels. For applications in geoscientific thematic mapping, the modes are

1) four multispectral bands plus two tilted stereo channels, and

2) four multispectral bands plus a high-resolution panchromatic channel.

Subject to the availability of the Space Transportation System, Spacelab D-2's launch into orbit is currently scheduled for spring 1992. As previously mentioned MOMS 02 realizes the three-line scanning concept that provides threefold coverage of objects on Earth in the course of a one-orbit track (see Fig. 6).

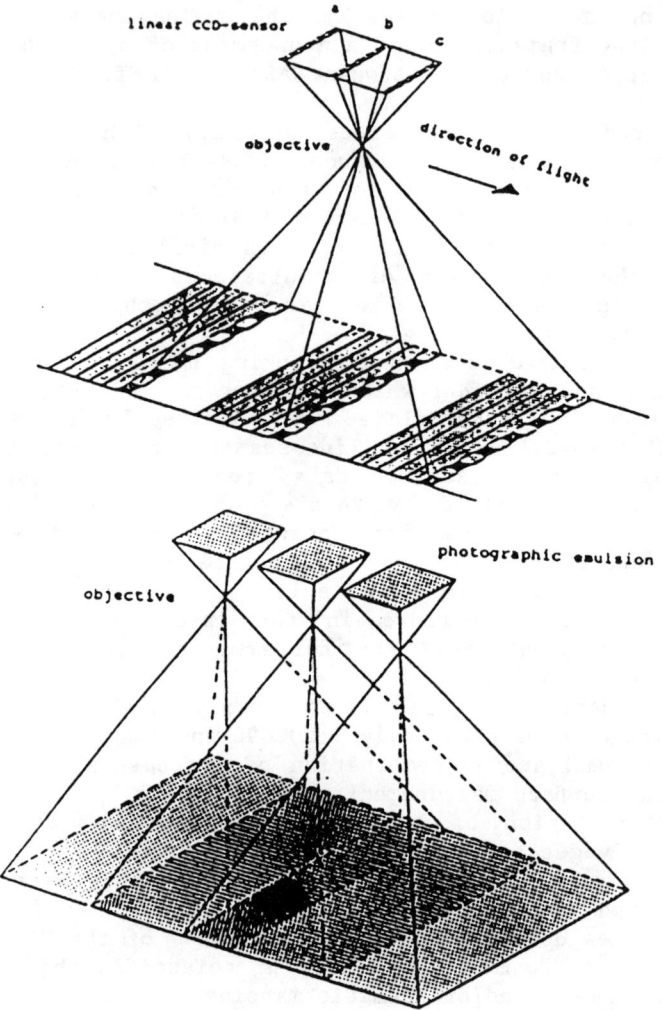

Fig. 6 The principle of stereoscopic data acquisition using a three-line scanner vs conventional photographic stereo image collection.

The concept provides for an automatic rigorous three-dimensional object-reconstruction with adequate algorithms, which will be developed during the scientific preparatory program.[6]

When the model reconstruction is performed, MOMS-02 imagery may be used for various tasks in photogrammetric applications:

1) Generation of digital terrain models.
2) Production of orthophotomaps.

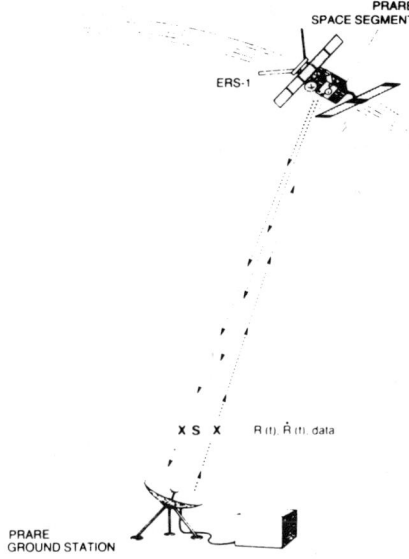

Fig. 7 Basic concept of PRARE.

3) Digital or analytical stereo and mono compilation.
4) Topographic map production.

Simulations on the accuracies in x, y, and z directions, which are envisaged with the geometrical parameters of MOMS-02, reach towards < 2 m for x and y and < 5 m for z, depending on the availability and accuracy of additional information such as attitude or position, density of homologous points, and existing DTMs. Maximum accuracy shall be obtained by "crossing orbit tracks" where sixfold surface coverage occurs.

In geoscientific thematic disciplines, the objectives are focussed on combination of three-dimensional information with multispectral measurements for:

1) Basic investigations in all disciplines by considering spectral-physical properties and simultaneously geometric relationships and morphology.

2) Improved visual interpretation and classification by taking into account relief-dependent variations of surface albedo and spectral signatures due to different sun expositions.

With these objectives, especially with the novel concept of photogrammetric data evaluation, MOMS-02 will demonstrate new application ranges, thus providing a significant milestone toward an operational use of remotely sensed data.

Presently, only the flight in connection with the German D-2 mission is assured. A similar instrument, MOMS-STEREO, has been proposed as a commercial payload for an ESA polar platform mission.

D. Precise Range and Range Rate Equipment/PRARE

This instrument is a national contribution to ESA's remote sensing satellite, ERS-1, which will be launched in September 1990.[9] The instrument will allow extension of the ERS-1 primary oceanographic mission to geodetic/geodynamic applications (i.e., support of the measurements of the altimeter) by providing very precise range and range rate data by two-way measurements between ground stations and the satellite. The expected accuracies will be in the dm and sub-dm range.

In this two-way microwave ranging system, the onboard equipment performs the measurements in X-band with some additional functions in S-band for ionospheric error correcting purposes. The ground stations are dedicated X-band regenerative transponders. Both units, the ground equipment and the space-borne one, are in many parts identical. One major difference is the S-band unit, which is a transmitter in the spacecraft and a receiver in the ground equipment. The S-band downlink is needed for the determination of the ionospheric effects (see Fig. 7).

The major technical features of the space and ground segments are given in Table 2. The procedure to perform range and range rate measurements is as follows:

Transmission of the PN-coded X-band signal starts aboard the program as soon as the satellite is within line of sight of the ground station (above 10° elevation angle). Precise ranging, however, will begin above 40° elevation angle. The PN-code signals are received by the ground station. The PN-codes are demodulated. The X-band PN-sequence (10 Mchips/s) remodulates the X-band ground transmitter (regenerative transponder). The cleaned signal then is sent up to the satellite. The transponder on ground is not only regenerative, but also coherent.

The carrier frequency of the uplink is, therefore, in a well defined relationship to the downlink carrier frequency. Onboard, it is thus possible to measure both the two-way range and the two-way Doppler very precisely.

On ground, the PN-sequences of the X-band and the S-band receivers are demodulated. The 10-MHz PN-code of the downlink is compared with the corresponding 1-MHz PN-code of the S-band. The time difference is a measure for the total electron content of the ionospheric path. The delay value is, for this reason, sent up to the satellite together with meteorological data and the housekeeping values of the ground station. The collected data are stored aboard and dumped to the PRARE command station during overflight via the PRARE X-band downlink (see Fig. 8).

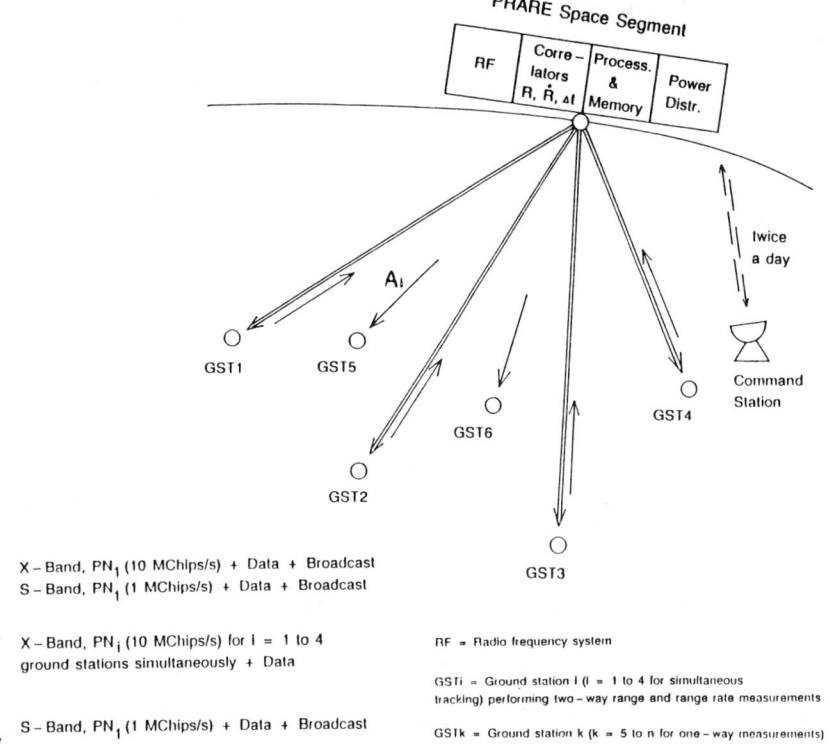

Fig. 8 Operational concept of PRARE.

Table 2 PRARE characteristics

Space segment

- electronic box size: 400 x 240 x 180 mm3
- mass: 17 kg
- power: 39 W (operating)
- antenna: 2 pairs of crossed dipoles

Ground segment

1 command station
2 ground stations
additional ground stations (15-20) provided by the users
steerable parabolic antenna with 0.6-m diam

Operating frequencies:
- at X-band (7.19-7.235 GHz), 10 MHz bandwidth
- at X-band (8.45-8.50 GHz), 10 MHz bandwidth in the downlink and uplink
- at S-band (2.20-2.29 GHz), 1 MHz bandwidth

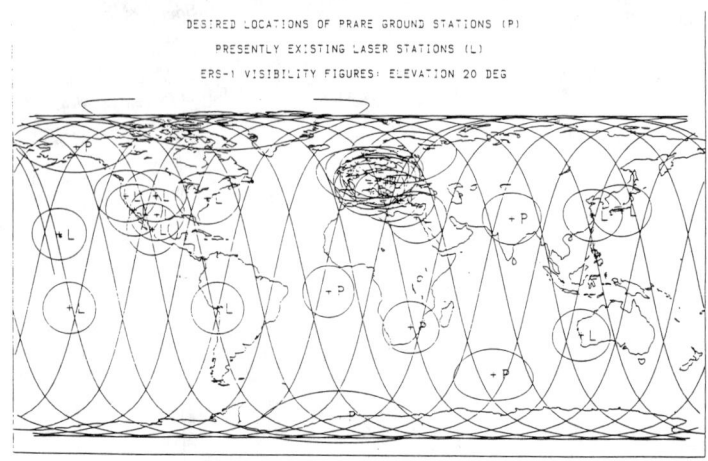

Fig. 9 Proposed locations of PRARE tracking stations complementing laser systems presently in operation.

The PRARE ground segment is made up of:
1) Primary ground stations that operate as regenerative coherent transponders and that contribute the space-ground-space range and Doppler data for precise orbit and station position determination. Minor data transmission capabilities are included.
2) Secondary ground stations that operate as "listen only" stations. These stations make use of the one-way Doppler signal in S- or S/X-band and the broadcast satellite ephemerides for on-site, on-line position determination. The S-band receive-only stations will be of particular interest to many users because they will be simple in design with an omnidirectional antenna and will be of low cost. The proposed locations of PRARE tracking stations (P) complementing laser systems (L) presently in operation are shown in Fig. 9. The overall data flow is described in Fig. 10.

IV. Activities Related to Remote Sensing Data Use

The German remote sensing activities and their promotion as part of the overall space program of the federal government cover a wide field of potential applications ranging from observation of land, ocean, sea and ice to physics of the solid earth, atmospheric physics/chemistry, climatology, etc.

Besides the German participation in ESA's Earth observation program and the development of selected space-borne instruments, including experimental flights, emphasis

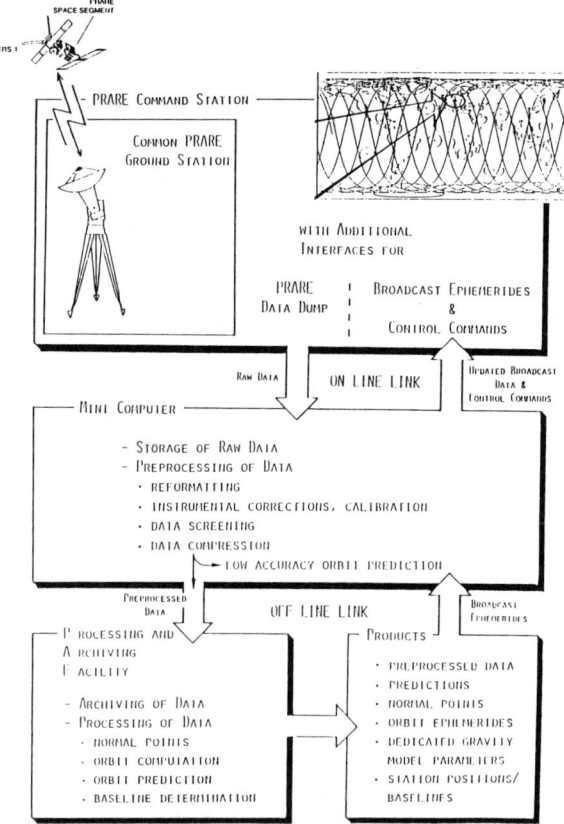

Fig. 10 PRARE overall data flow.

is also given to assure access by users to all kinds of data from foreign and international Earth observation satellites.

This is achieved by the establishment of an efficient and well-equipped national data center in the DLR. The following functions are carried out by the data center:

A. **Ground Segment**

The ground segment includes buildup and operation of a national payload ground segment for ESA and national projects.

B. **User Service**

User service consists of:
1) Standard data service, including a national point of

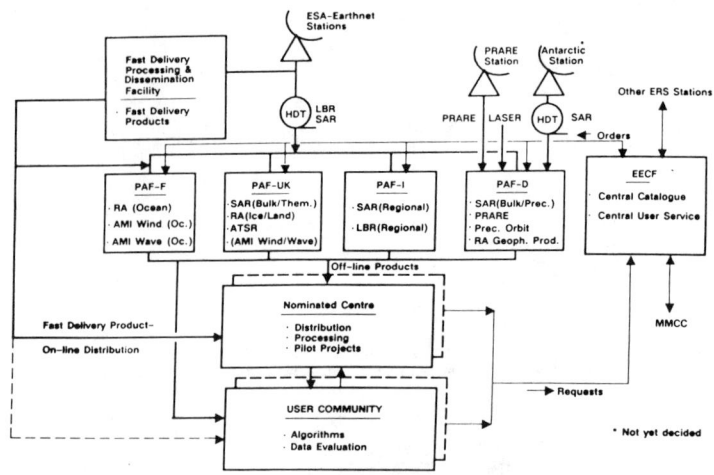

Fig. 11 The European ERS-1 ground segment for payload data.

contact function for ESA-EARTHNET, processing of raw data, geocoding, image generation, and archiving.

2) Image processing in the framework of cooperative pilot projects and on special request (on-line and off-line).

3) Promotion of application, i.e., user seminars, education, and demonstration projects.

A study was started at the end of 1988 to investigate the concept of an entire X-band SAR system, as a contribution of the FRG to the SAR payload of the second polar platform of NASA (NPOP-2) within the framework of EOS.

The EOS SAR design will be inherited from the SIR-C/X-SAR system, and its launch is planned for the end of the 1990s.

Significant changes between the present X-SAR design and the operational XEOS are based on changed system requirements and constraints, e.g.:

1) Orbit altitude of 700 km instead of 255 km.

2) Electronical beam steering in elevation and azimuth instead of mechanical beam steering.

3) Dual or quadro-polarization capability instead of single polarization.

4) Additional capability for Scan-SAR and high resolution mode operations.

5) Five-yr instrument life instead of three eight-day missions.

These new requirements and constraints are a challenge for the further development of the X-band SAR, and they require new technologies, especially for the realization of the SAR antenna.

The essential part of the present study is to compare system alternatives based on different technology approaches for the required "phased array X-band antenna" and to provide a baseline to initiate subsequent breadboard activities. This is necessary to meet the planned launch date, which requires feasibility studies (phase A) during 1991 and definition activities during 1992 parallel to technology developments. Then the XEOS development (Phase C/D) can start at the beginning of 1993 and will last 3-4 yr until the delivery for integration and test of the basic EOS SAR.

First estimates of the XEOS performance regarding the conventional side looking mode are as follows:
- ° Nominal orbit altitude   700 km
- ° Range of incidence angles   15-52 °C
- ° Swath width
  - Local high resolution   40-80 km
  - Regional medium resolution (Scan SAR)   80-600 km
- ° Resolution
  - Local high resolution   10 m
  - Regional medium resolution (Scan SAR)   10-30 m

The preceding figures are only a small sample of performance values but will provide a general overview of the essential user-oriented performance parameters estimated for XEOS.

C. Data Processing Technology

Data processing technology includes development of image processing systems for central and decentralized applications. An example of these functions is given in Fig. 11 for the ERS-1 ground segment. In addition to the promotion of a national data center, several federal ministries and the German Research Foundation sponsor activities within the user community to allow:

1) sensor calibration.
2) Verification of data.
3) Development of algorithms.
4) Development of new methods for data evaluation and interpretation.
5) Signature research.
6) Use of Earth observation data in research and application programs.

As far as data use for development aid is concerned, it has to be considered that government-sponsored activites are mainly executed in the framework of agreed projects that were proposed by the developing countries themselves. There are numerous examples where remote sensing data have been used to support projects between the FRG and developing

countries for applications such as:
1) Agricultural food production.
2) Early warning systems.
3) Land use.
4) Regional planning.
5) Development of infrastructure.
6) Exploration and management of natural resources.
7) Management of the environment.

In addition, significant training and user assistance to developing countries was given by the DLR and will continue in the future.

Mention should be made of the "International Conference on Remote Sensing for Development", in September 1986, which was hosted by the FRG in Berlin, and subsequent meetings.

The findings of those conferences are also considered by the German government to contribute to a better harmonization and coordination of training and user assistance activities in developing countries to apply remote-sensing techniques where they are regarded as an important tool for solving development problems.

V. Outlook: X-Band SAR for EOS (XEOS)

A logical step in the German microwave remote sensing program is the continuation of international cooperation with NASA and European partners and the development of the X-band SAR to an operational system.

References

[1] Wahl, M. and Ammendola, P., *Main Features of the X-SAR Project*, Proceedings of the International Geosciences and Remote Sensing Symposium '88, Edinburgh, Scotland, Sept. 1988, pp. 1018-1020.

[2] Hartmann, G.K., Künzi, K.F., and Schwartz, P.R., "Millimeter-Wave Atmospheric Sounder Instrument," Description for NASA ATLAS-1 Mission, March 1988.

[3] Ackermann, F., Bodechtel, J., and Lanzl, F., "MOMS-02 - A Sensor for Combined Stereoscopic and Multispectral Earth Observation", Experiment proposal in response to the announcement of opportunity for SPACELAB D-2 Experiments, 1987.

[4] Ackermann, F., Bodechtel,J, Meißner, D., Seige, P., Winkenbach, H., and Zilger, J., *MOMS-02 - A Sensor for Combined Stereoscopic and Multispectral Earth Observation*, Proceedings of the International Society of Photogrammetry and Remote Sensing-Congress, Kyoto, Japan, 1988.

[5] Bodechtel, J., Meißner, D., and Zilger, J., *The MOMS-01 Experiment on STS-7 and STS-11/41-B, First Results and Further Development of the Modular Optoelectronic Multispectral Scanner*, Proceedings of the

International Society of Photogrammetry and Remote Sensing Symposium, Rio de Janeiro, Brazil, 1984.

[6]Hofmann, O., Ebner, H., and Navé, P., "A Digital Photogrammetric System for Producing Digital Elevation Models and Orthophotos by Means of Linear Array Scanner Imagery", International Society of Photogrammetry Symposium, Commission III, Helsinki, Finland, 1982, and Photogrammetric Engineering and Remote Sensing, Vol. 5, pp. 1984, 1135-1142.

[7]Kaufmann, H., ét al., Design of MOMS-02 Spectral Bands, Proceedings of the Fourth International Symposium on Spectral Signatures of Objects in Remote Sensing, European Space Agency SP 287, Aussois, France, 1988.

[8]Salomonson, V. V., Nickeson, J. E., Bodechtel, J., and Zilger, J., Comparative Point-Spread Function Calculations for the MOMS-01, Thematic Mapper and SPOT-HRV Instruments, Proceedings of the Fourth International Symposium on Spectral Signatures of Objects in Remote Sensing, European Space Agency SP 287, Aussois, France, 1988.

[9]Hartl, P., and Reigber, C., "The Precise Range and Range Rate Equipment (PRARE)", Experiment proposal to European Space Agency for a supplementary payload on board ERS-1, 1981.

Author Index

Aikyo, K. 259
Argialas, D. P. 44
Bardelli, L. 132
Barsch, H. 214
Bescond, P. 11
Buxton, R. 77
Csornai, G. 205
Diaw, A. T. 170
Diop, N. 170
Dobashi, H. 259
Dodge, J. C. 18
Gebhardt, A. 214
Ghosh, S. K. 32
Hayashi, T. 271
Hayes, S. A. 70
Hung, R. J. 18
Huraib, F. S. 181
Kimura, E. 259
Krizek, M. 105
Langner, A. 296
Leckie, D. 77
Leese, J. A. 54
Marek, K.-H. 214
Martinino, F. 132
Maruo, K. 240
Mehmud, S. 94
Miller, J. 77
Morgan, K. M. 70
Murthy, S. N. 227
Newland, L. W. 70
Raney, R. K. 1
Rispoli, F. 132
Schüssler, H. 296
Sen, D. 109
Shahrokhi, F. 44
Shi, Z. 32
Strome, W. M. 77
Sultan, N. 150
Tarabzouni, M. A. 191
Thomas Y.-F. 170
von Noorden, W. D. 282
Weerakoon, W. T. 117
Weichelt, H. 214

PROGRESS IN ASTRONAUTICS AND AERONAUTICS SERIES VOLUMES

*1. **Solid Propellant Rocket Research** (1960)
Martin Summerfield
Princeton University

*2. **Liquid Rockets and Propellants** (1960)
Loren E. Bollinger
Ohio State University
Martin Goldsmith
The Rand Corp.
Alexis W. Lemmon Jr.
Battelle Memorial Institute

*3. **Energy Conversion for Space Power** (1961)
Nathan W. Snyder
Institute for Defense Analyses

*4. **Space Power Systems** (1961)
Nathan W. Snyder
Institute for Defense Analyses

*5. **Electrostatic Propulsion** (1961)
David B. Langmuir
Space Technology Laboratories, Inc.
Ernst Stuhlinger
NASA George C. Marshall Space Flight Center
J.M. Sellen Jr.
Space Technology Laboratories, Inc.

*6. **Detonation and Two-Phase Flow** (1962)
S.S. Penner
California Institute of Technology
F.A. Williams
Harvard University

*Out of print.

*7. **Hypersonic Flow Research** (1962)
Frederick R. Riddell
AVCO Corp.

*8. **Guidance and Control** (1962)
Robert E. Roberson,
Consultant
James S. Farrior
Lockheed Missiles and Space Co.

*9. **Electric Propulsion Development** (1963)
Ernst Stuhlinger
NASA George C. Marshall Space Flight Center

*10. **Technology of Lunar Exploration** (1963)
Clifford I. Cummings
Harold R. Lawrence
Jet Propulsion Laboratory

*11. **Power Systems for Space Flight** (1963)
Morris A. Zipkin
Russell N. Edwards
General Electric Co.

*12. **Ionization in High-Temperature Gases** (1963)
Kurt E. Shuler, Editor
National Bureau of Standards
John B. Fenn,
Associate Editor
Princeton University

*13. **Guidance and Control – II** (1964)
Robert C. Langford
General Precision Inc.
Charles J. Mundo
Institute of Naval Studies

*14. **Celestial Mechanics and Astrodynamics** (1964)
Victor G. Szebehely
Yale University Observatory

*15. **Heterogeneous Combustion** (1964)
Hans G. Wolfhard
Institute for Defense Analyses
Irvin Glassman
Princeton University
Leon Green Jr.
Air Force Systems Command

16. **Space Power Systems Engineering** (1966)
George C. Szego
Institute for Defense Analyses
J. Edward Taylor
TRW Inc.

17. **Methods in Astrodynamics and Celestial Mechanics** (1966)
Raynor L. Duncombe
U.S. Naval Observatory
Victor G. Szebehely
Yale University Observatory

18. **Thermophysics and Temperature Control of Spacecraft and Entry Vehicles** (1966)
Gerhard B. Heller
NASA George C. Marshall Space Flight Center

*19. Communication
Satellite Systems
Technology (1966)
Richard B. Marsten
*Radio Corporation
of America*

*20. Thermophysics of
Spacecraft and Planetary
Bodies: Radiation
Properties of Solids
and the Electromagnetic
Radiation Environment
in Space (1967)
Gerhard B. Heller
*NASA George C. Marshall
Space Flight Center*

21. Thermal Design
Principles of Spacecraft
and Entry Bodies (1969)
Jerry T. Bevans
TRW Systems

22. Stratospheric
Circulation (1969)
Willis L. Webb
*Atmospheric Sciences
Laboratory, White Sands,
and University of Texas
at El Paso*

23. Thermophysics:
Applications to Thermal
Design of Spacecraft
(1970)
Jerry T. Bevans
TRW Systems

24. Heat Transfer
and Spacecraft
Thermal Control (1971)
John W. Lucas
Jet Propulsion Laboratory

25. Communication
Satellites for the 70's:
Technology (1971)
Nathaniel E. Feldman
The Rand Corp.
Charles M. Kelly
The Aerospace Corp.

26. Communication
Satellites for the 70's:
Systems (1971)
Nathaniel E. Feldman
The Rand Corp.
Charles M. Kelly
The Aerospace Corp.

27. Thermospheric
Circulation (1972)
Willis L. Webb
*Atmospheric Sciences
Laboratory, White Sands,
and University of Texas
at El Paso*

28. Thermal
Characteristics
of the Moon (1972)
John W. Lucas
Jet Propulsion Laboratory

29. Fundamentals
of Spacecraft Thermal
Design (1972)
John W. Lucas
Jet Propulsion Laboratory

30. Solar Activity
Observations and
Predictions (1972)
Patrick S. McIntosh
Murray Dryer
*Environmental Research
Laboratories, National
Oceanic and Atmospheric
Administration*

31. Thermal Control
and Radiation (1973)
Chang-Lin Tien
*University of California
at Berkeley*

32. Communications
Satellite Systems (1974)
P.L. Bargellini
COMSAT Laboratories

33. Communications
Satellite Technology
(1974)
P.L. Bargellini
COMSAT Laboratories

34. Instrumentation
for Airbreathing
Propulsion (1974)
Allen E. Fuhs
Naval Postgraduate School
Marshall Kingery
*Arnold Engineering
Development Center*

35. Thermophysics and
Spacecraft Thermal
Control (1974)
Robert G. Hering
University of Iowa

36. Thermal Pollution
Analysis (1975)
Joseph A. Schetz
*Virginia Polytechnic
Institute*
ISBN 0-915928-00-0

37. Aeroacoustics: Jet
and Combustion Noise;
Duct Acoustics (1975)
Henry T. Nagamatsu,
Editor
*General Electric Research
and Development Center*
Jack V. O'Keefe,
Associate Editor
The Boeing Co.
Ira R. Schwartz,
Associate Editor
*NASA Ames
Research Center*
ISBN 0-915928-01-9

38. Aeroacoustics: Fan,
STOL, and Boundary
Layer Noise; Sonic
Boom; Aeroacoustics
Instrumentation (1975)
Henry T. Nagamatsu,
Editor
*General Electric Research
and Development Center*
Jack V. O'Keefe,
Associate Editor
The Boeing Co.
Ira R. Schwartz,
Associate Editor
*NASA Ames
Research Center*
ISBN 0-915928-02-7

SERIES LISTING

39. Heat Transfer with Thermal Control Applications (1975)
M. Michael Yovanovich
University of Waterloo
ISBN 0-915928-03-5

40. Aerodynamics of Base Combustion (1976)
S.N.B. Murthy, Editor
J.R. Osborn,
Associate Editor
Purdue University
A.W. Barrows
J.R. Ward,
Associate Editors
Ballistics Research Laboratories
ISBN 0-915928-04-3

41. Communications Satellite Developments: Systems (1976)
Gilbert E. LaVean
Defense Communications Agency
William G. Schmidt
CML Satellite Corp.
ISBN 0-915928-05-1

42. Communications Satellite Developments: Technology (1976)
William G. Schmidt
CML Satellite Corp.
Gilbert E. LaVean
Defense Communications Agency
ISBN 0-915928-06-X

43. Aeroacoustics: Jet Noise, Combustion and Core Engine Noise (1976)
Ira R. Schwartz, Editor
NASA Ames Research Center
Henry T. Nagamatsu,
Associate Editor
General Electric Research and Development Center
Warren C. Strahle,
Associate Editor
Georgia Institute of Technology
ISBN 0-915928-07-8

44. Aeroacoustics: Fan Noise and Control; Duct Acoustics; Rotor Noise (1976)
Ira R. Schwartz, Editor
NASA Ames Research Center
Henry T. Nagamatsu,
Associate Editor
General Electric Research and Development Center
Warren C. Strahle,
Associate Editor
Georgia Institute of Technology
ISBN 0-915928-08-6

45. Aeroacoustics: STOL Noise; Airframe and Airfoil Noise (1976)
Ira R. Schwartz, Editor
NASA Ames Research Center
Henry T. Nagamatsu,
Associate Editor
General Electric Research and Development Center
Warren C. Strahle,
Associate Editor
Georgia Institute of Technology
ISBN 0-915928-09-4

46. Aeroacoustics: Acoustic Wave Propagation; Aircraft Noise Prediction; Aeroacoustic Instrumentation (1976)
Ira R. Schwartz, Editor
NASA Ames Research Center
Henry T. Nagamatsu,
Associate Editor
General Electric Research and Development Center
Warren C. Strahle,
Associate Editor
Georgia Institute of Technology
ISBN 0-915928-10-8

47. Spacecraft Charging by Magnetospheric Plasmas (1976)
Alan Rosen
TRW Inc.
ISBN 0-915928-11-6

48. Scientific Investigations on the Skylab Satellite (1976)
Marion I. Kent
Ernst Stuhlinger
NASA George C. Marshall Space Flight Center
Shi-Tsan Wu
University of Alabama
ISBN 0-915928-12-4

49. Radiative Transfer and Thermal Control (1976)
Allie M. Smith
ARO Inc.
ISBN 0-915928-13-2

50. Exploration of the Outer Solar System (1976)
Eugene W. Greenstadt
TRW Inc.
Murray Dryer
National Oceanic and Atmospheric Administration
Devrie S. Intriligator
University of Southern California
ISBN 0-915928-14-0

51. Rarefied Gas Dynamics, Parts I and II (two volumes) (1977)
J. Leith Potter
ARO Inc.
ISBN 0-915928-15-9

52. Materials Sciences in Space with Application to Space Processing (1977)
Leo Steg
General Electric Co.
ISBN 0-915928-16-7

53. **Experimental Diagnostics in Gas Phase Combustion Systems** (1977)
Ben T. Zinn, Editor
Georgia Institute of Technology
Craig T. Bowman, Associate Editor
Stanford University
Daniel L. Hartley, Associate Editor
Sandia Laboratories
Edward W. Price, Associate Editor
Georgia Institute of Technology
James G. Skifstad, Associate Editor
Purdue University
ISBN 0-015928-18-3

54. **Satellite Communications: Future Systems** (1977)
David Jarett
TRW Inc.
ISBN 0-915928-18-3

55. **Satellite Communications: Advanced Technologies** (1977)
David Jarett
TRW Inc.
ISBN 0-915928-19-1

56. **Thermophysics of Spacecraft and Outer Planet Entry Probes** (1977)
Allie M. Smith
ARO Inc.
ISBN 0-915928-20-5

57. **Space-Based Manufacturing from Nonterrestrial Materials** (1977)
Gerard K. O'Neill, Editor
Brian O'Leary, Assistant Editor
Princeton University
ISBN 0-915928-21-3

58. **Turbulent Combustion** (1978)
Lawrence A. Kennedy
State University of New York at Buffalo
ISBN 0-915928-22-1

59. **Aerodynamic Heating and Thermal Protection Systems** (1978)
Leroy S. Fletcher
University of Virginia
ISBN 0-915928-23-X

60. **Heat Transfer and Thermal Control Systems** (1978)
Leroy S. Fletcher
University of Virginia
ISBN 0-915928-24-8

61. **Radiation Energy Conversion in Space** (1978)
Kenneth W. Billman
NASA Ames Research Center
ISBN 0-915928-26-4

62. **Alternative Hydrocarbon Fuels: Combustion and Chemical Kinetics** (1978)
Craig T. Bowman
Stanford University
Jorgen Birkeland
Department of Energy
ISBN 0-915928-25-6

63. **Experimental Diagnostics in Combustion of Solids** (1978)
Thomas L. Boggs
Naval Weapons Center
Ben T. Zinn
Georgia Institute of Technology
ISBN 0-915928-28-0

64. **Outer Planet Entry Heating and Thermal Protection** (1979)
Raymond Viskanta
Purdue University
ISBN 0-915928-29-9

65. **Thermophysics and Thermal Control** (1979)
Raymond Viskanta
Purdue University
ISBN 0-915928-30-2

66. **Interior Ballistics of Guns** (1979)
Herman Krier
University of Illinois at Urbana-Champaign
Martin Summerfield
New York University
ISBN 0-915928-32-9

*67. **Remote Sensing of Earth from Space: Role of "Smart Sensors"** (1979)
Roger A. Breckenridge
NASA Langley Research Center
ISBN 0-915928-33-7

68. **Injection and Mixing in Turbulent Flow** (1980)
Joseph A. Schetz
Virginia Polytechnic Institute and State University
ISBN 0-915928-35-3

69. **Entry Heating and Thermal Protection** (1980)
Walter B. Olstad
NASA Headquarters
ISBN 0-915928-38-8

70. **Heat Transfer, Thermal Control, and Heat Pipes** (1980)
Walter B. Olstad
NASA Headquarters
ISBN 0-915928-39-6

71. **Space Systems and Their Interactions with Earth's Space Environment** (1980)
Henry B. Garrett
Charles P. Pike
Hanscom Air Force Base
ISBN 0-915928-41-8

72. **Viscous Flow Drag Reduction** (1980)
Gary R. Hough
Vought Advanced Technology Center
ISBN 0-915928-44-2

73. Combustion
Experiments in a Zero-
Gravity Laboratory (1981)
Thomas H. Cochran
*NASA Lewis
Research Center*
ISBN 0-915928-48-5

74. Rarefied Gas
Dynamics, Parts I and II
(two volumes) (1981)
Sam S. Fisher
University of Virginia
ISBN 0-915928-51-5

75. Gasdynamics
of Detonations and
Explosions (1981)
J.R. Bowen
*University of Wisconsin
at Madison*
N. Manson
Université de Poitiers
A.K. Oppenheim
*University of California
at Berkeley*
R.I. Soloukhin
*Institute of Heat and Mass
Transfer, BSSR Academy
of Sciences*
ISBN 0-915928-46-9

76. Combustion in
Reactive Systems (1981)
J.R. Bowen
*University of Wisconsin
at Madison*
N. Manson
Université de Poitiers
A.K. Oppenheim
*University of California
at Berkeley*
R.I. Soloukhin
*Institute of Heat and Mass
Transfer, BSSR Academy
of Sciences*
ISBN 0-915928-47-7

77. Aerothermodynamics
and Planetary Entry (1981)
A.L. Crosbie
*University of Missouri-
Rolla*
ISBN 0-915928-52-3

78. Heat Transfer and
Thermal Control (1981)
A.L. Crosbie
*University of Missouri-
Rolla*
ISBN 0-915928-53-1

79. Electric Propulsion
and Its Applications
to Space Missions (1981)
Robert C. Finke
*NASA Lewis
Research Center*
ISBN 0-915928-55-8

80. Aero-Optical
Phenomena (1982)
Keith G. Gilbert
Leonard J. Otten
*Air Force Weapons
Laboratory*
ISBN 0-915928-60-4

81. Transonic
Aerodynamics (1982)
David Nixon
*Nielsen Engineering &
Research, Inc.*
ISBN 0-915928-65-5

82. Thermophysics of
Atmospheric Entry (1982)
T.E. Horton
University of Mississippi
ISBN 0-915928-66-3

83. Spacecraft Radiative
Transfer and Temperature
Control (1982)
T.E. Horton
University of Mississippi
ISBN 0-915928-67-1

84. Liquid-Metal Flows
and
Magnetohydrodynamics
(1983)
H. Branover
*Ben-Gurion University
of the Negev*
P.S. Lykoudis
Purdue University
A. Yakhot
*Ben-Gurion University
of the Negev*
ISBN 0-915928-70-1

85. Entry Vehicle
Heating and Thermal
Protection Systems: Space
Shuttle, Solar Starprobe,
Jupiter Galileo Probe
(1983)
Paul E. Bauer
*McDonnell Douglas
Astronautics Co.*
Howard E. Collicott
The Boeing Co.
ISBN 0-915928-74-4

86. Spacecraft Thermal
Control, Design, and
Operation (1983)
Howard E. Collicott
The Boeing Co.
Paul E. Bauer
*McDonnell Douglas
Astronautics Co.*
ISBN 0-915928-75-2

87. Shock Waves,
Explosions, and
Detonations (1983)
J.R. Bowen
University of Washington
N. Manson
Université de Poitiers
A.K. Oppenheim
*University of California
at Berkeley*
R.I. Soloukhin
*Institute of Heat and Mass
Transfer, BSSR Academy
of Sciences*
ISBN 0-915928-76-0

88. Flames, Lasers, and
Reactive Systems (1983)
J.R. Bowen
University of Washington
N. Manson
Université de Poitiers
A.K. Oppenheim
*University of California
at Berkeley*
R.I. Soloukhin
*Institute of Heat and Mass
Transfer, BSSR Academy
of Sciences*
ISBN 0-915928-77-9

89. **Orbit-Raising and Maneuvering Propulsion: Research Status and Needs** (1984)
Leonard H. Caveny
Air Force Office of Scientific Research
ISBN 0-915928-82-5

90. **Fundamentals of Solid-Propellant Combustion** (1984)
Kenneth K. Kuo
Pennsylvania State University
Martin Summerfield
Princeton Combustion Research Laboratories, Inc.
ISBN 0-915928-84-1

91. **Spacecraft Contamination: Sources and Prevention** (1984)
J.A. Roux
University of Mississippi
T.D. McCay
NASA Marshall Space Flight Center
ISBN 0-915928-85-X

92. **Combustion Diagnostics by Nonintrusive Methods** (1984)
T.D. McCay
NASA Marshall Space Flight Center
J.A. Roux
University of Mississippi
ISBN 0-915928-86-8

93. **The INTELSAT Global Satellite System** (1984)
Joel Alper
COMSAT Corp.
Joseph Pelton
INTELSAT
ISBN 0-915928-90-6

94. **Dynamics of Shock Waves, Explosions, and Detonations** (1984)
J.R. Bowen
University of Washington
N. Manson
Université de Poitiers
A.K. Oppenheim
University of California at Berkely
R.I. Soloukhin
Institute of Heat and Mass Transfer, BSSR Academy of Sciences
ISBN 0-915928-91-4

95. **Dynamics of Flames and Reactive Systems** (1984)
J.R. Bowen
University of Washington
N. Manson
Université de Poitiers
A.K. Oppenheim
University of California at Bereley
R.I. Soloukhin
Institute of Heat and Mass Transfer, BSSR Academy of Sciences
ISBN 0-915928-92-2

96. **Thermal Design of Aeroassisted Orbital Transfer Vehicles** (1985)
H.F. Nelson
University of Missouri-Rolla
ISBN 0-915928-94-9

97. **Monitoring Earth's Ocean, Land, and Atmosphere from Space — Sensors, Systems, and Applications** (1985)
Abraham Schnapf
Aerospace Systems Engineering
ISBN 0-915928-98-1

98. **Thrust and Drag: Its Prediction and Verification** (1985)
Eugene E. Covert
Massachusetts Institute of Technology
C.R. James
Vought Corp.
William F. Kimzey
Sverdrup Technology AEDC Group
George K. Richey
U.S. Air Force
Eugene C. Rooney
U.S. Navy Department of Defense
ISBN 0-930403-00-2

99. **Space Stations and Space Platforms — Concepts, Design, Infrastructure, and Uses** (1985)
Ivan Bekey
Daniel Herman
NASA Headquarters
ISBN 0-930403-01-0

100. **Single- and Multi-Phase Flows in an Electromagnetic Field: Energy, Metallurgical, and Solar Applications** (1985)
Herman Branover
Ben-Gurion University of the Negev
Paul S. Lykoudis
Purdue University
Michael Mond
Ben-Gurion University of the Negev
ISBN 0-930403-04-5

101. **MHD Energy Conversion: Physiotechnical Problems** (1986)
V.A. Kirillin
A.E. Sheyndlin
Soviet Academy of Sciences
ISBN 0-930403-05-3

102. Numerical Methods for Engine-Airframe Integration (1986)
S.N.B. Murthy
Purdue University
Gerald C. Paynter
Boeing Airplane Co.
ISBN 0-930403-09-6

103. Thermophysical Aspects of Re-Entry Flows (1986)
James N. Moss
NASA Langley Research Center
Carl D. Scott
NASA Johnson Space Center
ISBN 0-930403-10-X

104. Tactical Missile Aerodynamics (1986)
M.J. Hemsch
PRC Kentron, Inc.
J.N. Nielsen
NASA Ames Research Center
ISBN 0-930403-13-4

105. Dynamics of Reactive Systems Part I: Flames and Configurations; Part II: Modeling and Heterogeneous Combustion (1986)
J.R. Bowen
University of Washington
J.-C. Leyer
Université de Poitiers
R.I. Soloukhin
Institute of Heat and Mass Transfer, BSSR Academy of Sciences
ISBN 0-930403-14-2

106. Dynamics of Explosions (1986)
J.R. Bowen
University of Washington
J.-C. Leyer
Université de Poitiers
R.I. Soloukhin
Institute of Heat and Mass Transfer, BSSR Academy of Sciences
ISBN 0-930403-15-0

107. Spacecraft Dielectric Material Properties and Spacecraft Charging (1986)
A.R. Frederickson
U.S. Air Force Rome Air Development Center
D.B. Cotts
SRI International
J.A. Wall
U.S. Air Force Rome Air Development Center
F.L. Bouquet
Jet Propulsion Laboratory, California Institute of Technology
ISBN 0-930403-17-7

108. Opportunities for Academic Research in a Low-Gravity Environment (1986)
George A. Hazelrigg
National Science Foundation
Joseph M. Reynolds
Louisiana State University
ISBN 0-930403-18-5

109. Gun Propulsion Technology (1988)
Ludwig Stiefel
U.S. Army Armament Research, Development and Engineering Center
ISBN 0-930403-20-7

110. Commercial Opportunities in Space (1988)
F. Shahrokhi
K.E. Harwell
University of Tennessee Space Institute
C.C. Chao
National Cheng Kung University
ISBN 0-930403-39-8

111. Liquid-Metal Flows: Magnetohydrodynamics and Applications (1988)
Herman Branover,
Michael Mond, and
Yeshajahu Unger
Ben-Gurion University of the Negev
ISBN 0-930403-43-6

112. Current Trends in Turbulence Research (1988)
Herman Branover,
Michael Mond, and
Yeshajahu Unger
Ben-Gurion University of the Negev
ISBN 0-930403-44-4

113. Dynamics of Reactive Systems Part I: Flames; Part II: Heterogeneous Combustion and Applications (1988)
A.L. Kuhl
R & D Associates
J.R. Bowen
University of Washington
J.-C. Leyer
Université de Poitiers
A. Borisov
USSR Academy of Sciences
ISBN 0-930403-46-0

114. Dynamics of Explosions (1988)
A.L. Kuhl
R & D Associates
J.R. Bowen
University of Washington
J.-C. Leyer
Université de Poitiers
A. Borisov
USSR Academy of Sciences
ISBN 0-930403-47-9

115. Machine Intelligence and Autonomy for Aerospace (1988)
E. Heer
Heer Associates, Inc.
H. Lum
NASA Ames Research Center
ISBN 0-930403-48-7

116. Rarefied Gas Dynamics: Space-Related Studies (1989)
E.P. Muntz
University of Southern California
D.P. Weaver
U.S. Air Force Astronautics Laboratory (AFSC)
D.H. Campbell
University of Dayton Research Institute
ISBN 0-930403-53-3

117. Rarefied Gas Dynamics: Physical Phenomena (1989)
E.P. Muntz
University of Southern California
D.P. Weaver
U.S. Air Force Astronautics Laboratory (AFSC)
D. Campbell
University of Dayton Research Institute
ISBN 0-930403-54-1

118. Rarefied Gas Dynamics: Theoretical and Computational Techniques (1989)
E.P. Muntz
University of Southern California
D.P. Weaver
U.S. Air Force Astronautics Laboratory (AFSC)
D.H. Campbell
University of Dayton Research Institute
ISBN 0-930403-55-X

119. Test and Evaluation of the Tactical Missile (1989)
Emil J. Eichblatt Jr.
Pacific Missile Test Center
ISBN 0-930403-56-8

120. Unsteady Transonic Aerodynamics (1989)
David Nixon
Nielsen Engineering & Research, Inc.
ISBN 0-930403-52-5

121. Orbital Debris from Upper-Stage Breakup (1989)
Joseph P. Loftus Jr.
NASA Johnson Space Center
ISBN 0-930403-58-4

122. Thermal-Hydraulics for Space Power, Propulsion and Thermal Management System Design (1989)
William J. Krotiuk
General Electric Co.
ISBN 0-930403-64-9

123. Viscous Drag Reduction in Boundary Layers (1990)
Dennis M. Bushnell
Jerry N. Hefner
NASA Langley Research Center
ISBN 0-930403-66-5

124. Tactical and Strategic Missile Guidance (1990)
Paul Zarchan
Charles Stark Draper Laboratory, Inc.
ISBN 0-930403-68-1

125. Applied Computational Aerodynamics (1990)
P.A. Henne
Douglas Aircraft Company
ISBN 0-930403-69-X

126. Space Commercialization: Launch Vehicles and Programs (1990)
F. Shahrokhi
University of Tennessee Space Institute
J.S. Greenberg
Princeton Synergetics Inc.
T. Al-Saud
Ministry of Defense and Aviation Kingdom of Saudi Arabia
ISBN 0-930403-75-4

127. Space Commercialization: Platforms and Processing (1990)
F. Shahrokhi
University of Tennessee Space Institute
G. Hazelrigg
National Science Foundation
R. Bayuzick
Vanderbilt University
ISBN 0-930403-76-2

128. Space Commercialization: Satellite Technology (1990)
F. Shahrokhi
University of Tennessee Space Institute
N. Jasentuliyana
United Nations
N. Tarabzouni
King Abulaziz City for Science and Technology
ISBN 0-930403-77-0

(Other Volumes are planned.)